19th Annual Conference and 8th International Conference of Chinese Society of Micro/Nano Technology (CSMNT2017)

Journal of Physics: Conference Series Volume 986

Dalian, China
26 – 29 October 2017

ISBN: 978-1-5108-6068-1
ISSN: 1742-6588

Printed from e-media with permission by:

Curran Associates, Inc.
57 Morehouse Lane
Red Hook, NY 12571

Some format issues inherent in the e-media version may also appear in this print version.

Printed by Curran Associates, Inc. (2018)

For permission requests, please contact the Institute of Physics at the address below.

Institute of Physics
Dirac House, Temple Back
Bristol BS1 6BE UK

Phone: 44 1 17 929 7481
Fax: 44 1 17 920 0979

techtracking@iop.org

Additional copies of this publication are available from:

Curran Associates, Inc.
57 Morehouse Lane
Red Hook, NY 12571 USA
Phone: 845-758-0400
Fax: 845-758-2633
Email: curran@proceedings.com
Web: www.proceedings.com

Table of contents

Volume 986

The 19th Annual Conference and 8th International Conference of Chinese Society of Micro/Nano Technology (CSMNT2017)26–29 October 2017, Dalian, China

Accepted papers received: 9 March 2018Published online: 29 March 2018

Preface

Papers

IOP Conf. Series: Journal of Physics: Conf. Series **986** (2018) 011001 doi:10.1088/1742-6596/986/1/011001

Preface

This book collects papers selected from the 19th Annual Conference and 8th International Conference of the Chinese Society of Micro/Nano Technology (CSMNT2017).

As a newly emerging technology in the 21th century, Micro-nano has been used in various high-tech fields including electronics, machinery, materials, manufacturing. A lot of remarkable achievements have been made in the past few years. This technology is expected to change people's way of thinking. This conference gathers leading scientists and engineers in Micro-Nanotechnology to exchange information on their latest research progresses, and serves as a perfect forum for scientist and engineers of different disciplines to meet and discuss.

19th Annual Conference of the Chinese Society of Micro-Nano Technology have been held in Dalian, China. Dalian is located in the Liaodong peninsula. Dalian is a place where you can find natural scenes of stunning beauty, amiable climate, stylish architecture, comfortable beaches and summer tourist resorts. There are many interesting places in Dalian, such as Tiger Beach, Swallow Nest Hill, Star-Sea, Park Ice Valley and so on.

In the 19th conferences, Many top level MEMS and NANO technology researchers were invited to make plenary presentations. Including, Toshio Fukuda(Nagoya University), Evelyn L. Hu(Harvard University), Han Huang(The University of Queensland), Shoushan Fan(Tsinghua University), Ning Xi(The University of Hong Kong), Ai-Qun Liu(Nanyang Technological University), Feiyu Kang(Tsinghua University), Yufeng Jin(Peking University), Ning Gu(Southeast University), Suyuan Xie(Xiamen University), Wantai Yang(Beijing University of Chemical Technology), Junshan Liu(Dalian University of Technology).

Thanks the staff of Dalian university of technology and CSMNT Secretary for the efforts they have made, and thanks all honored attendees. Their contributions in all aspects are indispensable to the conference's success.

Thanks to IOP Publishing for their work on publishing the special volume for selected papers of this conference. We have a good experience for cooperation, and hope that we can continue this cooperation in coming conferences in the future.

The 20th Annual conference of the Chinese Society of Micro/Nano Technology will be held in October19-22, 2018, Zhengzhou, China.

See you in Zhengzhou!

Zheng You

Chairman of Organizing Committee

President of Chinese Society of Micro/Nano Technology

Peer review statement

All papers published in this volume of *Journal of Physics: Conference Series* have been peer reviewed through processes administered by the proceedings Editors. Reviews were conducted by expert referees to the professional and scientific standards expected of a proceedings journal published by IOP Publishing.

Research of temporary bonding for 3D integrational Microsystem

G X Shi[1,2,3], J Wu[1,2], K Q Qian[1,2], J Zhu[1,2]

[1]Nanjing Electronic Devices Institute, Nanjing 210016 China
[2]Science and Technology on Monolithic Integrated Circuits and Modules Laboratory Nanjing 210016 China
[3]E-mail:xiong525625@163.com

Abstract. Due to the thickness of current interposer must be very small, the thickness of active and passive wafers are no more than 200um, even less than 100um, sometime even less than 50um, So handing the thin wafer is the key problem for 3D stack, in this paper designed and experimented the temporary bonding for 3D stack and substrate thinning , analyzed the result of the experiment , and the foundation for the 3D integration technology in the future.
Keywords: Three dimension integration, Temporary bonding, Interposer.

1. Introduction

The temporary bonding is important for 3D stack and thin wafer handing, the results analysis of the different temporary bonding material. The temporary bonding between the device wafer and the carry wafer used a special polymer material as an intermediate layer to bonded the two wafers together, then the carry wafer support of device wafer complete at the process such as the thinning of the device wafer, for this technology 3D integration was widely used in microsystem, such as multi-wafers permanent stacked[3]. After the permanent stacked the temporary bonding between the device wafer and the carry wafer will be released. The polymer bonding layer on the surface of carry wafer removed by chemical. The temporary bonding of polymer materials and equipment must be fully compatible with later process in the device manufacture, for example, the de-bonding temperature is preferably lower than 200 ℃ and the temporary material is required to have high stability of thermal and chemical, the temporary bonding also required sufficient adhering strength and lower bond stress. After the process completed, the temporary bonding material must be able to removed by the method of laser, heat or chemical etching, easy to de-bonding and surface cleaning. The basic process of temporary bonding shown in Figure 1, first step is through spin coating the temporary material on the surface of carry wafer, in some cases the device can also be coated the temporary bonding material , then handing temporary bonding material at 60℃ for 2 min before baking, then handing the temporary bonding material at 200℃ to remove the solvent in temporary bonding material , and reached the initial curing; then the carry wafer and device wafer attached and aligned together then heated temperatures material at 200℃ to completely bonding the two wafer together. Mostly temporary bonding material is in liquid form, By spin coating and spray coating method, the thickness is usually 10 ~ 50μ m, the uniformity of the TTV is less than 10%, during bonding in order to reduce surface bubbles requires a certain vacuum conditions, such as 5 * 10 $^{-2}$ mbar; bonding temperature varies depending on the material, but the temperature uniformity requires less than 1%, the interfacial energy of temporary bonding can generally up to 2J / m^2, completely reach to the requirements of the follow-thinning and other process.

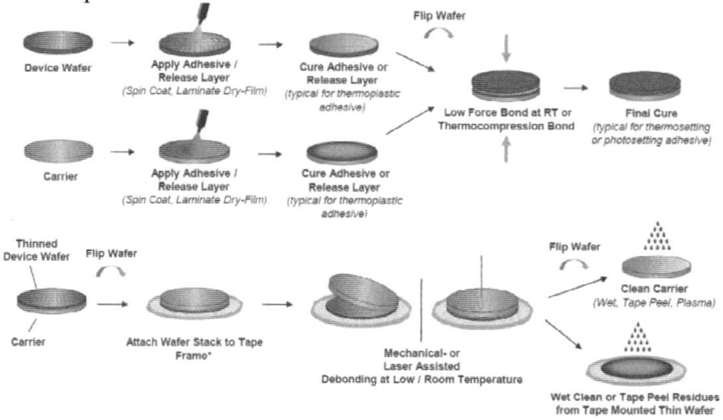

Figure.1 General process flow for temporary bonding

2. Temporary bonding material

Research applied to the application solution of temporary bonding and de-bonding material system by three-dimensional integrated. Temporary bonding is one of the key steps in the three-dimensional integration, there are many materials and equipment suppliers to supply mass type production of temporary bonding materials and equipments. Temporary bonding material eventually require removal of the auxiliary wafer, so it is necessary to use biodegradable polymer material as the temporary bonding material. The thermoplastic and thermosetting of temporary bonding of materials for bonding and de-bonding properties has a significant impact , the characteristics of thermosetting temporary bonding material is cross linking after heated, can no longer be cleaved or softening by heating . The thermoplastic polymer material can be repeatedly softened by heating and solidified after cooled. Such materials are solid polymer at room temperature, no intermolecular cross-linking, with only by Vander Waals or hydrogen bonds attract each other. This material is heating softens and flows after heating and pressure , chemical cross linking does not occur, cure after cooled . Using temporary bonding material must be properly selected according to the subsequent process. Temporary bonding material mainly applied is in Table 1 [1].

Table 1. Temporary bonding material

Material name	Material type	Tolerable temperature/℃	Thickness /μm	carrier	Coating		Debond	
					Device wafer	Carrier	Type	temperature
TMAT	Slilia gel	～250	20～200	silicon/ glass	Spin-coating	Spin-coating	Mechanical debond	Room Temperature
Brewer HT10.10	resin	～220	＜100	silicon/ glass	Spin-coating	-	Mechanical debond	180℃
Brewer Zonebond	-	～250	＜120	silicon/ glass	Spin-coating	-	Mechanical debond	Room Temperature
3M	resin	～250	＜125	glass	Spin-coating	Spin-coating	Laser debond	Room Temperature
HD/DuPont	polymde	＞350	2～20	glass	Spin-coating	-	Laser debond	Room Temperature
TOK	-	～250	＜130	Silicon/ glass	Spin-coating	-	Mechanical debond	Room Temperature

3. Temporary bonding process

Because of the high strength of the temporary bond, the gap of the device wafer and carrier (the thickness of temporary bonding layer) is very small, coupled with restrictions on the compatibility of materials and temperature , how to remove the temporary bonding material to achieve de-bonding is the main difficulties, different methods for removing of different temporary bonding material varies. The project intends to adopt the HT10.10 temporary bonding material, by the temporary bonding material is heated and softened, use the edge of the opened (peeled) or transverse passage way for the detachment between device wafer and slide. After spin-coating, requiring to baking the film and remove the solvent. The process of before baking is generally at 120 ~ 180 ℃ and during 3 ~ 5min [2]. The temperature of before drying has a very big influence on the bonding parameters and bonding quality. At the same time the temperature of before baking is higher, the temperature of bonding is more higher. While the main drawback of a lower temperature of before baking is not fully volatilize the solvent, not completely volatile solvent during the high temperature process of bonding temperature is released, resulting in the presence of air bubbles at the surface of bonded, affecting the bond quality. The topic by the temporary bonding of silicon and glass Pyrex7740, observe the bonding interface quality to analysis the effect of temporary bonding, preliminary results is mainly existence of bubble figure 2. By improving the bonding temperature and pressure and the chamber pressure to avoid the cavity on the surface.

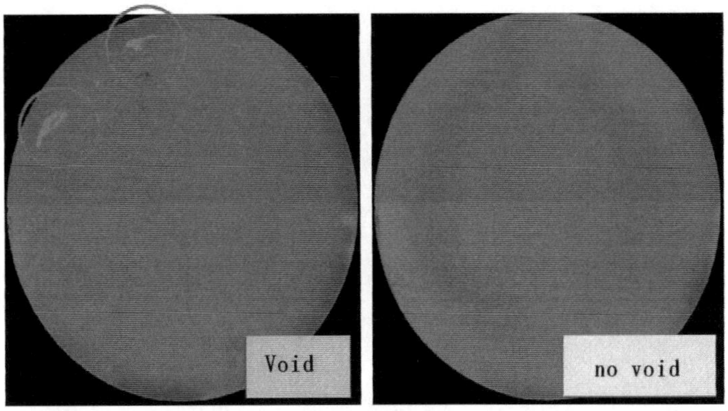

Figure.2 Temporary bonding between silicon and glass

4. Thinning process

Thinning process is mainly to achieve precise thickness by removal of substrate material, in the integrated circuit on thinning process on the substrate is only for chip packaging needs. But for the microsystems, 3D stack height requirements the interposer thickness extended from a few microns to several hundred microns[5], for the microwave and radio frequency device signals transmission, the thickness of interposer has been desired exactly to less than 1μ m, therefore, for the thickness of different device layers requires very precise nano-thinning process. The interposer thickness is generally less than 100μ m in this 3D integration, it is need to use carry wafer handing thinned interposer, therefore, the first step of interposer thinning process is temporary bonding the device wafer to the carry wafer. Mechanical grinding include the coarse grinding and fine grinding two steps. The particle on coarse grinding wheel is relatively large, so it can be thinning very fast to the target thickness. Because coarse grinding can damage wafer surface and produce residual stress, so we must use fine grinding to thin 10 ~ 20μ m [8] on the wafer surface after coarse grinding. After grinding mechanical grinding, there will be some lattice defects and damage on the wafer show in the figure 3, so the surface damage layer can be removed by chemical mechanical polishing or wafer dry and wet etch.

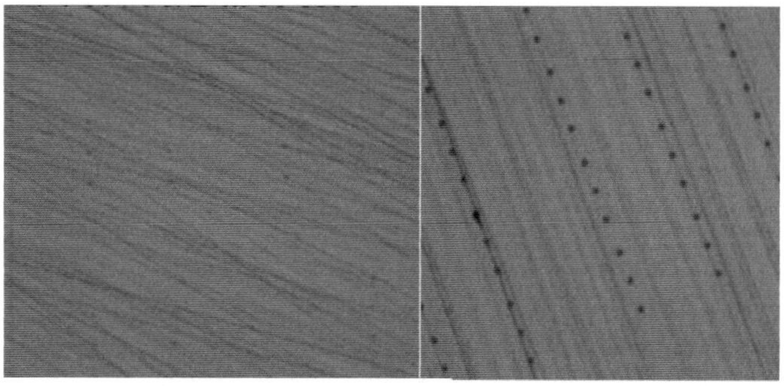

Figure.3 Result of interposer grinding

CMP can obtain a very good surface, the surface damage layer can be removed completely, but the cost of the process is very high. Wet chemical etching method can be used to remove the surface damage layer and is the most effective method, but the after wet-etch, the surface roughness is a little poor, in this paper, HF/HNO_3 mixed solution was used for surface wet-etch[4] . Dry-etch was able to completely remove the surface damage layer, and relatively easy to control, so in this paper ,we used dry- etch to remove the damage layer.

Also after the thinning process ,the interposer surface cleaning is a key step, since the grinding process [6] will be left a lot of silicon cutting and polishing on the surface of the silicon interposer, H_2O_2,DHF,SC1 are used to clean the interposer [7]. Figure 3 show the surface after the debond and cleaning, the surface is smooth and the Cu pillar for interconnect are bring out show in figure 4.

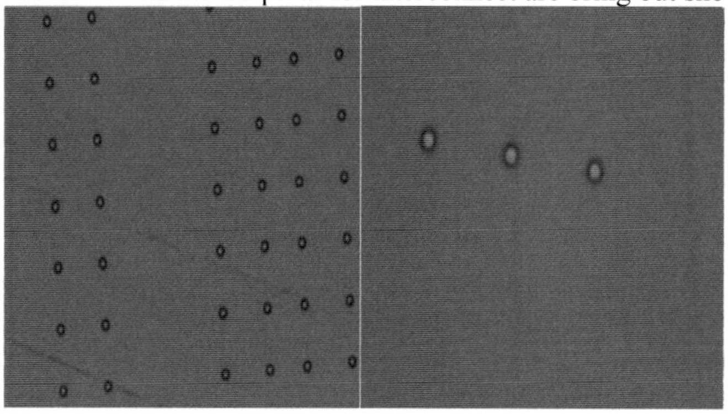

Figure.4 The smooth surface and the Cu pillar

5. Summary

In the paper we introduced a temporary bond method for 3D integrational interposer and thinning the wafer to 200um. Also introduced the wafer surface treatment and cleaning after grinding and debond. This experimental results can be used for 3D integration.

References

[1] Wang Zheyao 2014 *Three-Dimensional Integration Technology* (Beijing: Tsinghua Press) P173-290
[2] Peter Ramme 2012 *Handbook of Wafer Bonding* (New York: WILEY-VCH)
[3] Reza Ghodssi 2011 *MEMS Materials and Processes Handbook* (New York: Springer)
[4] Rao R. Tummala 2008 *Introduction to System-on-package* (New York: Mc Graw Hill)
[5] John H. Lau 2010*Advanced MEMS Packaging* (New York: WILEY-VCH)
[6] Wen Liang and Wang Jiayou 2004 *Micro&nano electronic technology* vol41（6）pp: 30-34
[7] Liu Huan and Zhou Zhen 2011 *Journal of sensing technology* vol24（2）pp: 200-203
[8] Wang Zheyao 2008 *Microsystem design and Fabrication* (Beijing: Tsinghua Press) P 103-110

Improvement of absolute positioning of precision stage based on cooperation the zero position pulse signal and incremental displacement signal

H H Wang, Y P Shi, X H Li, K Ni, Q Zhou, and X H Wang

Division of Advanced Manufacturing, Graduate School at Shenzhen, Tsinghua University, Shenzhen 518055, China

E-mail: li.xinghui@sz.tsinghua.edu.cn, wang.xiaohao@sz.tsinghua.edu.cn

Abstract. In this paper, a scheme to measure the position of precision stages, with a high precision, is presented. The encoder is composed of a scale grating and a compact two-probe reading head, to read the zero position pulse signal and continuous incremental displacement signal. The scale grating contains different codes, multiple reference codes with different spacing superimposed onto the incremental grooves with an equal spacing structure. The codes of reference mask in the reading head is the same with the reference codes on the scale grating, and generate pulse signal to locate the reference position primarily when the reading head moves along the scale grating. After locating the reference position in a section by means of the pulse signal, the reference position can be located precisely with the amplitude of the incremental displacement signal. A kind of reference codes and scale grating were designed, and experimental results show that the primary precision of the design achieved is 1μ m. The period of the incremental signal is 1μ m, and 1000/N nm precision can be achieved by subdivide the incremental signal in N times.

1. Introduction.

Linear encoder is a kind of linear distance measuring instrument with nanometer resolution and micrometer precision [1-3], applied in various precision positioning fields [4], the range of which depends on the length of the scale grating (tens of centimeters). Linear encoder transfers light intensity of the moire fringe or interference fringe of the grating into electrical signal, and displacement of the encoder will be achieved after sampling and processing the electrical signal [5]. Among linear displacement transducer with nanometer precision are linear encoder, laser interferometer which achieves a resolution of 0.1-10nm over a range of 1m [6], atom force microscope (AFM) which achieves a resolution of 0.1nm over a range of 30-50μ m and scanning electron microscope (SEM) which achieves a resolution of 1 nm over a range of 3-5mm [7]. Thus it can be seen that only linear encoder and laser interferometer provide a nanometer resolution and a relative large range at the same time. Compared with the laser interferometer, the linear encoder shows advantages in optical structure, circuit and data processing, which makes it more simple and compact. In addition, the linear encoder has prominent advantage over anti-jamming, cost and reliability, and is employed for various precision measurement and positioning applications such as closed loop CNC machine [8].

According the type of the encoder, there are incremental type and absolute type linear encoder. The incremental type linear encoder has to find a reference as zero before measurement while the absolute type achieve the reading head position directly when star-up [9]. In the research, a two-probe optical encoder [10], consists of a zero position interpretation system and an incremental displacement system, is designed. The zero position interpretation system generates pulse signal to find reference positions while the incremental displacement system transfers interference intensity into elective signal to achieve the displacement. The positioning precision of the zero position interpretation is related to the width of pulse signal directly.

A pulse signal, in condition of a kind of reference codes, 150 bits, each width 10μ m, is simulated by using angular spectrum theory [11-13]. Figure 1(a) shows the simulation waveform of the zero position pulse signal, where the width at half-peak is 23μ m. Figure1(b) the zero position pulse signal, achieved from experiment which performed with the designed codes as above, where the width at half-peak is 29μ m, is consistent with the simulated pulse signal in waveform. A flat region of 0.5μ m wide, caused by diffraction effect produced by the slit on mask and the installation error, appears at the bottom of the pulse signal when amplify it. It limits the positioning accuracy of reference zero to 0.5μ m, and thus constricts the measurement and positioning accuracy of the linear encoder.

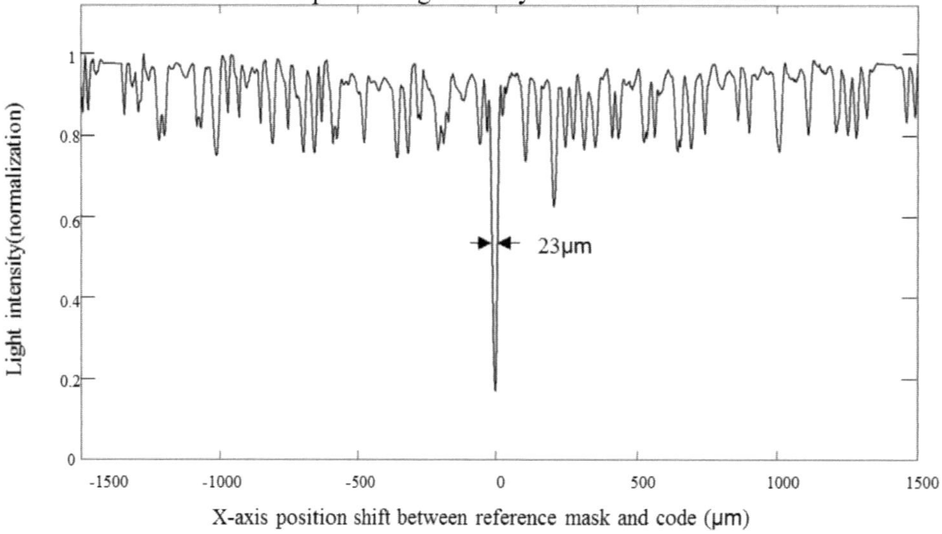

(a)

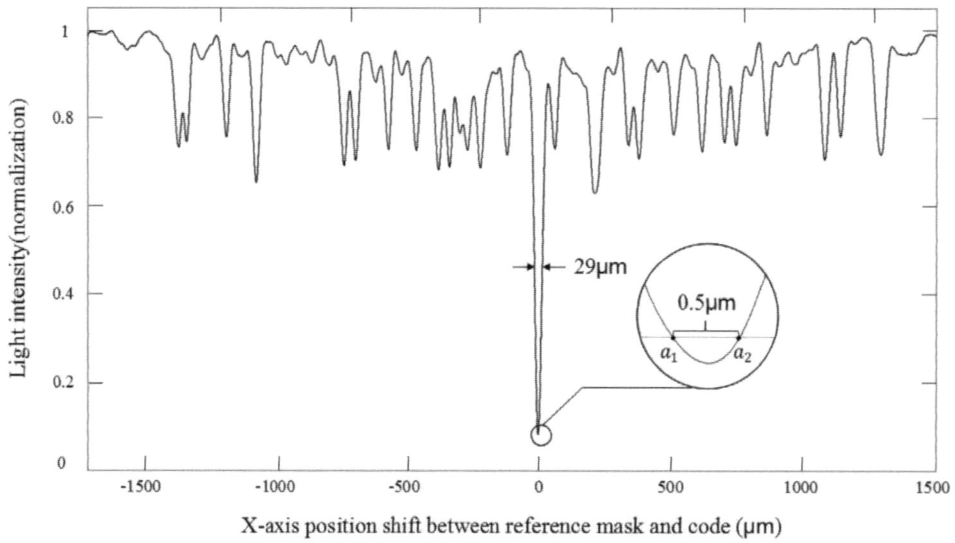

(b)

Figure 1. The simulate pulse signal (a) and the experimental pulse signal (b)

The positioning accuracy need to be further improved so that the linear encoder can apply in ultra-high precision measurement, such as lithography machine and closed loop CNC machine. In order to promote the positioning accuracy, in this study, a new method to find the reference positions, in aid of subdivision of the periodic incremental signal which is used to aligning reference zero, is proposed.

2. Principle and Experiments.

The reference positions of the linear encoder are spread on grooves of scale grating in a mode of different spacing. For example, there are 13 reference positions on the scale grating, the distances between adjacent references codes are: 10.02mm, 9.98mm, 10.04mm, 9.96mm, 10.06mm, 9.94mm, 10.08mm, 9.92mm, 10.10mm, 9.90mm, 10.12mm, and 9.88mm. The distances between each reference codes are unique, and so are the positions in incremental signal within a period, where each peak of zero position pulse corresponding.

When linear encoder moves along X-axis, the phase of the incremental signal has a relationship with the displacement of the encoder as following.

$$\Omega_X = \Omega_{X_0} + \frac{2\pi \cdot \Delta X}{g} \qquad (1)$$

Where Ω_X is the phase of the incremental signal, Ω_{x0} is the initial phase of the incremental signal, ΔX is displacement of the encoder, g is grating constant.

The phase of the incremental signal corresponds to the displacement of encoder, as show in Figure 2, and the reference positions can be marked on the incremental signal within a period. When located in a period on the incremental signal with the zero position pulse, the reference positions can be found precisely, which will improve the positioning accuracy of the zero position pulse signal significantly and make the positioning accuracy of linear encoder related to degree of subdivision of the incremental signal directly.

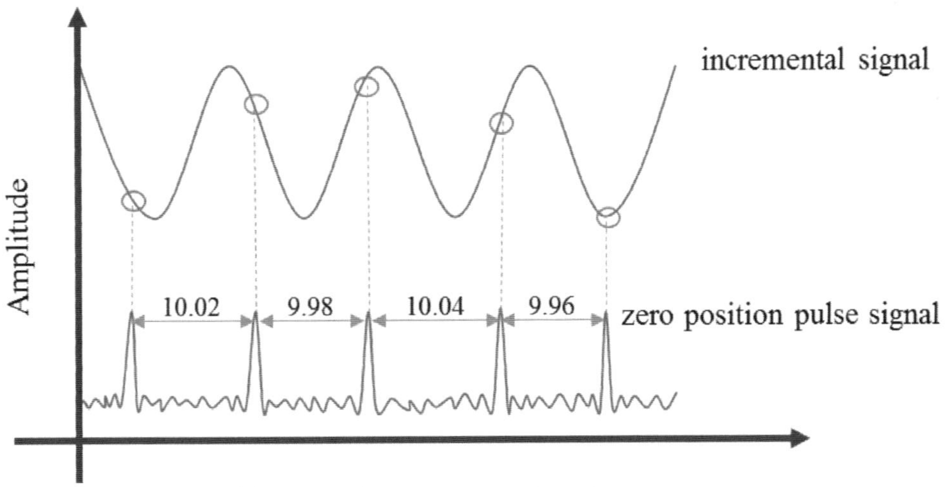

Figure 2. The relationship of incremental signal and zero position pulse signal

Steps of positioning the reference position based on cooperation the zero position pulse and incremental displacement signal are as follows.

First, we locate the zero position pulse signal in a period of the incremental signal by set a threshold.

$$I_{th} = I_{Max} \cdot k \qquad (2)$$

$$I_{x1} = I_{x2} = I_{th} \qquad (3)$$

$$|x_2 - x_1| \le P_s \qquad (4)$$

Where I_{th} is the threshold of zero position pulse signal, I_{Max} is the maximum amplitude of zero position pulse signal, x_1 and x_2 is corresponding position when the amplitude of zero position pulse signal achieve the threshold (suppose $x_1 < x_2$); P_S is periodic constant of the increment signal.

According (2)-(4), pulse interval $[x_1, x_2]$ is achieved when the width of pulse signal equal to periodic constant of the incremental signal by adjust K to set a proper threshold, where the reference position stay. In this step primary positioning is realized as the reference position is located in incremental signal within a period.

Second, searching the reference position marked with phase previously in $[x_1, x_2]$ to complete the precisely positioning.

In order to verify the process, experiments were done to prove the position the peak of the zero position pulse signal corresponding to in the incremental displacement signal is unique. We did two group of experiments, stop the linear encoder at several different positions during measurement, and achieved the incremental displacement signal waveform as below.

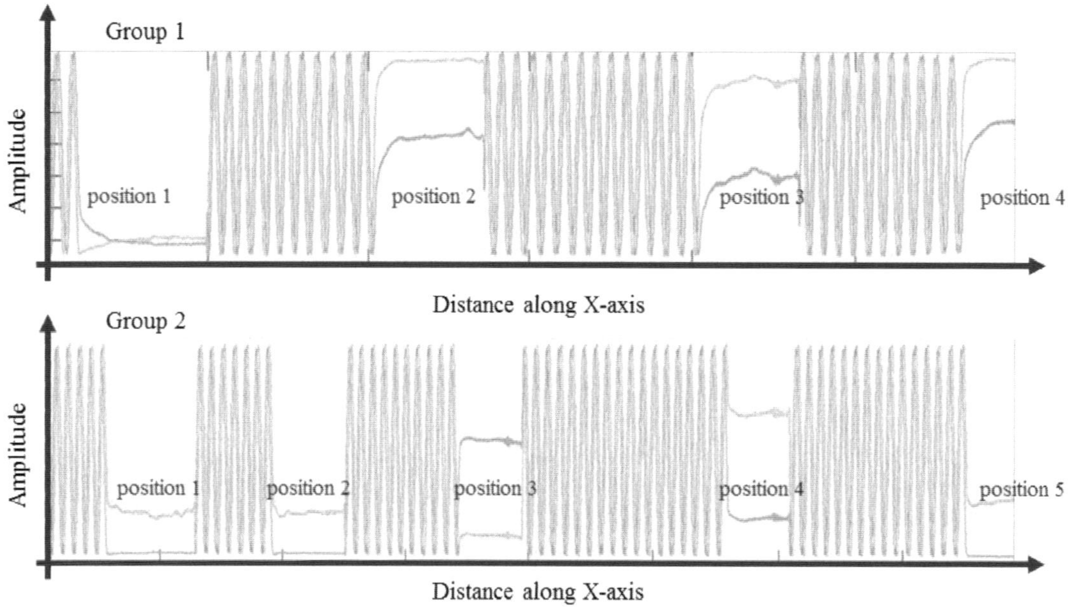

Figure 3. The waveform of incremental displacement signal

Two channels interference signal is extract from encoder in each group show in Figure 3, we can see that the two channels incremental displacement signal have different light intensity when laser beam from reading head shoot at different position of the scale grating, which proves the incremental displacement signal is unique at different positions. Therefore the position in the incremental displacement signal the peak of the zero position pulse signal corresponding to is unique too, which can be used to locate the reference positions.

To confirm the result proved above even further, we did another 4 groups of experiments, moving reading head from position 1 to position 2 along the scale grating, passing the same reference position every time. The positions in the incremental displacement signal, which the zero position pulse corresponding to, are show in Figure 5.

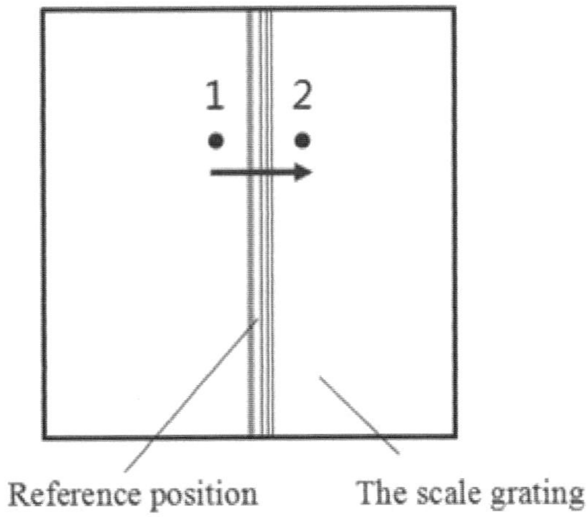

Figure 4. The reference position on the scale grating

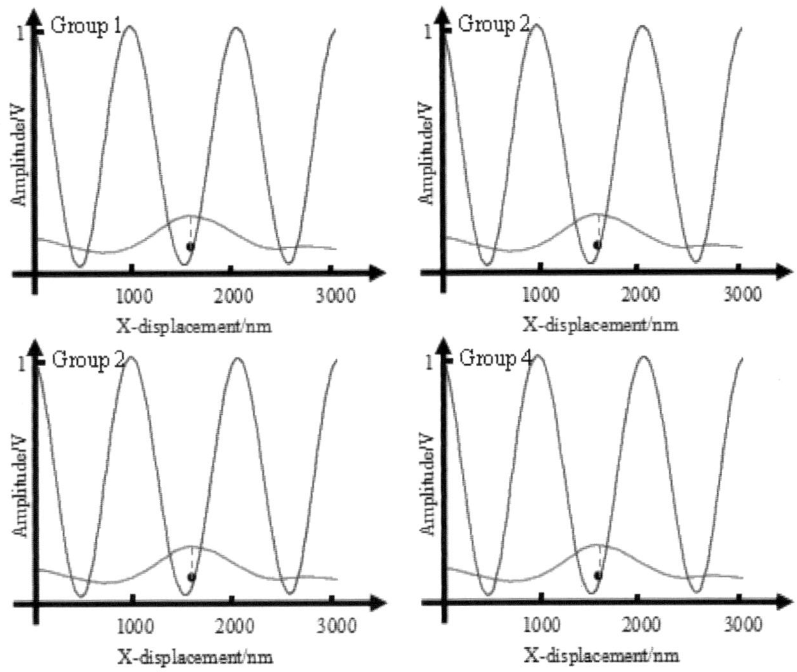

Figure 5. The result of positioning the reference position with the incremental signal

In the four sets of experiments positions in incremental displacement signal which the zero position pulse corresponding to are recorded when reading head move from position 1 to position 2.The point as show in figure 5 is the peak position of pulse signal. It is obvious that each peak of the zero position pulse signal corresponds to the same position in the incremental signal period, which further validates the feasibility of the scheme proposed before.

3. Conclusion.

A new scheme is proposed in this paper to align the reference positions from zero position interpretation system, cooperating with the incremental sinusoidal periodic signal. The experiments results show the positions in incremental sinusoidal signal the peaks of the zero position pulse signal corresponding to are unique. Because of the distances between each adjacent reference positions on the scale grating are different and determined by range coding, the positions in incremental sinusoidal signal the peaks of the zero position pulse signal corresponding to are different. Therefore subdivision of the incremental displacement signal [14-16] within a period is a powerful aid in location of the reference positions, which further improves the positioning accuracy.

References

[1] Engelhardt K and Seitz P 1996 Absolute, high-resolution optical position encoder Appl. Opt. **35**（1）, 201-208
[2] Yao R, Li X, Jiang L and Yao P 2010 Miniaturized large-grating-clearance linear encoder based on beam-splitting optical system with MOEMS LD and wireless moving part for rebound nondestructive testing in Proceedings of International Nanoelectronics Conference (IEEE), pp.1315-16
[3] Eom T B, Jeong D Y, Kim M S, Kim J W and Kim J A 2013 Tape measuring system using linear encoder and digital camera Proc.SPIE 8788,87883P
[4] Yan R, Ma X, Li G and Ying Z 2008 Research on the nano measurement technology of linear encoder Journal of Anhui Vocational College of Electronics & Information Technology,**7**(1),70-71
[5] Zhou K 2016 Research on the Zero Alignment Technology for Plane Grating Measurement System in dissertation of Harbin Institute of Technology
[6] Xian Y 2015 Applications of Laser Interferometer Tool Engineering **48**(2), 79-85
[7] Zhu J and Sun R 2005 Research on Principle and Methodology of Atomic Force Microscopy Life Science Instruments **3**(1),22-26
[8] Si J and Duan L 2012 Application of Linear Grating in Numerical Control Machine Tool Plant Maintenance Engineering (6), 40-41
[9] .Shi K, Gao Y, .Zhao Y and Qian B2011 Comparison and analysis of incremental encoder and absolute encoder MACHINERY DESIGN & MANUFACTURE(1),96-97
[10] Li X, Wang H, Ni K, Zhou Q, Mao X, Zeng L, Wang X and Xiao X 2016 Two-probe optical encoder for absolute positioning of precision stages by using an improved scale grating Opt. Express **24**(19), 21378-91
[11] Saez-Landete J, Salcedo-Sanz S, Rosa-Zurera M, Alonso J and Bernabeu E 2007 Generation of Optical Reference Signals Robust to Diffractive Effects IEEE PHOTONIC TECH L.**19**(15),1133-35
[12] Li C 2008 Digital diffraction optics based on the MATLAB in dissertation of Shandong Normal University
[13] Cai Q 2015 Research on integrated operational amplifier circuit Journal of Harbin Institute of Vocational Technology(1),152-153
[14] Xia H 2006 Research on Precise 2-D Plane Grating Measurement System and Key Technology in Doctoral dissertation of Hefei University of Technology
[15] Li X, Huang S, Zhang Y and Feng Z 2004 Study on the subdividing technigue for morie fringes of grating MODERN MANUFACTURING ENGINEERING(6),83-85
[16] Liu W and Fei Y 1999 A High-Accuracy Algorithm for the Interpolated Displacement of Grating Signal Journal of Applied Sciences(1),70-74

Characterisation of the Microstructure of Fe–Al/Cr_3C_2 Composite Coatings

Xiaoming Liu[1,3]*, JunhuiDong[2], Yuehong Yang [1], Changming Sun [1], Ya Tuo[1], Yanwei Li [1]

1. Inner Mongolia Electric Power Science and Research Institute, Hohhot, 010020, China
2.School of Materials Science and Engineering, Inner Mongolia University of Technology,Hohhot,010051,China
3.Inner Mongolia Key Laboratory of Thin Film and Coatings, Hohhot,010051,China

Corresponding author's e-mail: 31535029@qq.com

Abstract. An Fe–Al/Cr_3C_2 composite coating is investigated to assess its suitability for treating high-temperature components in a power plant. The coating exhibits excellent high-temperature properties including good corrosion, erosion and friction-wear resistance at high temperatures. To deduce the formation of the Fe–Al/Cr_3C_2 composite coating and to provide an adequate theoretical basis for its extensive application, its structures and microstructures are investigated. Scanning electronic microscopy (SEM)is used along with energy-dispersive X-ray analysis (EDAX) to analyse the surface of the coating. Energy-dispersive spectroscopy (EDS) is used to analyse the cross-section of the coating. Further, X-ray diffraction (XRD) and transmission electron microscopy (TEM) are used to analyse the phases and micro structural features within the coating. The results reveal that the basic phases are two orderly inter metallic compounds (Fe_3Al and FeAl) and that the reinforcement includes two oxides (Al_2O_3 and Cr_2O_3) as well as substantial quantities of Cr_3C_2. $Al_2O_3$is formed using two mechanisms: oxidation of aluminium in the coating and separation of Al_2O_3crystals from Fe_3Al and FeAl. The grain size of Al_2O_3 and Cr_2O_3 in the coatings is nanometric. These two oxides may increase the corrosion–erosion and wear resistances of the coating when they are used as reinforcements.

1. Introduction

Iron aluminide intermetallics (including FeAl and Fe_3Al) are intermetallic compounds that depictexcellent anti-erosion properties at high temperatures. Therefore, these materials can be potentially applied to manufacture cheap high-temperature structural materials. However, the room-temperature brittleness of iron aluminide intermetallics results in shaping difficulties and limits their industrial application[1–4]. High-velocity flame spraying (HVFS), which is a flame-spraying technology developed in the 1980s, is especially beneficial for spraying composite coatings because of its high spray velocity and diminished temperature effect on the substrate[5–7]. HVFS technology has been further used to produce FeAl intermetallic coatings containing powder materials on AISI 1020 steel. This technology not only eliminates the problems that are encountered while fabricating these alloys into useful shapes but also enables the effective use of their outstanding high-temperature characteristics. Adding Cr_3C_2 as are inforcement to iron aluminide intermetallic coatings can yield an Fe–Al/Cr_3C_2 composite coating, which exhibits several excellent properties. Chromium is the only element that is renowned to improve the room-temperature mechanical properties of iron aluminide intermetallics[8–10].

Further, an Fe–Al/Cr_3C_2 composite coating was prepared using high-velocity arc spraying (HVAS) technology. However, the sprayed materials were observed to be cored wires. The Cr_3C_2 content was limited to 20 wt% because of the coating rates of these cored wires. However, the authors could not exploit the complete potential of the material properties of Cr_3C_2in order to influence the properties of the as-fabricated material. In contrast, the spraying materials that were used in HVFS technology were powders. Thus, the Cr_3C_2 content can be adjusted in accordance with the necessary requirement.

In the current study, we sprayed an Fe–Al/Cr_3C_2 alloy powder onto AISI 1020 steel using HVFS technology, resulting in an Fe–Al/Cr_3C_2 composite coating containing a Cr_3C_2 content of45wt%.We subsequently used scanning electron microscopy (SEM) along with energy-dispersive X-ray analysis (EDAX) to analyse the surface of the Fe–Al/Cr_3C_2 composite coating. Further, energy dispersive spectroscopy (EDS) was used to analyse the cross-section of the Fe–Al/Cr_3C_2 composite coating, whereas X-ray diffraction (XRD) and transmission electron microscopy (TEM) were used to analyse the phase composition and microstructural features of the composite coating.

2. Materials and Methods

2.1. Experimental Materials

The substrate material that was used in this study was an AISI 1020 steel. The sprayed material was an Fe–Al/Cr_3C_2composite powder composed of Fe, Al, Cr_3C_2 and a small amount of Ni. Overcoming the limit of a reinforcement-phase content of 15wt% that was caused due to a low cladding ratio, we increased the content of the Cr_3C_2reinforcement phase in this experimental material to reach 45wt%. The contents of the basic phases Fe,Al andNi were observed to be 35,15 and 5wt%, respectively. The powder was prepared using water atomising technology, and the particle diameters were observed to be sufficiently small for all of the particles to pass through a 200-mesh sieve.

2.2. Experimental Methods

The materials were sprayed onto the surface of the AISI 1020 carbon substrate using an HVFS system that was equipped with a CP-5000 high-velocity spray gun. The specimen dimensions were 25 mm (diameter) × 5 mm (length), and the materials were sprayed on one end-face. The process parameters were as follows: an oxygen pressure of 0.8 MPa, anacetylene pressure of 1.1 MPa, an air pressure of 0.4 MPa and a spray stand-off distance of300 mm.

The cross-section microstructure of the Fe–Al/Cr_3C_2composite coating was observed using a Philips Quant200 scanning electron microscope (Field Electron and Ion Company, USA). The experimental parameters included a resolution of 3.5nm, an amplification multiple of 7×10^6,anaccelerating voltage of 30–200kV,amaximum beam current of 2A and a sample-chamber pressure ranging from 6×10^{-6} to 2600Pa.

The phase compositions of the Fe–Al/Cr_3C_2composite coating were analysed usingaD8 Advance X-ray diffractometer (Bruker AXS Company, Germany) that was equipped with a Cu Kα radiation source, having a wavelength, λ, of 0.154157nm. The source was operated at a voltage of 40kV and a current of 20mA. Further, samples were scanned using a step length of 0.02°.

The microstructures of coatings were observed usingaH800 transmission electron microscope (Hitachi Ltd., Japan), which exhibits high resolution and can provide information about the microstructure and crystal defects of the Fe–Al/Cr_3C_2composite coating. Selected-area electron diffraction(SAED) was used to simultaneously characterise the crystal structure and orientation of the Fe–Al/Cr_3C_2composite coating. The main technical parameters include a point resolution of 0.45nm, a lattice resolution of 0.204nm, an acceleration voltage of 35, 75,100,150 or 200kV, amagnification of 100 to 500,000×and an electron-diffraction camera length of 0.4,0.8, 1.2,1.6or2m.The process of sample thinning was divided into three steps. First, the sample was cut from the substrate using a linear cutting machineandmechanically thinned to 300μm. The sample was further thinned to 120μm using a metallographic sand paper. AGL-6960ion-beam thinner was subsequently used to thin the perforated sample. Further, the angle between the surface of the sample and the ion beam was gradually decreased from 30°to 15°.

3. Results

3.1. Structure and Microstructure of the Coating

Figure 1depictsthe microstructure of the Fe–Al/ Cr_3C_2 composite coating. Figure 1(a) illustrates the even distribution of the Cr_3C_2. Figure 1(b) exhibits that the binding method that was applied between the coating and substrate involved mechanical jointing. Mechanical jointing can be explained as follows: a molten metal drop impacts the treated substrate surface where it spreads out as a thin liquid film and adheres to the rough substrate surface. The drops further penetrate and form a mechanical joint with the rough substrate surface.

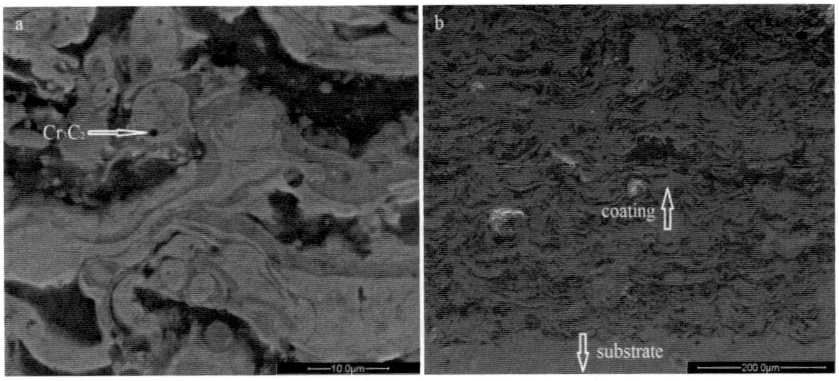

Figure 1. Microstructure of the Fe–Al/Cr_3C_2 composite coating: (a) cross-section and (b) junction of the coating.

Figure 2 depicts the cross-sectional morphology and EDS spectra of the Fe–Al/Cr_3C_2 composite coating. The coating was observed to be stratified or lamellar. The oxyacetylene heat source that is used during the HVFS process partially oxidised the sprayed materials to form ferrous, chromic and aluminium oxides.

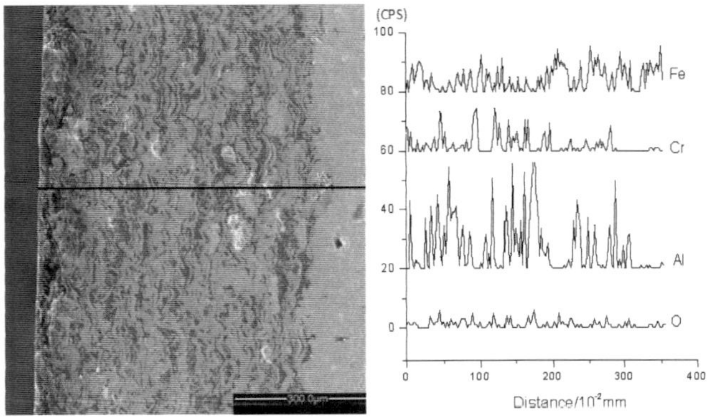

Figure 2. The cross-sectional morphology of the Fe–Al/Cr_3C_2 composite coating and the corresponding results of EDS analysis.

IOP Conf. Series: Journal of Physics: Conf. Series **986** (2018) 012003 doi:10.1088/1742-6596/986/1/012003

Figures 3 and 4 present an SEM micrograph depicting the morphology of the Fe–Al/Cr_3C_2 composite coating and the corresponding EDAX spectrum. Figure 4 illustrates the XRD pattern of the coating. The surface of the coating depicts dense aggregation, as evident in Figure 3. EDAX analysis illustrates that the coating contained Fe, Al, Cr, C, O and Ni. Further, the XRD pattern in Figure 4 indicates the presence of α-Fe, FeAl, Fe_3Al and Cr_3C_2in the coating. The grey basic phases in Figure 1 are observed to be iron aluminide intermetallics (FeAl and Fe_3Al),whereas the white phases are reinforcements. Further, we observe that the black Al_2O_3 phase is embedded in the tabular grains. The diffraction peaks of the phases containing elemental C were unnoticeable in these XRD patterns because the Cr_3C_2 content in the coating was relatively high and this phase underwent less decomposition.

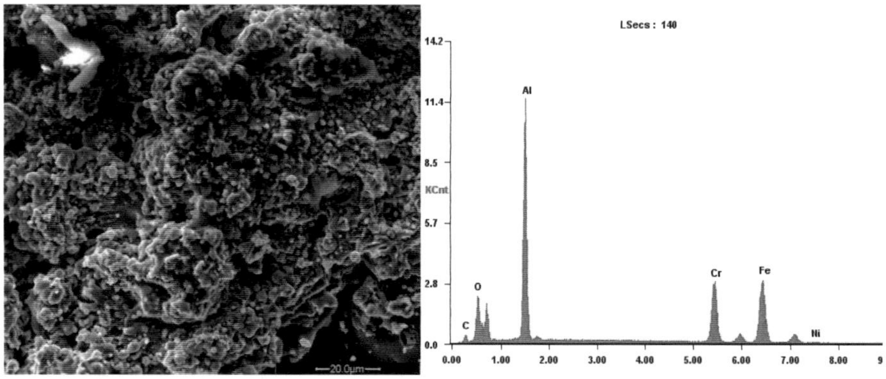

Figure 3.SEM micrograph depicting the morphology ofthe Fe–Al/Cr_3C_2 composite coating; its corresponding EDAX spectrum is also depicted.

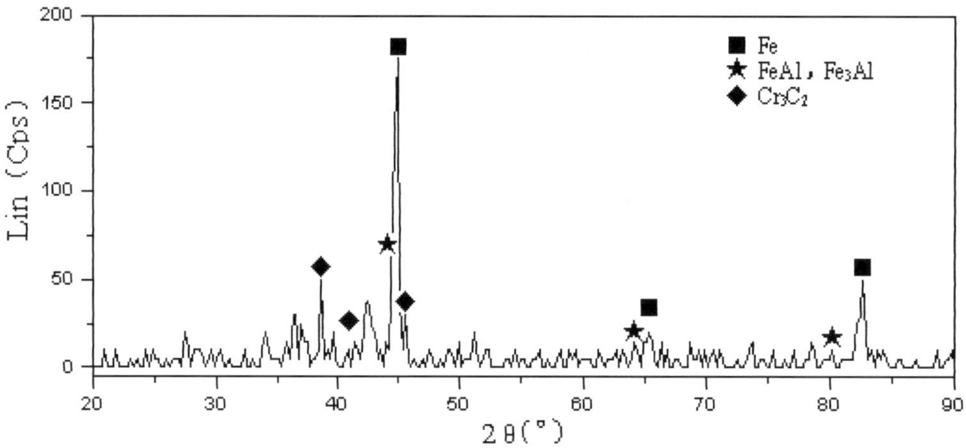

Figure 4.XRDpattern of the Fe–Al/Cr_3C_2 composite coating.

Several phases were formed because HVFS imposes a rapid cooling rate. Further ,numerous chemical reactions occur causing in homogeneity during molten-drop alloying. The relations governing the behaviour of interactions between phases were observed to be complicated.
Fe–Al/Cr_3C_2 composite coatings are typically used in industrial applications. Therefore, this investigation about the microstructure of an Fe–Al/Cr_3C_2 composite coating was deemed to be important.

3.2. Experimental Results and Microstructural Analysis of the Composite Coating

IOP Conf. Series: Journal of Physics: Conf. Series **986** (2018) 012003 doi:10.1088/1742-6596/986/1/012003

Figures5 and 6 depict the TEM micrographs and SAED patterns that are used for crystallographic phase matching of the Fe_3Al and FeAl phases, which formed the main basic phases of the Fe–Al/Cr_3C_2 composite coating. In Figure 5, an amorphous structure is indicated by the diffraction ring, which was caused due toFe_3Al. Nano particles that precipitated from the Fe_3Al are further observed in the TEM micrograph. The FeAl content was determined by calibrating against the aforementioned diffraction ring, as depicted in Figure 6.Similarly, the FeAl phase additionally contained nanoparticles formed by rapid cooling. By analysing the crystallographic phase matches, we observed that two types of orderly intermetallics were inside the flat granules (Fe_3Al and FeAl), among which the basic phases were observed to be anisotropic.

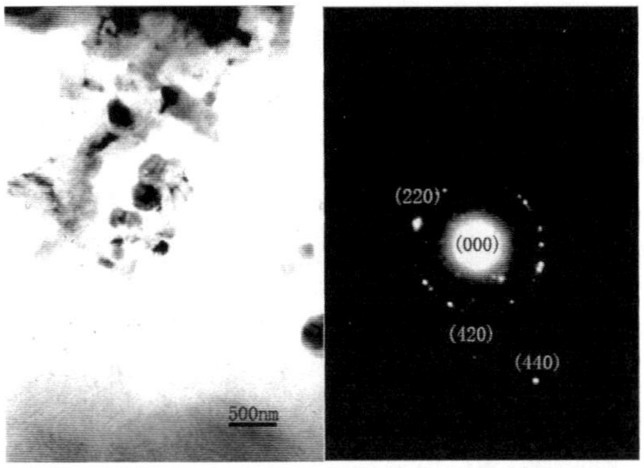

Figure 5.TEM micrograph and SAED pattern depicting the Fe_3Al and corresponding crystallographic phases.

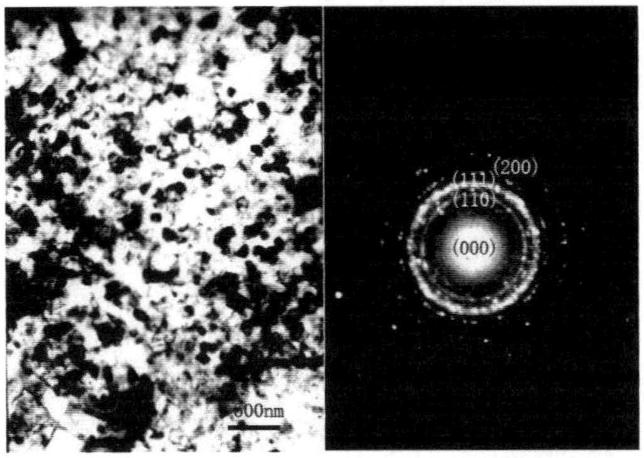

Figure 6.TEM micrograph and SAED pattern depicting the FeAl and corresponding crystallographic phases.

Figures7 and 8 depict the TEM micrographs and results of crystallographic phase matching between the formation and precipitation of oxides during the coating process.

In Figure 7, an amorphous diffraction ring can be observed in the Fe_3Al phase. After calibration, it was determined that Al_2O_3wasprecipitated from Fe_3Al.

In Figure 8, the Al_2O_3 content was determined by calibration, which separated the crystalline phases from the Fe_3Al, exhibiting high hardness and good high-temperature stability. The grain size in the Al_2O_3 phase was observed to be nanometric. As a reinforcement, Al_2O_3exhibitedhigh compatibility with the basic phase that was observed, thereby further improving the performance of the Fe–Al/Cr_3C_2 composite coating.

IOP Conf. Series: Journal of Physics: Conf. Series **986** (2018) 012003 doi:10.1088/1742-6596/986/1/012003

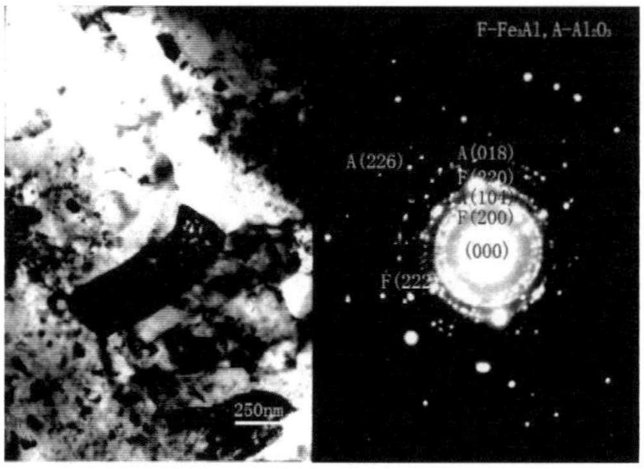

Figure 7.TEM micrographs depicting the results of Fe_3Al/Al_2O_3 and crystallographic phase matching.

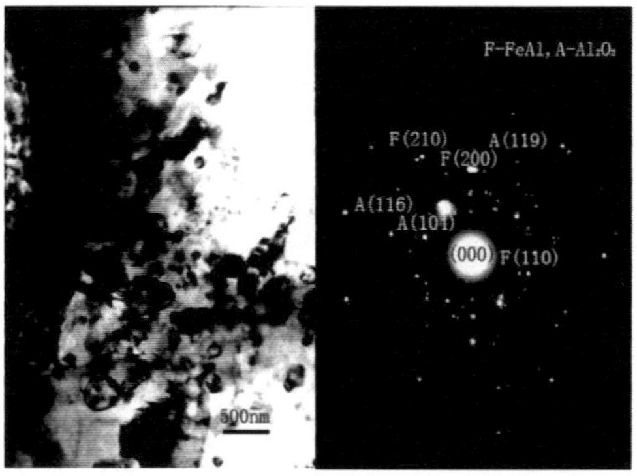

Figure 8.TEM micrographs depicting the results of $FeAl/Al_2O_3$ and crystallographic phase matching.

In Figure 9, a lath structure is observed in the basic phase. After calibration, it was observed to be Al_2O_3. Its formation mechanism can be described as follows: Relatively small droplets of aluminiumare oxidised to form Al_2O_3.When the droplets collide with the base metal during spraying, the solidified droplets becomes pherical, whereas those that are semi-molten become flat, as depicted in Figure 5. The melting point of Al_2O_3 is 2030°C,and it crystallises in a trigonal system. Al_2O_3exhibits excellent high-temperature oxidation and sulphur corrosion resistance. Further, it does not produce volatiles at high temperatures because of its high thermal stability. The oxidation process for aluminium is observed to be the same as that for iron. First, the oxygen gas molecules are adsorbed onto the surface of the composite coating and are decomposed into oxygen atoms. The oxygen atoms are further ionised, and the oxygen ions, which are the product of this ionisation process, react with the aluminium ions to form Al_2O_3. At the start of the oxidation process, no diffusion of cations, anions or electrons was observed to occur. Thus, the oxidation process is controlled solely by the rate of the chemical reaction. After the formation of theAl_2O_3 oxide film, which exhibits good adhesion, diffusion of Al^{3-} and O^{2-}becomethe main factors that drive the oxidation process.

In case of the high-temperature oxidation behaviour of Fe–Cr–Al alloy, researchers assumed that oxides with high iron and chromium contents were present on the surface of the alloy and that Al_2O_3 was present within these oxides at the onset of oxidation. Thus, there arrangement of the oxides was observed to occur in accordance with their thermodynamic stability over time. Huntz speculated that the formation of Al_2O_3 was driven by the short-range diffusion of aluminium and that the short-range diffusion of oxygen was a key factor along with the nature of the crystal boundaries, microscopic cracks and shrinkage.

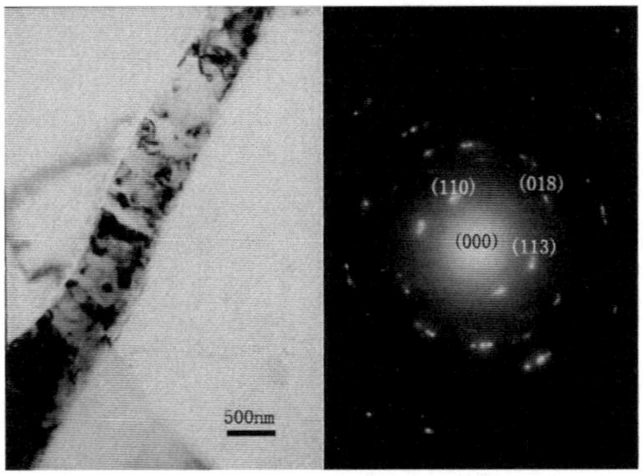

Figure 9.TEM micrographs depicting the results of Al_2O_3 and crystallographic phase matching.

In Figure 10, the oxide is observed to be an amorphous basic phase. However, after calibration, it was observed to be Cr_2O_3. Meanwhile, nanoparticles were observed to precipitate from Cr_2O_3. Further, we observe that the melting point of Cr_2O_3 is 2035°C and thatits crystalline structure is trigonal. In the crystalline structure of Cr_2O_3,each Cr^{3+} is surrounded by six O^{2-}ions,and each $O^{2-}$is bonded to four Cr^{3+} in fixed positions. The fine granular structure enhances the combination of the coating materials, thereby improving the performance of the coating. Aluminium and chromium could prevent the basic metal (elemental Fe) from oxidising. We observe that aluminium and chromium are more easily oxidised than iron. Thus, the basic metal can be prevented from corroding. However, iron oxide can be reduced by aluminium and chromium, enabling proper control of the iron content. When adense oxide film forms on the metal surface, the corrosion medium can be isolated from the base metal so that the corrosion–erosion resistance of the coating can be enhanced. During the growth of the oxide films of Al_2O_3 and Cr_2O_3, outward diffusion of aluminium and chromium and inward diffusion of oxygen can both occur. Additionally, boundary diffusion also plays a key role in this process by forming new oxides on the crystal boundaries. Because of the high concentration of oxygen in the outer layer, the diffusion effects of aluminium and chromium are enhanced, resulting in the growth of new oxides near the outer surface of the oxide film. Thus, we observe that the film becomes more uniform and can better protect the base metal.

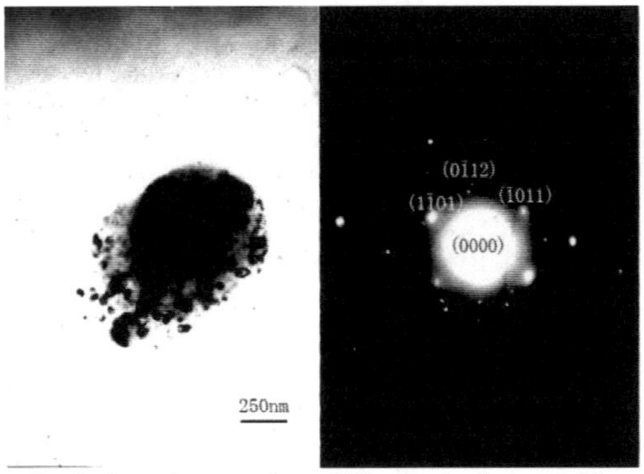

Figure 10.TEM micrographs depicting the results of Cr_2O_3 and crystallographic phase-matching.

4. Conclusions

The Fe–Al/Cr_3C_2 composite coating was mechanically bonded with its substrate. The compound phases in the Fe–Al/ Cr_3C_2 composite coating were Fe_3Al, FeAl and Cr_3C_2. Further, we observed that the coating had a lamellar structure and that the Cr_3C_2reinforcement was uniformly distributed. The finer-grained Fe–Al/Cr_3C_2 composite coating was amorphous. The interior of the coating was composed of two types of orderly intermetallics (Fe_3Al and FeAl) and anisotropic basic phases.

$Al_2O_3$was formed using two mechanisms that can be given as follows: the oxidation of aluminium in the coating and precipitation ofAl_2O_3 from Fe_3Al and FeAl. The formation of Cr_2O_3 was observed to be beneficial for the formation of an Al_2O_3 phase with a compact structure. The grain size of the Al_2O_3 and Cr_2O_3 particles in the coating was observed to be nanometric. These two oxides acted as reinforcements in the coating, which may have improved its properties, including its corrosion–erosion and wear resistances.

References

[1] Zhang Z B,Liang X B,Xu B S,Chen Y X. Microstructure and Performance of Al-Based Amorphous and Nanocrystalline Composite Coatings Prepared by High Velocity Arc Spraying. *Rare Metal Materials and Engineering*, **2012**,5,872-876.

[2] Xiang J,Zhang X H,Ye Q,Li J L,Shen X Q. In-situ Preparation and Microwave Absorption Performances of Fe-Ni/C Composite Nanofibers[J]. *Chemical Journal of Chinese Universities*. **2014**, 7,1379-1387.

[3] Yang S F,Zhang Y,Cheng J L,Zhu C,Chen W P. Microstructure and Properties of Al0.4FeCrNiCo1.5Ti0.3 High Entropy Alloy Prepared by MA-HP Technique. *Rare Metal Materials and Engineering.***2014**,43,2948-2952.

[4] Song Z Y,Wang Y,Zhang J,Sun X L,Zhao S S,Song D. Preparation and Characterization of Al/Cr_2O_3/ETN Nanocomposite Energetic Materials. *Science Technology and Engineering.***2016**, 16,62-65.

[5] Xu B S,Dong S Y,Shi P J. States and Prospects of China Characterised Quality Guarantee Technology System for Remanufactured Parts. *Journal of Mechanical Engineering.***2013**,49,84-89.

[6] Yu Y C,Chen W Q,Zheng H G. Research on the Hot Ductility of Fe-36Ni Invar Alloy.*Rare Metal Materials and Engineering.***2014**,43,2969-2973.

[7] Qin Y J,Wu Y P,Zhang J F,Guo W M,Hong S,Chen L Y. Long-term corrosion behavior of HVOF sprayed FeCrSiBMn amorphous/nanocrystalline coating. *Transactions of Nonferrous Metals Society of China*. **2015**, 25,1144-1150.

[8] Wang H X,Mao X Y,Shen T. Effect of nano-SiC particles on properties of Ni-SiC micro-composite coatings. *The Chinese Journal of Nonferrous Metals.***2015**, 25,1560-1566

[9] Wu Y M,Zhao J,Chen X M,Yang X,Mao P Z,Zhao P. Microstructure and properties of high velocity sprayed nano WC composite coatings and chromium platings. *Transactions of Materials and Heat Treatment*. 2015, 36,171-176

[10] Lin C C,Kong M G,Zhu H Y,Huang L P,Zheng X B,Zeng Y. Tribological Behavior of Vacuum Plasma Sprayed B4C-Mo Composite Coating. *Journal of Inorganic Materials*. **2016**, 31,100-106

Study on film resistivity of Energy Conversion Components for MEMS Initiating Explosive Device

Wei Ren[1, 2], Bin Zhang[2], Yulong Zhao[1], Enyi Chu [2], Ming Yin[2], Hui Li [2], Kexuan Wang[2]

[1] The State Key Laboratory for Manufacturing Systems Engineering, Xi'an Jiaotong University, Xi'an, 710049, China;
[2] Science and Technology on Applied Physical Chemistry Laboratory, Shaanxi Applied Physics-Chemistry Research Institute, Xi'an, 710061, China
E-mail address: rw0192@163.com (Ren Wei)

Abstract: Resistivity of Plane-film Energy Conversion Components is a key parameter to influence its resistance and explosive performance, and also it has important relations with the preparation of thin film technology, scale, structure and etc. In order to improve the design of Energy Conversion Components for MEMS Initiating Explosive Device, and reduce the design deviation of Energy Conversion Components in microscale, guarantee the design resistance and ignition performance of MEMS Initiating Explosive Device, this paper theoretically analyzed the influence factors of film resistivity in microscale, through the preparation of Al film and Ni-Cr film at different thickness with micro/nano, then obtain the film resistivity parameter of the typical metal under different thickness, and reveals the effect rule of the scale to the resistivity in microscale, at the same time we obtain the corresponding inflection point data.

Key Words: MEMS energy conversion component; Scale effect; resistivity

1 Introduction

MEMS initiating explosive device which base on the MEMS technology or compatible with MEMS process, it has some size characteristics: the structure of energy conversion components and the scale of energy materials in micro, the scale of core device in the sub-millimeter level, and the scale of system in millimeter level. This kind of initiating explosive device is integrated with information control, safe and actuator and detonating unit, bring it has the characteristics of information, small and multi-function, and it's a key basic technology of informatization, smart ammunition[1-3]. Energy conversion components as the core device in the MEMS initiating explosive device, which realize the energy conversion function, energy conversion components generally are the surface resistance that made of metal thin film material, its realize the heat transfer to energy material and also further amplify the energy.

Lower ignition energy and the high energy conversion efficiency are the main characteristics of ideal energy conversion components of MEMS initiating explosive device, and it's mainly affected by resistance and bridge structure of energy conversion components[4]. Therefore, we must pay more attention to the ignition resistance and bridge structure when design MEMS energy conversion components.

At present, the research mainly focused on the influence of substrate[5-6] and bridge structure[7-9] on energy conversion components performance, but the influence research of ignition resistance is less[10], and it's a matter with material and its micro structure scale. Thus, this paper theoretical analyzed the influence factors of thin film energy conversion components, prepared the different material and different thickness metal film by magnetron sputtering technology, through analyzed their morphology and electrical parameters, we obtained the basic data and test results, and verified the micro-nano scale effect rule of the resistivity, which would provide important support to optimal design and preparation of MEMS energy conversion components.

2 Theoretical analysis

Material resistivity is a physical quantity that characterization of material resistance properties, which is the electrical properties of the material itself. The product of bulk resistivity of metal materials and the mean free path of a electron, and this has no relationship with geometric dimensions of materials. With the development of micro/nano technology, geometric dimensions or microstructure scale of metal materials in the energy conversion components of initiating explosive device decreased gradually from the macro scale to micron, sub-micron or even nanometer level. And the resistivity at the room temperature tends to show obvious size effect, namely when the structure of the metal materials reaches sub-micron or nanometer, its resistivity increased dramatically. If we still use the bulk resistivity of metal materials for the design of MEMS energy conversion components, it would lead to large deviation between design resistance and test resistance, which would affect the design resistance accuracy of the MEMS energy conversion components.

According to the metal material resistivity calculation theory, Fuch-Sondheimer(F-S)[11-12], Mayadas-Shatzke(M-S)[13] model gives a detailed explanation. Among them, the calculation formula is given by the F-S theory, which mainly combined with the statistical distribution of the mean free path (λ),

$$\frac{\rho_0}{\rho_{FS}} = \frac{k}{\varphi_P(k)} \tag{1}$$

$$\frac{1}{\varphi_P(k)} = \frac{1}{k} - \frac{3}{2k^2}(1-P)\int_1^\infty \left(\frac{1}{t^3} - \frac{1}{t^5}\right)\frac{1-e^{-kt}}{1-Pe^{-kt}}dt \tag{2}$$

Where $k = t/\lambda_0$, and λ_0 is the mean free path, and t is the thickness of metal film, and ρ_0 is the bulk resistivity, and P is thin film surface reflect electronic score, and its value between 0 and 1, and P=0 for the diffuse the situation, and P=1 for all of specular reflection.

Another calculation formula is given by the M-S theory, which mainly combined with materials internal microstructure parameters, especially grain boundary in the process of film deposition and the average grain size of the difference, then put forward with the reflection coefficient R to show the effect of grain boundaries on the resistivity, namely

$$\frac{\rho_0}{\rho_{MS}} = \left[1 - \frac{2}{3}\alpha + 3\alpha^2 - 3\alpha^2 \ln\left(1 + \frac{1}{\alpha}\right)\right]^{-1} \quad (3)$$

$$\alpha = \frac{\lambda_0}{D}\left(\frac{R}{1-R}\right) \quad (4)$$

Where λ_0 is the mean free path as defined above, and D is the average grain size, and R is the reflection coefficient of grain boundary, and its value between 0 and 1, and R=0 for completely mirror reflection and R=1 for the diffuse the situation. For the block metal, $R_{Al} = 0.17$.

All the above theory, show that the resistivity of metal film is higher than the corresponding bulk materials, the main reason is the role of film surface and grain boundary scattering. Metal film showed obvious scale effect when they reach micron/nanometer lever: the smaller film thickness, as its average grain size decrease, the film resistivity greatly increase. When metal film thickness is less than a certain critical size (usually within 100 nm), the increase of resistivity is mainly due to the interface and grain boundary scattering at the same time; and when metal film thickness is greater than a certain critical size, the interface for free electron scattering effect is abate and the increase of resistivity is mainly due to the grain boundary scattering.

The range of thin thickness for energy conversion components of MEMS initiating explosive device is usually in 0.2~2.0μm, so in this paper we focus on the effects of grain boundaries on the film resistivity. The effect of grain boundaries on the resistivity is mainly the increase of membrane residual tensile stress, and membrane residual tensile stress is closely related to the 111 orientation[14]. For the surface energy, the close surface's (111) is lowest; for the strain energy, 111 orientation of grain in the strain energy density is the largest. That's to say, with the increase of the residual tensile stress, the membrane (111) orientation was enhanced. At this point, the strain energy of thin film can gather more, and it caused grain deformation, which increased the distortion degree of grain boundary. Hence, the increase of film resistivity can be characterized by grain boundary distortion (111 orientation).

3 The experiment and test

We use 76mm*25mm*1mm glass as base layer in the process of preparing metal film, and use polyimide tape to keep out in the middle of the glass before sputter deposition. Then, we remove the tape after sample preparation to keep the sample step. Last, we test the thickness and electrical of the sample. We ignored the substrate impact on the resistivity of metal film with the same substrate.

Using DISCOVERY635 magnetron sputtering machine that made by Denton companies in the United States for film preparation, and using Al target (purity 99.99%). According to the early stage on the process, we determine the best sputtering process of Al that is 0.5A/20SCCM, and thickness range from 0.2μm to 2.0μm with 0.2μm gradient. Also, we do the same research for metal material such Cr、W、Ni-Cr, which are the most frequently used material for energy conversion components.

3.1 characterization of surface topography

Frist of all, using the SEM for grain size and uniformity test of the different thickness Al films, and partial results obtained are shown in Figure 1. Through the comparative analysis shows that within the scope of the research method to the scale, with the film thickness smaller, the grain size to be smaller, and the film surface smoother.

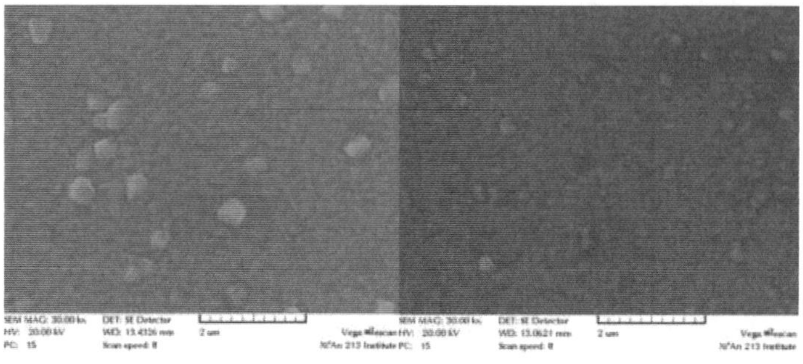

（a）200 nm thick （b）600 nm thick

Figure 1. Test result of SEM of Al film

Then, using the AFM for surface roughness test of the different thickness Al films, and partial results obtained are shown in Figure 2. And the Al surface roughness(R_q) is 15.2nm and 33.5nm, which show that thin film surface consistency control is better.

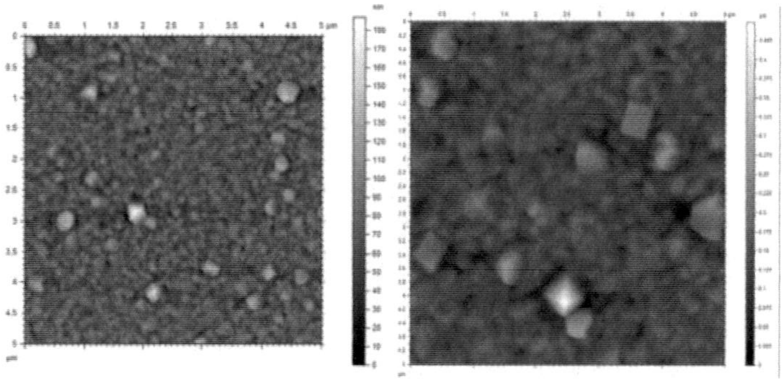

（a）200 nm thick （b）600 nm thick

Figure 2. Test result of AFM of Al film

Last, using X-ray diffraction analysis the Al thin film, and its diffraction pattern shown in Figure 3. Compared with standard atlas of Al, the results show that Al mainly tetragonal crystal system exists in the film, and Al_2O_3 was not detected, and it may exist in less amorphous or polycrystalline state. Al (111) orientation in the film are significantly higher than the strength of the standard atlas of Al (111) orientation, namely strain energy accumulation is more in the film, and increasing the grain boundary distortion degree to lead to increase grain boundary scattering of electrons, which made the film resistivity increases exceedingly.

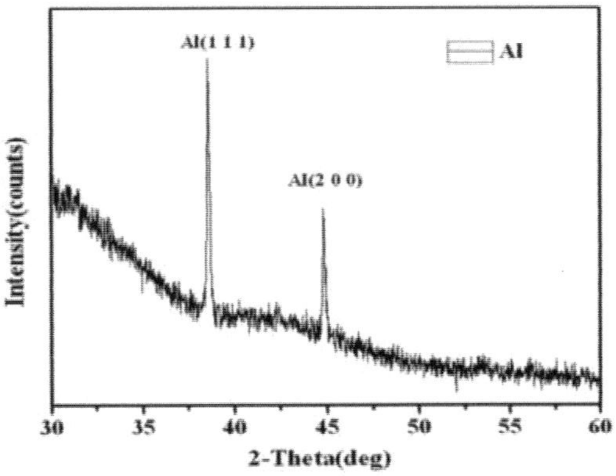

Figure 3. Resistivity fitting curve of Al film under the different thickness

3.2 electrical performance analysis

Upon completion of the metal film grain size, surface roughness and XRD diffraction spectrum analysis, using the steps apparatus to measure the thickness for the different thickness of Al film, and taking three times the average of the measurement for thin film thickness. Then, using the four point probe to measure the square resistance for the different thickness of Al film, and the results as shown in Figure4, which is the square resistance and resistivity parameter under different thickness of Al films.

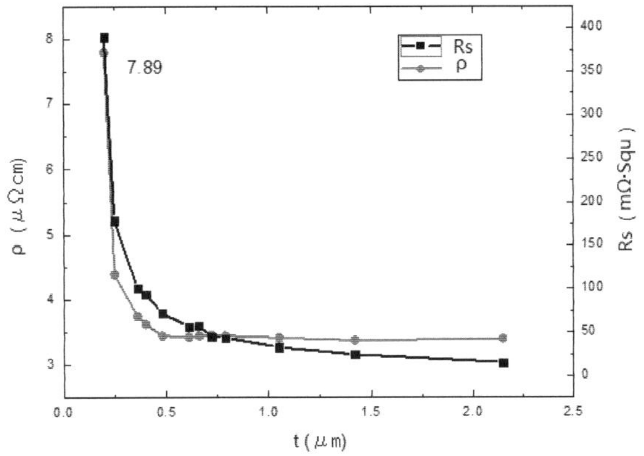

Figure 4. Square resistance and resistivity of Al film with different thickness

Figure 4 is the square resistance and resistivity test curve under different thickness of Al films. The figure shows that the square resistance (R_s) and the change trend of resistivity (ρ) is consistent, namely decreasing as the film thickness increase, and gradually become metal Al volume resistivity (2.75μΩ·cm). When the thickness is less than 300nm, the films resistivity begin to steep, and up to 7.89μΩ·cm in 200nm with an increase of nearly 200%. Then, we had the Boltzmann fitting for the test of Al film resistivity data, and obtained the relation equation between the film resistivity (ρ) of Al film and film thickness (t) in a specific scale range (0.2~2.0μm), and fitting degree R^2 is 0.99188. The fitting curve is shown in Figure5.

$$\rho = A_2 + (A_1 - A_2) * (1 + e^{-\frac{t-t_0}{dt}})^{-1} \tag{5}$$

Where is A_1=2647.97615, A_2=3.47471, t_0=-0.01142, $dt = 0.03295$.

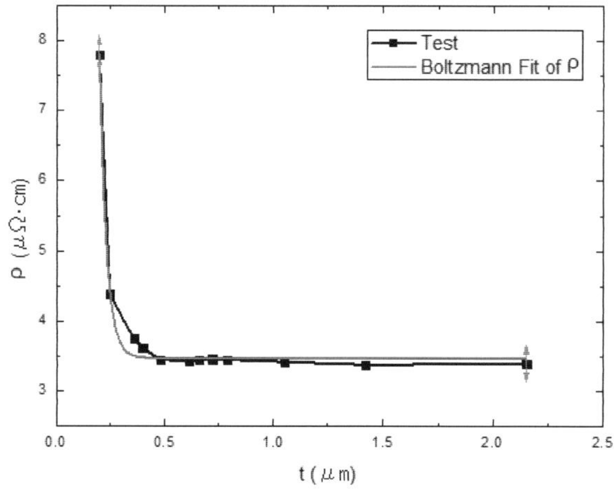

Figure 5. Resistivity fitting curve of Al film with different thickness

Similarly, we do the same work for the different thickness of the Cr, W, and Ni - Cr film, and the data obtained as shown in Figure 6, Figure 7, Figure 8.

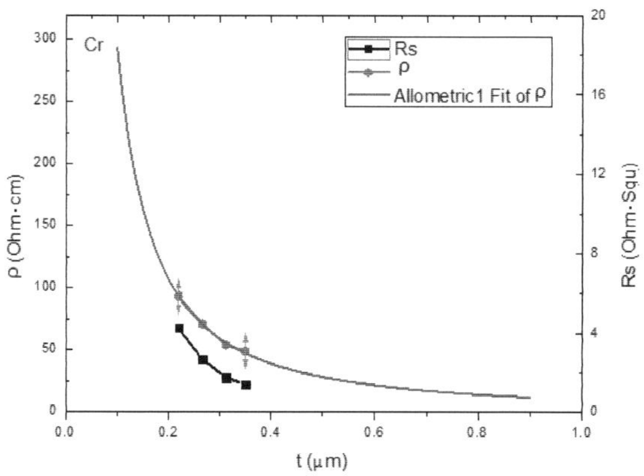

Figure 6. Square resistance and resistivity of Cr film with different thickness

Figure 6 is the square resistance, resistivity test curve and fitting curve under different thickness of Cr film. The figure shows that the square resistance (R_s) and the change trend of resistivity (ρ) is consistent, namely decreasing as the film thickness increase, and gradually become metal Cr volume resistivity (12.9μΩ·cm). Because Cr film data collection point is relatively small, and fitting R^2 based on the Boltzmann fitting is poorer, so choosing the Allometric function to fitting for Cr film resistivity, and fitting degree R^2 is 0.99328. Fitting equation is

$$\rho = a * t^b \tag{6}$$

Where is a=10.33677, b=-1.45296.

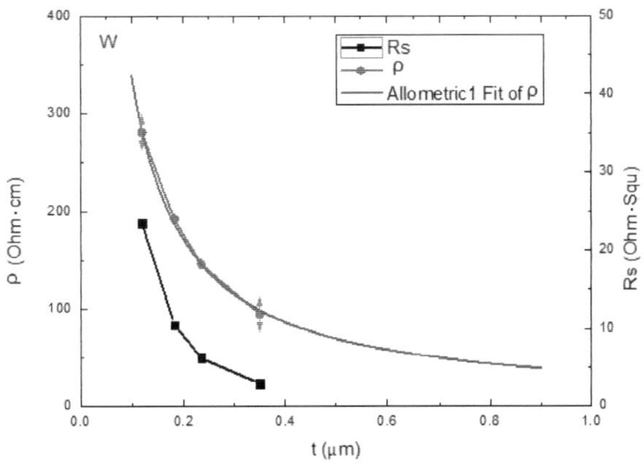

Figure 7. Square resistance and resistivity of W film with different thickness

Figure 7. is the square resistance, resistivity test curve and fitting curve under different thickness of W film. The figure shows that the square resistance (R_s) and the change trend of resistivity (ρ) is consistent, namely decreasing as the film thickness increase, and gradually become metal Cr volume resistivity (5.5μΩ·cm). Choosing the Allometric function to fitting for Cr film resistivity, and fitting degree R^2 is 0.99644. Fitting equation is

$$\rho = a * t^b \tag{7}$$

Where is a=35.14768, b=-0.98384.

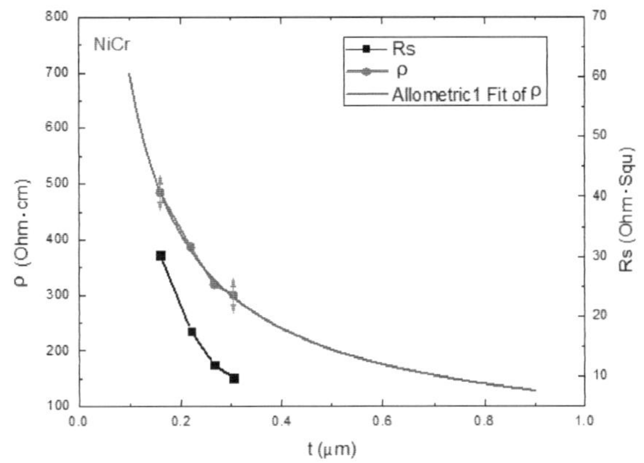

Figure 8. Square resistance and resistivity of W film with different thickness

Figure 8 is the square resistance, resistivity test curve and fitting curve under different thickness of Ni-Cr film. The figure shows that the square resistance (R_s) and the change trend of resistivity (ρ) is consistent, namely decreasing as the film thickness increase, and gradually become metal Ni-Cr volume resistivity (90μΩ·cm). Choosing the Allometric function to fitting for Ni-Cr film resistivity, and fitting degree R^2 is 0.99013. Fitting equation is

$$\rho = a * t^b \tag{8}$$

Where is a=118.83507, b=-0.76887.

From the above results, we known that the metal film thickness less than 1μm, and the film resistivity as rapidly increases with the decrease of the thickness, when the thickness of the metal film is greater than 1μm, the film resistivity tends gradually metal volume resistivity. When the metal started to deposit, the metal existed as main island shaped structure, and at this point, the conductive mechanism of thermionic emission tunnel movement, so the film resistivity is greater than volume resistivity. With the increase of film thickness, metal film slowly by island structure into a mesh structure, and at this point, the electrons through the conductive path and form seepage conductive, which increases greatly the conductivity of the metal film, so the resistivity decreases sharply as the film thickness increases. When the metal form a continuous film, it's mainly presents the nature of the metal, and the film resistivity gradually tends to the metal volume resistivity with the increase of film thickness. The main reason is that with the increase of film thickness, film grain increase, which decrease the grain boundary electron scattering, thus the resistivity reduce. The film resistivity will always be greater than the metal volume resistivity, because the film structure than the block of metal structure is loose, and has many defects or containing impurities.

Base on the study results, we design and manufacture MEMS energy conversion components shown in Figure 9, which with plane-film structure and use the Ni-Cr material. The resistance value shown in Tab.1, its deviation less than 0.35Ω, and indicate that we can accurately control the MEMS energy conversion components resistance value design.

V-100 V-50 F-150 S-100 L-1 Snake-1

Figure 9. Samples of energy conversion components

Table 1. design and Test resistance of the typical MEMS energy conversion components

Structure	V-50	V-100	F-150	S-100	L-1	Snake-1
Design resistance/Ω	6.5	5.5	5	9	8.5	31
Test resistance Average/Ω	6.63	5.6	5.28	9.22	8.68	31.32
Deviation	0.13	0.1	0.28	0.22	0.18	0.32
Variance	0.296	0.293	0.255	0.331	0.071	0.343

4 Conclusion and perspectives

The film resistivity of energy conversion components of MEMS initiating explosive device is the key parameter to influence the resistance performance and combustion, and it has important relationship with the preparation of thin film technology, and scale, and material and etc. The paper theoretical analyzed influence factors of metal film resistivity, and tested grain size and the electrical performance for Al、Cr、W and Ni-Cr under the micro-scale, and obtained the scale effect of partial metal film.

Metal film resistivity is higher than the corresponding bulk materials, and the main reason is that a combined action of thin film surface and grain boundary scattering. When the thickness of Al is less than 300nm, the films resistivity begin to steep, and up to 7.89μΩ·cm in 200nm. When the thickness of Cr、W and Ni-Cr is less than 1μm, the increase trend of film resistivity is more apparent with the decrease of film thickness. With the further decrease of the material microstructural scale and geometrical dimensions, it's a major challenge for the basic theory of initiating explosive device ender microscale, and it needs to think a lot about film resistivity scale effect.

References

[1] Yang W, Liu Y, Xu L, Leng Y and Huang N 2005 Review of film growth by sputtering technology. *Chinese Journal of Vacuum Science & Technology, 25*(3), 204-210.

[2] KONG J F and LI B 2010 A new generation of initiation technique and application. *Technology Foundation of National Defence*, (7):40-43.

[3] SHEN R Q, YE Y H and DAI S Z 2000 Conception and Application of Digital Initiation Technique[J]. *Initiators & Pyrotechnics*, (2):37-39.

[4] Hu Y, Ye Y H, Shen R Q and Dong N F 2010 Preparation and characterization of Al-CuO reactive bridge film. *Chinese Journal of Energetic Materials, 18*(3), 339-342.

[5] Zhang K L, Chou S K and Ang S S 2007 Fabrication, modeling and testing of a thin film Au/Ti microheater. *International Journal of Thermal Sciences, 46*(6), 580-588.

[6] Chopin H 2010 Low-cost MEMS initiators and Micro Assembly Technologies. *NDIA 54th Annual Fuze Conference*.

[7] Neha Mehta 2003 Chemical Engineer, Design and development of micro energetic initiators. *NDIA 47th Annual Fuze Conference*.

[8] Xie R Z, Xue Y Ren X M, Liu L and Peng Z M 2012 Effect of bridge parameters on the firing sensitivity of Ni-Cr alloy thin film transducer elements. *Initiators & Pyrotechnics*.

[9] Lee J and Kim T 2013 Mems solid propellant thruster array with micro membrane igniter. *Sensors & Actuators A Physical, 190*(2), 52-60.

[10] Zhang B, Chu E Y, Ren W, Wang K X, Li H and Yin M 2017 Research progress in energy conversion components for mems initiating explosive device. *Hanneng Cailiao/chinese Journal of Energetic Materials, 25*(5), 428-436.

[11] Fuchs K and Mott N F 1938 The conductivity of thin metallic films according to the electron theory of metals. *Mathematical Proceedings of the Cambridge Philosophical Society, 34*(1), 100-108.

[12] E H Sondheimer 1952 The mean free path of electrons in metals. *Advances in Physics, 50*(1), 499-537.

[13] Mayadas A F and Shatzkes M 1970 Electrical-resistivity model for polycrystalline films: the case of arbitrary reflection at external surfaces. *Physical Review B, 1*(4), 1382-1389.

[14] Wu T, Deng L, Xu K and Lu J 2008 Relationship between resistivity of metallic film and its surface roughness, residual stress. *Rare Metal Materials & Engineering, 37*(4), 617-620.

IOP Conf. Series: Journal of Physics: Conf. Series **986** (2018) 012005 doi:10.1088/1742-6596/986/1/012005

Design and Characterization of a Novel Bio-inspired Hair Flow Sensor Based on Resonant Sensing

X Guo [1,2], B Yang [1,2], Q H Wang [1,2], C F Lu [1,2], and D Hu [1]

[1]School of Instrument Science & Engineering, Southeast University, Nanjing 210096, China
[2]Key Laboratory of Micro-Inertial Instrument and Advanced Navigation Technology, Ministry of Education, Nanjing 210096, China

Correspondence: 101011019@seu.edu.cn

Abstract. Flow sensors inspired by the natural hair sensing mechanism have great prospect in the research of micro-autonomous system and technology (MAST) for the three-dimensional structure characteristics with high spatial and quality utilization. A novel bio-inspired hair flow sensor (BHFS) based on resonant sensing with a unique asymmetric design is presented in this paper. A hair transducer and a signal detector which is constituted of a two-stage micro-leverage mechanism and two symmetrical resonators (double ended tuning fork, DETF) are adopted to realize the high sensitivity to air flow. The sensitivity of the proposed BHFS is improved significantly than the published ones due to the high sensitivity of resonators and the higher amplification factor possessed by the two-stage micro-leverage mechanism. The standard deep dry silicon on glass (DDSOG) process is chosen to fabricate the proposed BHFS. The experiment result demonstrates that the fabricated BHFS has a mechanical sensitivity of 5.26 Hz/(m/s)^2 at a resonant frequency of 22 kHz with the hair height of 6 mm.

1. Introduction

Nature has gradually become a significant source of inspiration for engineering since it has evolved to make the optimal design. Therefore, many innovative solutions can be inspired by imitating natural engineering designs. A large quantity of natural creatures such as arthropods, aquatics and mammals are equipped with extraordinary sensitive and robust hair structures to help them assess the surrounding environment parameters and rapidly changing inertia parameters for their survival and locomotion [1]. The bio-inspired hair flow sensor (BHFS) based on natural hair structures is a potential biomimetic flow sensor combining the micro-electro-mechanical system (MEMS) technology.

Based on the sensing principle, a few published bio-inspired hair flow sensors can be classified into several categories, namely, piezo-resistive, capacitive, piezo-electric and optical sensors [2]. Hair flow sensors based on the lateral line perception system of aquatics and piezo-resistive detection have been published in the literature [3]. Reference [4] presented a BHFS with a capacitance signal detector and a micro-hydraulic amplification mechanism. A hair flow sensor based on piezo-electric transductor with an array of micro-pillars was developed for monitoring turbulent flow [5]. Reference [6] developed a BHFS adopted a photo-electronic sensing mechanism for dual-axis air flow and dynamic pressure detecting.

Compared with the traditional sensing principles, there is a great potential for improving the sensitivity of BHFS based on the principle of resonant sensing since the resonators are highly sensitive, highly stable, and able to output quasi-digital signal. In this paper, a novel bio-inspired hair flow sensor based on resonant sensing with a unique asymmetric design is presented to achieve a better performance on the sensitivity, structural stability and durability.

2. Device Description and Simulation analysis

2.1. Device design

The structure scheme of proposed bio-inspired hair sensor is shown in Figure 1. The device adopts a unique asymmetric design. A hair transducer is attached to a planar silicon micro-detector which is constituted of a two-stage micro-leverage mechanism and two symmetrical resonators. Then the signal detector is bonded to a Pyrex glass wafer by bonding anchors. A novel lever pivot which is constituted of a rotational anchor and eight rotational beams is adopted to ensuring the rotational movements of main frame in the plane, and four groups of swing suppression beams are adopted to suppress structural swing of main frame for accuracy and device safety. The drag force derived from the external fluid interact with the hair transducer is amplified by the two-stage micro-leverage mechanism and applied to the resonators to induce the natural frequency drifts of resonators. Then, the input air flow velocity can be calculated by detecting the shifts of natural frequencies of resonators.

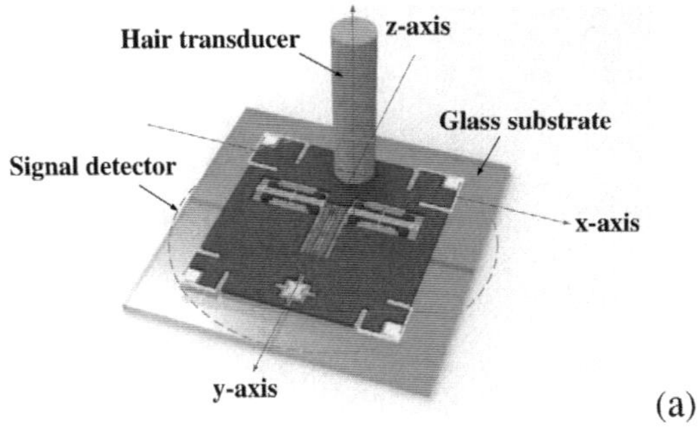

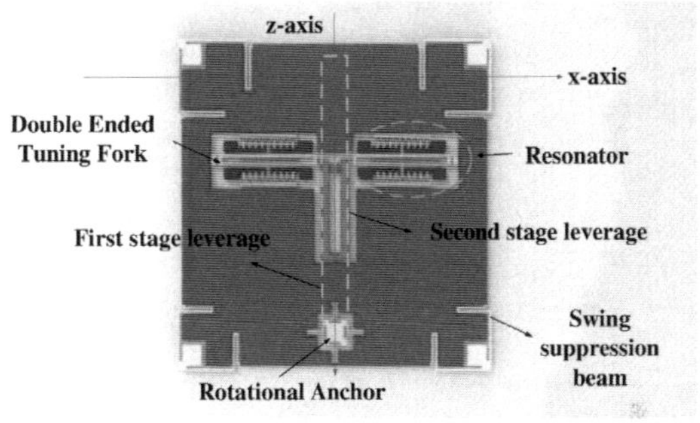

Figure 1. Schematic of bio-inspired hair flow sensor. (a) Three layers; (b) Signal detector.

2.2. Theoretical description

Hypothesize the flow velocity around the hair post is approximately constant. The drag force caused by air flow exerted on the hair post can be expressed as [7]

$$F_f = \frac{1}{2} C_D \rho D L_H u^2 \tag{1}$$

where u is the flow velocity, C_D and ρ are the drag coefficient and density of the specified fluid, D and L_H are the diameter and length of the hair post. The moment applied to the hair post can be simplified as

$$M_f = \frac{1}{4} C_D \rho D u^2 L_H{}^2 \tag{2}$$

The theoretical mechanical sensitivity of the proposed BHFS for air flow velocity measurement can be approximately expressed as

$$S_f = \frac{df}{du^2} \approx \frac{\lambda L^2 \eta A C_D \rho D L_H f_0}{2Ehw^3} \tag{3}$$

IOP Conf. Series: Journal of Physics: Conf. Series **986** (2018) 012005 doi:10.1088/1742-6596/986/1/012005

where λ is the attenuation coefficient of resonator, f_0 is the natural frequency of resonator without axial force F_f, h and w are the thickness and width of resonant beam, L is the length of coupling combs. The frequency variation is theoretically linear with the square of flow velocity.

2.3. Simulation and Optimization

The natural frequency and natural modes of proposed BHFS can be analyzed by modal analysis. The vibration characteristics determine the response of structure to various input loads. A modal simulation is implemented by commercially available Finite Element Method (FEM) program ANSYS to determine the geometric properties. The first twelve mode frequencies are shown in Table 1 and the three effective modes are shown in Figure 2. The simulation results comfirm that the effective modes and the interference modes are effectively isolated, the feasibility of structural design is verified.

Table 1. The first twelve mode frequencies of bio-inspired hair flow sensor.

Modal	1	2	3	4	5	6
Frequency (Hz)	471	565	1364	1624	11251	12068
Modal	7	8	9	10	11	12
Frequency (Hz)	15257	18892	18905	19889	19891	20963

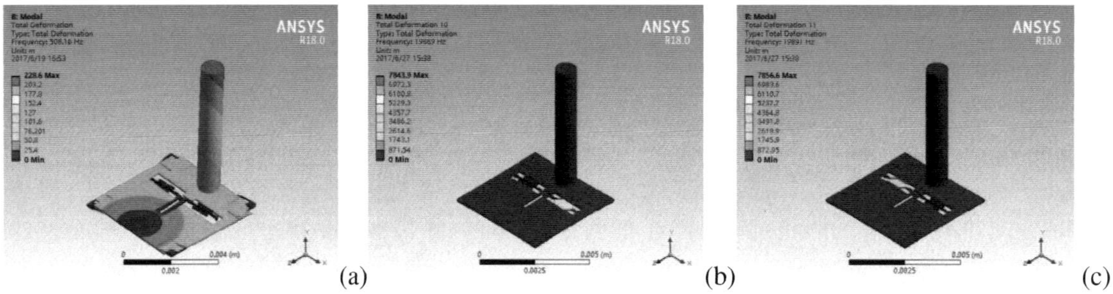

Figure 2. Effective modes of bio-inspired hair flow sensor. (a) Sensing mode; (b) Resonator working mode 1; (c) Resonator working mode 2.

A fluid-structure interaction simulation is implemented to evaluate the performance of the proposed BHFS. Figure 3 shows the simulation of structural deformation at an input air flow of 10 m/s. The sensitivities of proposed BHFS are 3.48 Hz/(m/s)^2 in height of 3 mm and 5.88 Hz/(m/s)^2 in height of 6 mm and 8.11 Hz/(m/s)^2 in height of 9 mm, respectively. The simulation result demonstrates that the increase of the hair height will improve the sensitivity which is consistent with the theoretical analysis since the drag force is positively related to the surface area of hair post.

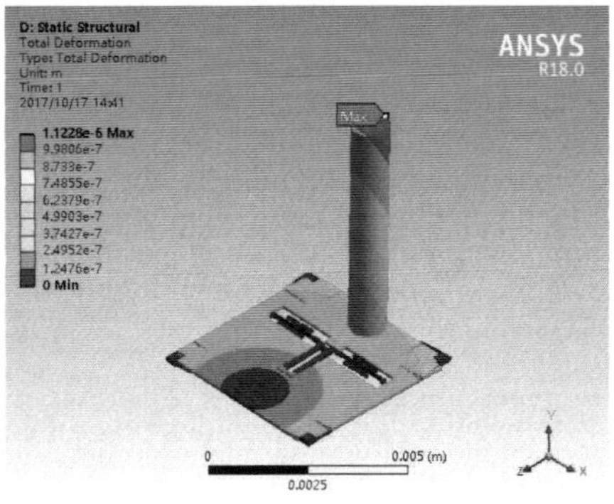

Figure 3. Fluid-structure interaction simulation of bio-inspired hair flow sensor.

IOP Conf. Series: Journal of Physics: Conf. Series **986** (2018) 012005 doi:10.1088/1742-6596/986/1/012005

3. Fabrication and Experiments

3.1. Fabrication
The standard deep dry silicon on glass (DDSOG) is chosen to fabricate the novel BHFS. As shown in Figure 4, the DDSOG process starts with a 200 μm thick monocrystalline and a 500 μm thick Pyrex glass substrate. The photoresist is patterned and etched via the deep reactive ion etching (DRIE) to define the bonding anchors. Then a Cr/Ti/Au stack layer is sputtered to define the electrode wires and pads in the Pyrex glass wafer. The silicon wafer and the Pyrex glass wafer are connected via an electrostatic anodic bonding process. A wet etching process is adopted to decrease the thickness of the silicon wafer. Subsequently, the silicon wafer is lithographically patterned and etched to release the mechanical structures via the DRIE. Finally, a micro-assembly process is adopted to install the metal hair. The optical micrograph of BHFS is shown in Figure 5. The fabricated bio-inspired hair flow sensor has a size of 5100 μm (length) × 5100 μm (width) × 100 μm (height) with a hair diameter of 1 mm and a height 6 mm.

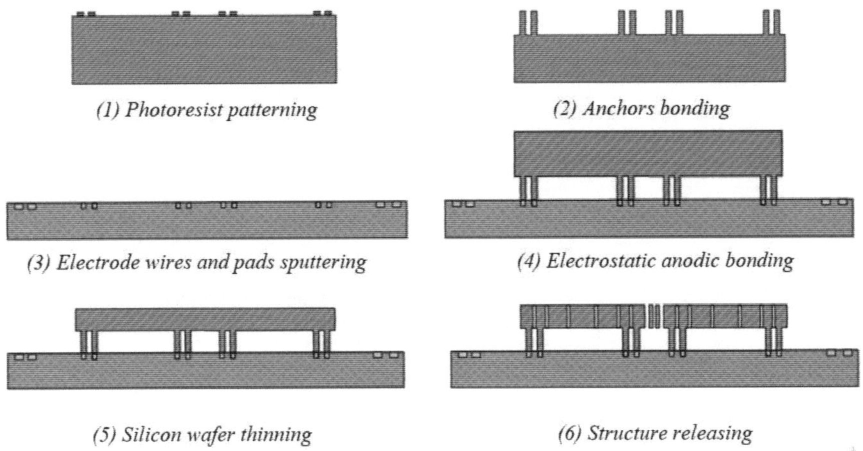

Figure 4. DDSOG process flow of BHFS.

Figure 5. Optical micrograph of BHFS

3.2. Performance experiments
After the hair post assembly and encapsulation, the performance of BHFS was preliminarily tested. DC power supply, spectrum analyzer, digital multimeter, DC brushless fan and wind speed sensor are

used to evaluate the electromechanical characterization of the fabricated prototypes. The amplitude-frequency response curves of DETFs are obtained by the spectrum analyzer, the subsequent experiments demonstrate that the symmetrical resonators have natural resonant frequencies of 22.280 kHz and 21.960 kHz with quality factors of 40.5 and 40.7, respectively. The quality factors of two resonators are smaller than conventional ones due to the large damping coefficient in the air.

The air flow measurement can be implemented by using the anemometer to calibrate the wind speed produced by the DC brushless fan. The output frequency difference of prototype is shown in figure 6(a). Figure 6(b) demonstrates the linearity of the square of input air flow rate and the output frequency difference, which is also consistent with the previous theoretical analysis. The air flow measurement result demonstrates that the fabricated prototype with a hair height of 6 mm has a mechanical sensitivity of 5.26 Hz/(m/s)^2.

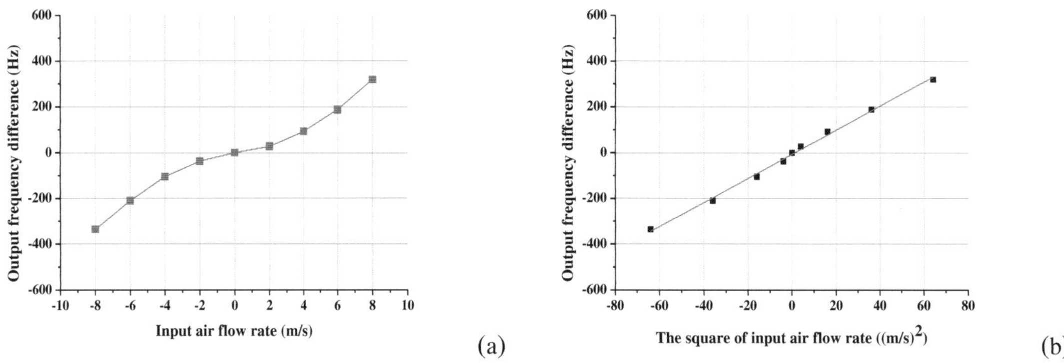

Figure 6. Relationship between input air flow and output frequency difference. (a) Output frequency difference with input air flow. (b) Output freqency difference with the square of input air flow.

4. Conclusions

This paper presents a novel bio-inspired hair flow sensor with a unique asymmetric design. The mechanical sensitivity of the proposed BHFS is improved significantly by adopting a signal detector which is constituted of a two-stage micro-leverage mechanism and two symmetrical resonators. The DDSOG process is chosen to fabricate the proposed BHFS. The fabricated bio-inspired hair flow sensor has a size of 5100 μm (length) × 5100 μm (width) × 100μm (height) with a hair diameter of 1 mm and a height 6 mm. The priliminary experiment result demonstrates that the fabricated BHFS has a mechanical sensitivity of 5.26 Hz/(m/s)^2 at a resonant frequency of 22 kHz.

5. References

[1] Rizzi F, Qualtieri A, Dattoma, T, Epifani G, and de Vittorio M 2015 Biomimetics of underwater hair cell sensing *Microelectronic Engineering* 132(c) pp 90-97

[2] Tao J and Yu X 2012 Hair flow sensors: From bio-inspiration to bio-mimicking—A review *Smart Mater. Struct.* 21(11) pp 1451-1466

[3] Chen N, Tucker C, Engel J, Yang Y, Pandya S and Liu C 2007 Design and characterization of artificial haircell sensor for flow sensing with ultrahigh velocity and angular sensitivity *J. Microelectromech. S* 16(5) pp 999-1014

[4] Sadeghi M, Peterson R and Najafi K 2011 Micro-hydraulic structure for high performance bio-mimetic air flow sensor arrays *In Proceedings of 2011 IEEE International Electron Devices Meeting* 47(10) pp 29.4.1–29.4.4

[5] Tao J, Yu X, and Berilla J 2011 Micropillar sensing element for bio-inspired flow sensors *In Proceedings of 8th Int. Workshop on Structural Health Monitoring* vol 2 pp 1732-1739

[6] Huang K Y and Huang C T 2011 Hair sensor using a photoelectronic principle for sensing airflow and its direction *Optical Engineering* 50(1) pp 123-128

Acknowledgments

This work is supported by the National Natural Science Foundation of China (Grant No. 61571126), the Equipment pre-research field foundation (Grant No. 6922001217), the NSAF (Grant No. U1230114), the Aviation Science Foundation (Grant No. 20150869005), and the China Academy of Space Technology Innovation Foundation and the Eleventh Peak Talents Programme Foundation in the Six New Industry Areas.

IOP Conf. Series: Journal of Physics: Conf. Series **986** (2018) 012006 doi:10.1088/1742-6596/986/1/012006

Study on Silicon Microstructure Processing Technology Based on Porous Silicon

Yingqi Shang [1], Linchao Zhang [1], Hong Qi [1], Yalin Wu [1,2], Yan Zhang [1], Jing Chen [1]

[1] The 49th Research Institute, China Electronics Technology Group Corporation, Harbin 150001, China;

[2] School of Mechatronics Engineering, Harbin Institute of Technology, Harbin 150001, China

shangyingqi@163.com ;zhanglinchao88@gmail.com ;qihong156@gmail.com ;wuyalin@cetc.com.cn ;zhangyan1376@gmail.com ;chenjing1884@gmail.com

Abstract. Aiming at the heterogeneity of micro - sealed cavity in silicon microstructure processing technology, the technique of preparing micro - sealed cavity of porous silicon is proposed. The effects of different solutions, different substrate doping concentrations, different current densities, and different etching times on the rate, porosity, thickness and morphology of the prepared porous silicon were studied. The porous silicon was prepared by different process parameters and the prepared porous silicon was tested and analyzed. For the test results, optimize the process parameters and experiments. The experimental results show that the porous silicon can be controlled by optimizing the parameters of the etching solution and the doping concentration of the substrate, and the preparation of porous silicon with different porosity can be realized by different doping concentration, so as to realize the preparation of silicon micro-sealed cavity, to solve the sensor sensitive micro-sealed cavity structure heterogeneous problem, greatly increasing the application of the sensor.

1. Introduction

The traditional sensor seal cavity is achieved by the bonding of silicon glass, which makes the sensor sealed cavity heterogeneous structure, resulting in the sensor by stress, reduce the performance of the sensor and limit the use of the sensor environment. Therefore, it is proposed to use porous silicon as the sacrificial layer to realize the processing of silicon microstructure. Solve the heterogeneous problem of micro-sealed cavity. Porous silicon is a unique nano-silicon cluster skeleton of the "quantum sponge" or "branch" microstructure silicon-based materials, is an important micro-machining material, the tower has a large surface area, good biocompatibility, room temperature can be Luminous and refractive index adjustable features, in the micro-sensor and actuator manufacturing has been widely used. Porous silicon due to its pore structure characteristics, has a very large surface area, its surface activity is high, the hole surface adsorption of organic gases, inorganic gases and water vapor. So it can be used for sensitive materials of gas and humidity sensors. Porous silicon, although filled with porous silicon thickness, but its structure is still showing a single crystal, the use of its single crystal properties, it can grow epitaxial silicon on it. The thickness of porous silicon is larger and can be precisely controlled, and the removal rate is high, so it can be used as a sacrificial layer material in surface micromachining technology. At the same time, porous silicon is also a good insulation material, rich holes can absorb air to reduce thermal conductivity, improve the insulation performance, it can be used as a thermal sensor base material. Porous silicon was discovered as early as 1956[1], but in 1990 it made much progress in its research[2-5]. With the rapid development of MEMS theory and the rapid development of process technology, porous silicon in the field of micro-machining has shown great prospects for development. So far, the theoretical study on porous silicon has focused on the preparation methods, formation mechanism, microstructure, luminescence mechanism, biotechnology, photo catalysis and super capacitor[6]. In the field of sensors, the use of porous silicon as a sacrifice layer has great advantages. The traditional sacrificial layer material thickness is only a few microns, limiting the fabrication of many microstructures and microsensors. The formation rate of the porous silicon film is faster than the conventional material, and can reach several microns per minute and has a thickness of up to 100 μ m. The film or epitaxial silicon film can be prepared on porous silicon, which can form porous silicon within a particular region by a suitable mask, KOH at 1% concentration at room temperature can rapidly etch the porous silicon layer, and the preparation process can be compatible with the CMOS process[7,8].

2. Preparation principle of porous silicon

Preparation of stable porous silicon is the basis and key of the study. There are many ways to prepare porous silicon at home and abroad, which are mainly chemical corrosion, electrochemical corrosion, photochemical corrosion, hydrothermal corrosion, laser ablation. Chemical etching method for the preparation of porous silicon is the simplest method, which is prepared by the appropriate composition of the etching solution directly on the base monocrystalline silicon wafer for general chemical corrosion and preparation of porous silicon. However, the preparation of porous silicon in this process is difficult to control the process, the prepared porous silicon has poor homogeneity and has a limited thickness. Photochemical etching is a kind of porous silicon which reacts with the etching solution under the action of light, however, the light-assisted preparation of porous silicon requires very high wavelengths for incident light wavelength is too long to produce electron-hole pairs, Light waves too short will affect the electron-hole pair generation rate, porous silicon must be prepared using incident light of the appropriate wavelength. Hydrothermal corrosion method needs to be carried out in high pressure liquid system, and the preparation process temperature is higher. Electrochemical corrosion method has the advantages of simple operation and controllable preparation, and it is a relatively common preparation method. The use of electrochemical corrosion method to prepare porous silicon, not only to maintain the advantages of anisotropic corrosion, but also to avoid the high concentration of corrosive materials doped and high stress and other shortcomings. The reaction equation for the silicon surface is:

$$Si+2HF+(2-n)e^+ \rightarrow SiF_2+2H^++ne^- \quad (1)$$

$$2SiF_2 \rightarrow Si+SiF_4 \quad (2)$$

$$SiF_4+2HF \rightarrow H2SiF_6 \quad (3)$$

Where n <2, e + denotes a hole, and e- denotes an electron.

In the pores of the porous silicon, the reaction is carried out in two ways, respectively, for the divalent and tetravalent reactions. The bivalent reaction is:

$$Si+2HF+(2-n)e^+ \rightarrow SiF_2+2H^++ne^- \quad (4)$$

$$2SiF_2+HF \rightarrow SiF_4 \quad (5)$$

$$SiF_4+H_2+2HF \rightarrow H_2SiF_6 \quad (6)$$

The tetravalent reaction is:

$$Si+4HF+(4-n)e^+ \rightarrow SiF_4+4H^++ne^- \quad (7)$$

IOP Conf. Series: Journal of Physics: Conf. Series **986** (2018) 012006 doi:10.1088/1742-6596/986/1/012006

$$SiF_4 + 2HF \rightarrow H_2SiF_6 \tag{8}$$

3. Preparation of porous silicon

There are two factors that affect the structure of porous silicon, one is the silicon itself, including the silicon type, doping concentration and crystal orientation, the second is the process parameters, including the composition of the corrosive liquid, current density and corrosion time. In this paper, the electrochemical corrosion method was used to prepare porous silicon, and the relationship between different solutions, different types of silicon wafers, different current densities, different etching times and the rate, porosity, thickness and morphology of prepared porous silicon was studied. Experiments were carried out using a <100> crystal wafer with a thickness of 400 μ m. The preparation of porous silicon is accomplished by double-channel electrochemical etching. Experiments were carried out with different corrosion solutions (HF and H_2O mixed solution, HF and C_2H_5OH mixed solution and HF, C_2H_5OH and H_2O_2 mixed solution), different substrates (p^+ silicon and p-type silicon), different current densities (5 mA / cm^2, 10 mA / cm^2, 15 mA / cm^2, 20 mA / cm^2, 25 mA / cm^2), different etching time (40min, 80min, 90min, 100min, 140min) for porous silicon. Finally, the porous silicon was etched with 1% KOH solution, the effects of different solutions, different substrate doping concentration, different current density and different etching time on the rate, porosity, thickness and morphology of the prepared porous silicon were analyzed.

4. Results and discussion

4.1. Effect of Corrosion Solution on Microstructure, Porous Silicon Thickness and Porosity of Porous Silicon

Aiming at the preparation principle of porous silicon, the surface morphology of porous silicon and the surface state of silicon wafer after etching were studied. The main components of the corrosion solution are HF, using different mixed solution (HF and H_2O mixed solution, HF and C_2H_5OH mixed solution and HF, C_2H_5OH and H_2O_2 mixed solution) as the etching solution. Porous silicon was prepared on an undoped p-type silicon wafer using a different ratio, corrosion time is 5min, current density 10mA / cm^2. After the process is completed, the porous silicon microstructure is tested. The prepared porous silicon was removed by 1% KOH solution, the surface morphology of the silicon wafer was tested, and the porous silicon thickness and porosity were calculated. Corrosion solution ratio is shown in Table 1. Preparation of porous silicon microstructure is shown in Figure 1. The surface state of the silicon wafer after removing the porous silicon is shown in Figure 2. Corrosion solution and porous silicon thickness and porosity relationship is shown in Figure 3.

Table 1. Corrosion solution ratio.

Corrosion solution	Corrosion solution ratio				
HF: H_2O	3:1	2:1	1:1	1:2	1:3
HF: C_2H_5OH	3:1	2:1	1:1	1:2	1:3
HF: C_2H_5OH: H_2O_2	1:1:1	1:1:2	1:1:3	2:2:1	3:3:1

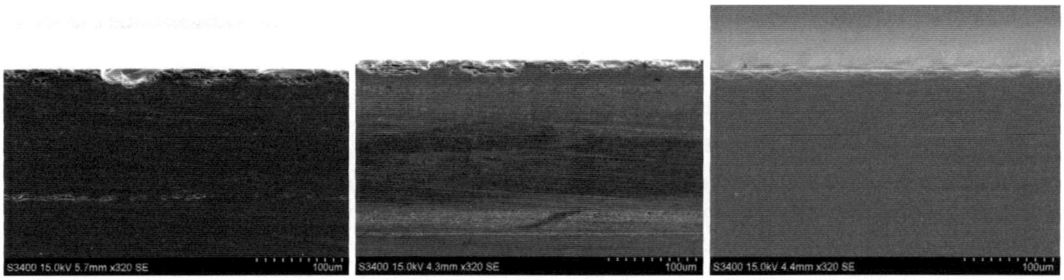

a) HF: H_2O=2:1 b) HF: C_2H_5OH=2:1 c) HF: C_2H_5OH: H_2O_2=2:1:1

Figure 1. The same ratio of different corrosive solutions prepared porous silicon microstructure.

a) HF: H_2O=2:1 b) HF: C_2H_5OH=2:1 c) HF: C_2H_5OH: H_2O_2=2:1:1

Figure 2. The surface state of the silicon after removing the porous silicon.

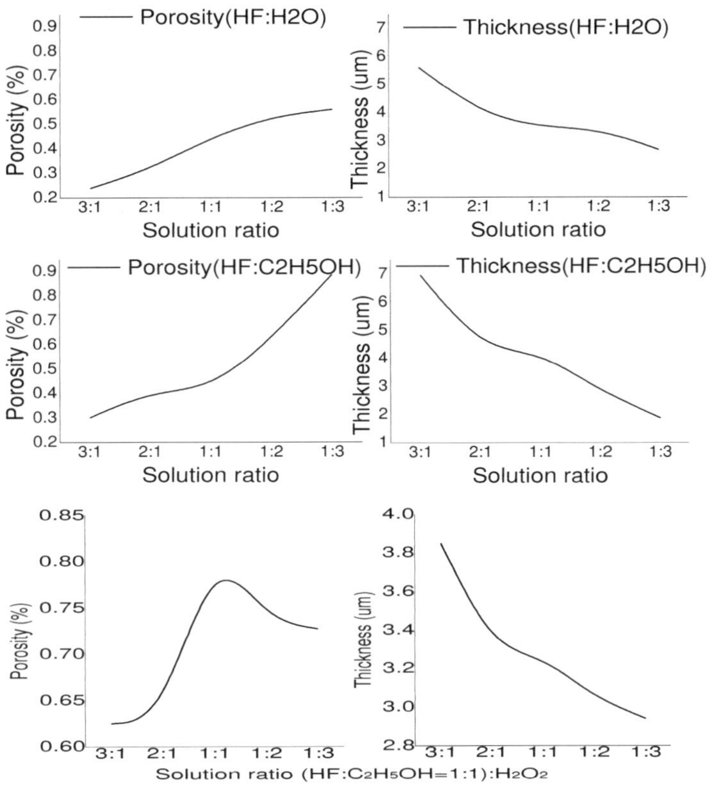

Figure 3. The relationship between the different corrosion solutions of the same ratio and the thickness and porosity of the porous silicon.

The above test results show that HF and H_2O mixed solution prepared porous silicon pore size is inconsistent, the surface is not uniform, HF, C_2H_5OH and H_2O_2 mixed solution prepared porous silicon, the silicon surface has obvious hills after remove the porous silicon, HF and C_2H_5OH mixed solution prepared the porous silicon effect is relatively good, and the silicon surface smooth after remove the porous silicon. As a result of the preparation of porous silicon from different solutions, the thickness of porous silicon decreases with the decrease of HF concentration, and the porosity increases with the decrease of HF concentration. When H_2O_2 was added to the solution, the porosity increased first and then decreased with the increase of H_2O_2 concentration.

4.2. Effect of Silicon Type on Porous Silicon Microstructure, Porous Silicon Thickness and Porosity

Experiments were carried out using the mixed solution of etchant HF and C_2H_5OH, the etching time was 10min, and the current density was 10mA / cm^2 as the preparation method of porous silicon. The preparation of porous silicon was carried out using p-type and p^+ -type silicon as substrates. After the process is completed, the microstructure of the porous silicon is tested, as shown in Figure 4. The prepared porous silicon was removed by 1% KOH solution. The surface morphology of the tested silicon wafer was shown in Figure 5, and the calculated porous silicon thickness and porosity, the relationship between the silicon type and the porous silicon thickness and porosity is shown in Figure 6.

a)p b)p^+

Figure 4. Preparation of porous silicon microstructure by different types of wafers.

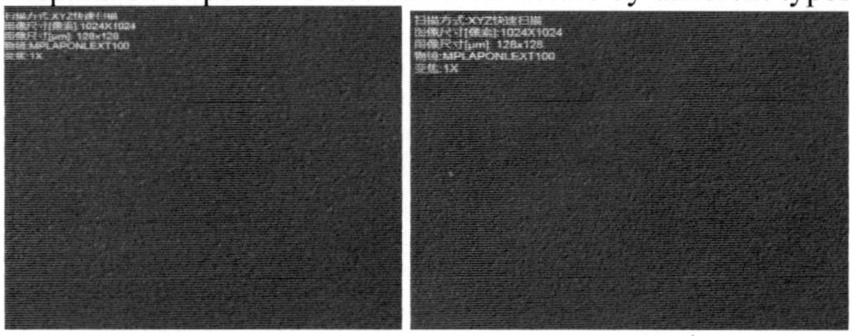

a)p b)p^+

Figure 5. The surface state of the silicon after removing the porous silicon.

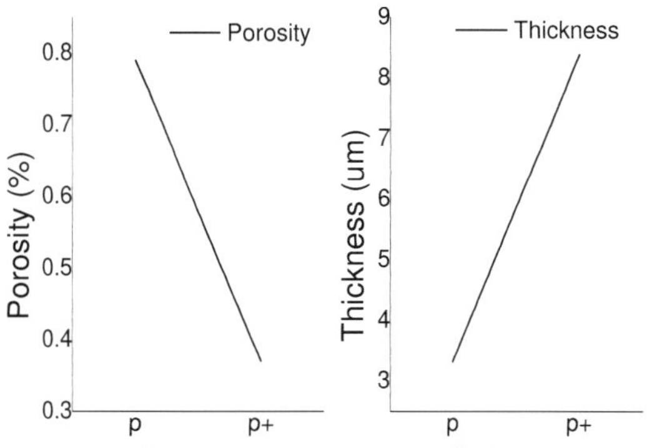

Figure 6. The relationship between different types of silicon and porous silicon thickness and porosity.

The above test results show that the porous silicon of p^+ -type silicon is uniform. Compared with p^- type silicon, porous silicon is prepared by p^+-type silicon that porosity is low, the thickness of porous silicon is thick.

4.3. Influence of Current Density on Microstructure, Porous Silicon Thickness and Porosity of Porous Silicon

Experiments were carried out using p-type silicon wafers, the etching solution was mixed with HF and C_2H_5OH, and the etching time was 10min. The current densities were set to 5 mA / cm^2, 10 mA / cm^2, 15 mA / cm^2, 20 mA / cm^2 and 25 mA / cm^2, respectively. After the completion of the process, the microstructure of the porous silicon is shown in Figure 7. The prepared porous silicon was removed by using 1% KOH solution to test the surface morphology of the silicon wafer. As shown in Figure 8, the porous silicon thickness and porosity were calculated. The relationship between current density and porous silicon microstructure, porous silicon thickness and porosity is shown in Figure 9.

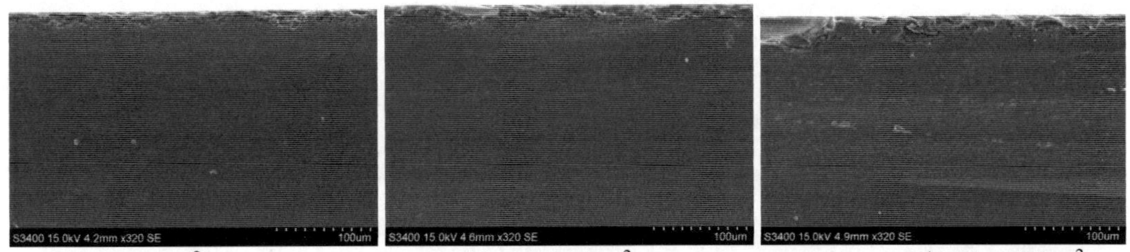

a) 5mA/cm^2 b) 15mA/cm^2 c) 25 mA/cm^2

Figure 7. Preparation of porous silicon microstructure by different current densities.

a) 5mA/cm^2 b) 15mA/cm^2 c) 25 mA/cm^2

Figure 8. The surface state of the silicon after removing the porous silicon.

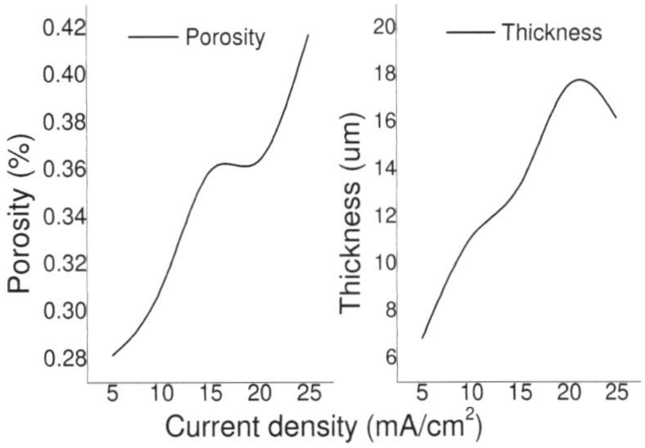

Figure 9. The relationship between different current densities and porous silicon thickness and porosity.

The above test results show that the pore size of porous silicon increases with increasing current density. The thickness and porosity of the porous silicon increased first and then decreased with the current density. This is because the current density exceeds the critical current density, the occurrence of electrochemical polishing phenomenon.

4.4. Effects of Corrosion Time on Microstructure, Porous Silicon Thickness and Porosity of Porous Silicon

Experiments were carried out using p-type silicon wafers, the etching solution was mixed with HF and C_2H_5OH, and the current density was 10min. The etching times were set to 40min, 80min, 90min, 100min, and 140min, respectively. After the completion of the process, the microstructure of the porous silicon is shown in Figure 10. The prepared porous silicon was removed by using 1% KOH solution to test the surface morphology of the silicon wafer. As shown in Figure 11, the porous silicon thickness and porosity were calculated. The relationship between etching time and porous silicon microstructure, porous silicon thickness and porosity is shown in Figure 12.

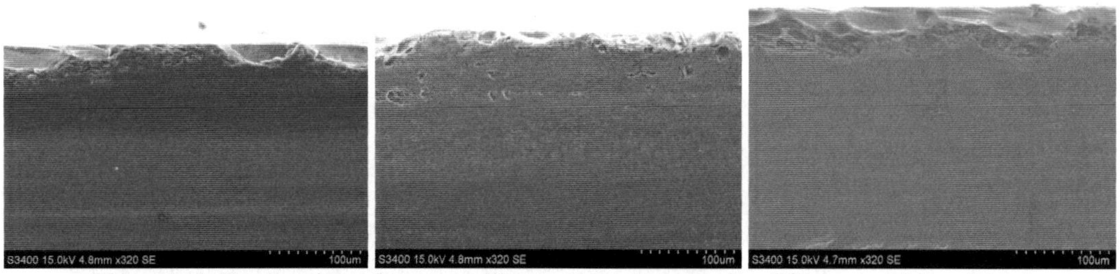

Figure 10. Preparation of porous silicon microstructure by different etching times.

a）40min b）90min c） 140min

Figure 11. The surface state of the silicon after removing the porous silicon.

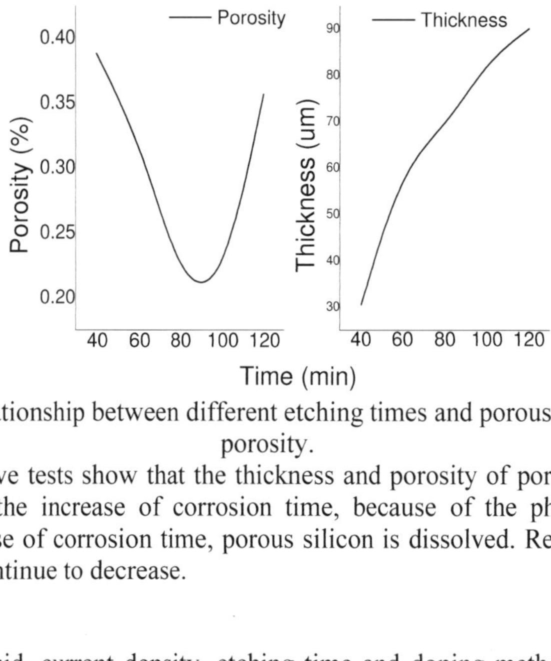

Figure 12. The relationship between different etching times and porous silicon thickness and porosity.

The results of the above tests show that the thickness and porosity of porous silicon increase first and then decrease with the increase of corrosion time, because of the phenomenon of excessive corrosion with the increase of corrosion time, porous silicon is dissolved. Resulting in porous silicon thickness and porosity continue to decrease.

5. Conclusions

The effects of etching fluid, current density, etching time and doping method on the preparation of porous silicon were studied. The effects of different process parameters were tested and tested. The relationship between the process parameters and the microstructure of porous silicon, the thickness of porous silicon and porosity was obtained. Optimize the process parameters, select different doping forms, can control the porous silicon pore size and porous silicon thickness. The use of porous silicon properties, the use of porous silicon as a sacrifice layer, silicon microstructure can be achieved, greatly increasing the application of the sensor.

References

[1] J.López-García,R.J.Martín-Palma,et al.Porous silicon based structures for the electrical biosensing of glucose,Sensors and Actuators B:Chemical,2006, 45(9):336~341.
[2] A Kanjilal,M Song,K Furuya,et al.Observation of room-temperature resonant photoluminescence in porous silicon,Physica E:Low-dimensional Systems and Nanostructures,2006,35(1):151~156.
[3] L.T.Canham.Silicon quantum wire array fabrication by electrochemical and chemical dissolution of wafers[J].Applied Physics Letters,1990,57(10):1046-1048.
[4] D.R.Turner.Electropolishing silicon in hydrofluoric acid solutions[J].Journal of The Electrochemical Society,1958,105(7):402-408.
[5] C.Pickering,M.I.J.Beale,D.J.Robbins,P.J.Pearson,R.Greef.Optical studies of the structure of porous silicon films formed in p-type degenerate and non-degenerate silicon[J]. Journal of Physics C:Solid State Physics,1984,17:6535-6552.
[6] C.S.Fuller,J.A.Ditzenberger.Diffusion of donor and accepter elements in silicon[J].Journal of Applied Physics,1956,27(5):544-553.
[7] Lai Chuan. Study on Corrosion Behavior of Porous Silicon in Alkaline Solutions[M].2014.
[8] J.N.Chazalviel,R.B.Wehrspohn,et al.Electrochemical preparation of porous semiconductors:from phenomenology to understanding,Materials Science and Engineering B,2000,69(14):1~10.

Reliability analysis of different structure parameters of PCBA under drop impact

P S Liu[1,2], G M Fan[1] and Y H Liu[1]

[1]Jiangsu Key Laboratory of ASIC Design, Nantong University, Nantong 226019, China
[2] Tongke School of Microelectronics, Nantong University, Nantong 226019, China
**E-mail: psliu@ntu.edu.cn*

Abstract. The establishing process of PCBA is modelled by finite element analysis software ABAQUS. Firstly, introduce the Input-G method and the fatigue life under drop impact are introduced and the mechanism of the solder joint failure in the process of drop is analysed. The main reason of solder joint failure is that the PCB component is suffering repeated tension and compression stress during the drop impact. Finally, the equivalent stress and peel stress of different solder joint and plate-level components under different impact acceleration are also analysed. The results show that the reliability of tin-silver copper joint is better than that of tin-lead solder joint, and the fatigue life of solder joint expectancy decrease as the impact pulse amplitude increases.

Key words: Board assembly; PBGA package; finite element analysis; drop impact; reliability

1. Introduction

It is well known that Drop impact test contains plate drop and product drop, and the plate drop is more maneuverable. Handheld electronic products tend to be miniaturized and versatile, which makes electronic packaging design to be of high density and small size. Compact handheld electronics often fall unexpectedly in transit, storage, and use, which make the printed circuit board subject to transient impact. The transient bending of the plate causes transient dynamic stresses at the solder joint and other joints, which brings about mechanical and electrical failure. The small solder joints between the circuit board and the chip become the most critical parts of the damage. The mechanical behaviour in the drop process becomes an important subject of the reliability research of electronic products.

2. PBGA solder joint structural parameter

The PBGA packaging model consists of a chip, a package body, a substrate, solder balls, copper pads, and PCB. The only heat source of the PBGA comes from the chip. The welding ball is a uniform array of 8×6, in which 0.05mm thickness pad is attached on upper and lower surfaces respectively. The pitch between the solder balls is 1mm. As can be seen from Table 1 and 2, the PBGA structure parameters and component parameters are given.

Table 1. Structural parameters of PBGA

component	Dimensions/mm
Solder joint	Diameter:0.56 ×Height:0.42
Copper pad	Height:0.05
Package	12×8×0.625-8×6×0.275
Substrate	12×8×0.25
Chip	8×6×0.275
PCB	90×70×0.675

Table 2. Material parameters

Material	Elastic modulus/ GPa	Density/ Kg/m^3	Poisson's ratio
FR-4	22	2680	0.28
Cu	120	8930	0.34
Substrate	22	1800	0.38
Si	131	2990	0.23
Modelling compound	28	1890	0.35
Pb90Sn10	22.5	10800	0.40
Sn63Pb37	31.7	8420	0.35
Mix	33.5	7531	0.35
SAC387	38	7500	0.38
SAC305	54	7400	0.40

3. Input-G method

Figure 1 shows the PCB under the impact pulse under the condition that the four screws are fixed. Establishing PCB installation model by ABAQUS software is a useful way. The simplicity of the Input-G allows it to calculate three times faster than the free fall model. Therefore, the Input-G method is an effective way to optimize reliability of PCB drop, and it is used by many scholars to analyse the reliability of solder joints under drop impact [1-3].

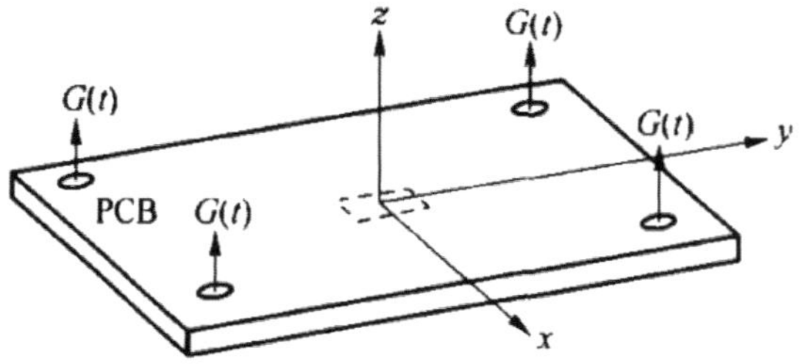

Figure 1. Input-G method of four fixed screws

4. Life model of drop impact

To predict the fatigue failure of solder joints,damage parameter is a scalar of solder joint failure that should be first taken into considered. Generally, pressure and strain parameters are two damage parameters for drop test failure criteria. Peeling stress is the dominant factor in the reliability analysis of drop [4]. Therefore, the impact life model can be determined according to the simulated results of peeling stress and the actual measured life span. Numerical analysis can calculate the maximum peeling stress of the solder joint, when predict the life of the product or evaluate the performance of the product. Considering the fatigue life model with the maximum peel stress and fatigue life of the solder joints, the power criterion combines can be described as:

$$N_{50} = c_1 \sigma_Z^{C_2} \tag{1}$$

N_{50} is the average drop impact life, σ_z is the critical delamination stress of the solder joint, C_1 and C_2 are correlation coefficients with values of 9.045×10^8 and -3.1485[5].

5. Finite element analysis

5.1 Failure mechanism analysis under drop impact

In this paper, the conduct explicit transient dynamic analysis is based on ABAQUS software. In the process of free fall, the PCB board is suffering repeated tension and stress, and it will lead to solder joint fail. From the distribution of stress, the maximum stress of the whole component is 1595.9 MPa.S33 is the peeling stress that perpendicular to the PCB plate. The S33 stress is also to predict the solder joint failure life. From the distribution nephogram of peeling stress, the maximum peeling stress of PCB component is 554.35 MPa.

Firstly, the semi sinusoidal acceleration is simulated under the condition that the acceleration amplitude of 1500G and the pulse duration of 0.5ms. The half sine acceleration is the boundary condition that applied to the four screw holes of the PCB, and the direction is perpendicular to the PCB plate. In addition, apply the horizontal symmetry constraint to the symmetry surface of the model. The drop impact analysis of the model shows that the stress concentration location of the solder joint is located at four vertex angles, and the stress at the side of the PCB plate is the maximum. Because the model is symmetrical, select the point of maximum stress in one of the solder joints, and the distribution of Mises stress is in Figure 2. From Figure 3, we can see the distributions of the peel stress. As the solder joint is suffering a semi sinusoidal acceleration, the peeling stress and the Mises stress of the solder joint propagate in a sinusoidal form. Figure 4 is a curve of the critical solder joint stress with time, wherein S11, S12, S13, S22, S23 and S33 represent stress in each direction and each plane respectively. S33 represents the stress direction different from other stresses, and its amplitude is the largest. Therefore, the peeling stress is the main cause of crack initiation, propagation and full fracture of solder joints, and it is the main basis for solder joint failure. The peeling stress of the spot leads to the growth of the internal cracks in the BGA solder ball, and the crack growth is the root cause of the solder ball failure in the drop reliability test [6]. S_{13} represents the shear stress in the X-Z direction with a peak value of 639 MPa.

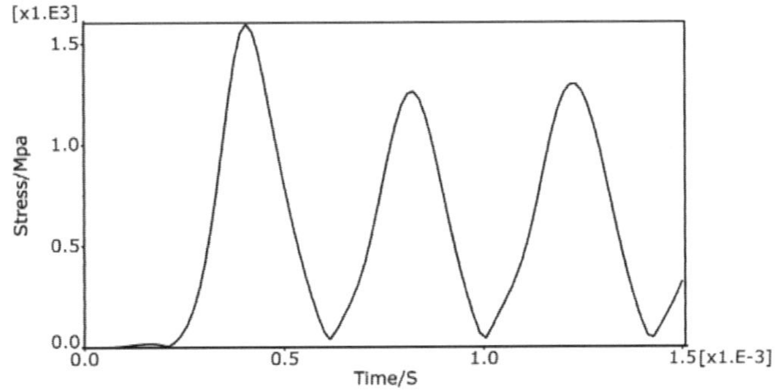

Figure 2. Mises stress curve of critical solder joint

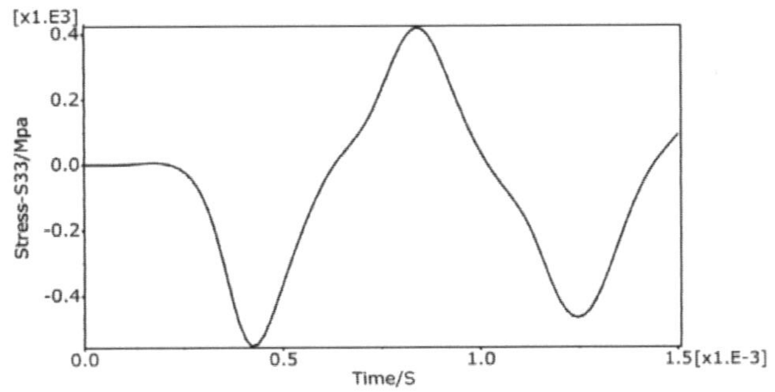

Figure 3. Peel stress distribution of critical solder joint

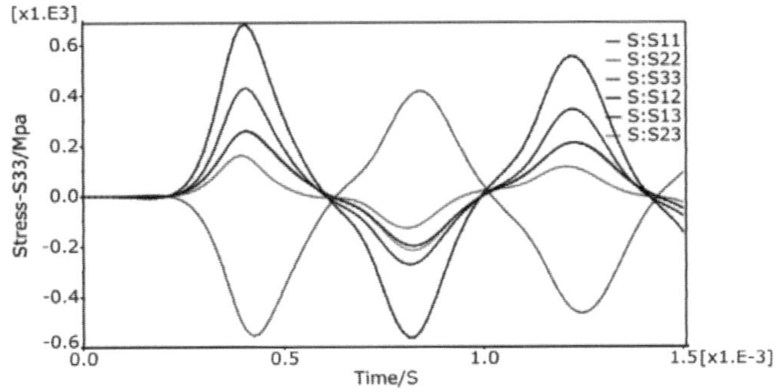

Figure 4. Stress of critical solder joint

One of the four vertex angles of the PCB board is selected. The displacement change is also shown in Figure 5. In 1.5ms, the displacement at the screw hole of the PCB plate is shifted in the negative direction of Z, and the maximum displacement change is 5.857mm. The displacement changes at the centre of the PCB are shown in Figure 6, the selected node is node 30257, and the maximum displacement is 6.099 mm. The deflection of PCB is the displacement of its apex angle relative to the centre displacement, shown as Figure 7. At 0.42ms, the deflection of PCB reaches the maximum value of 0.638 mm, and the deflection of PCB reaches the maximum value of -0.621mm at 0.82ms. The vibration of PCB changes periodically and decays gradually over time, the size and trend are close to the existing experimental results, which proves that the simulation results are true.

The comparison of the stress curve and deflection curve of PCB is in Figure 8. It is found that the general trend of the two is the same, and the peak value of the amplitude appears approximately the same. However, the peak value of stress lags slightly behind the peak of deflection, indicating that the stress in the solder joint is due to the deflection of the PCB. The displacement of PCB leads to tension and compression stress acting on the solder joint, which makes the solder joint become the key solder joint.

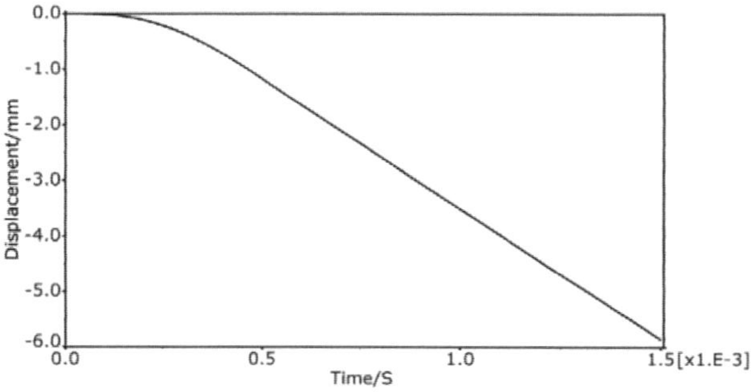

Figure 5. Displacement of PCB vertex

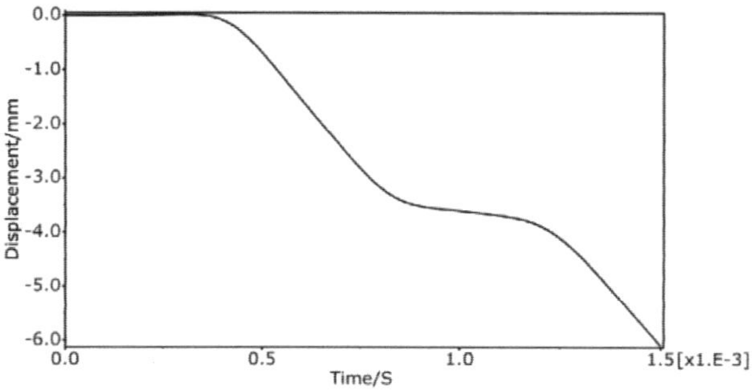

Figure 6. Displacement curve of PCB center

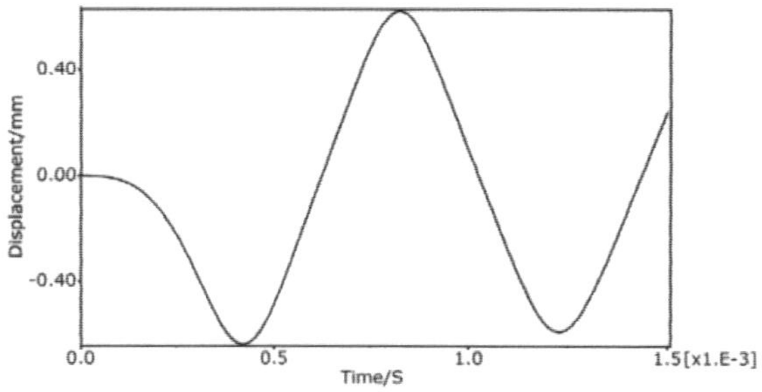

Figure 7. Displacement curve of PCB center point

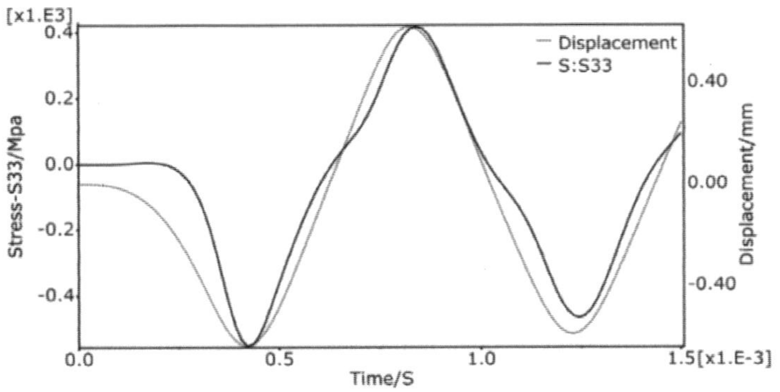

Figure 8. Peeling stress curve of solder joint and PCB deflection curve

From the deflection curve and the peeling stress curve of the solder joint, we can know that the joints with four vertex angles are the key parts of the subject. In the solder joint, an 8x6 array is established with two paths, each of which starts from a vertex angle of the solder joint array and ends at the other corner. The two paths of X and Y are shown in Figure 9. The nodes selected on the X path are the 8 nodes of 1430, 107, 41, 110, 38, 53, 95, and 56. The nodes selected on the Y path are the 6 nodes of 1433, 1476, 1390, 1519, 1562, and 1347.The two paths are shown in Figure 10 and Figure 11 respectively.

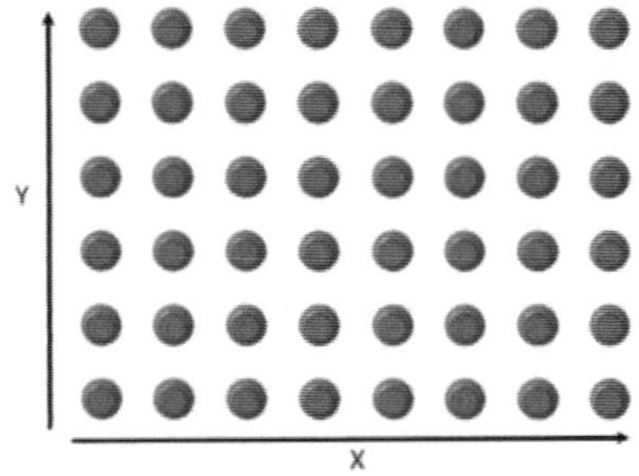

Figure 9. Illustration of X direction and Y direction

From Figure 10 and Figure 11, it is found that the stress of the solder joint in the Y axis is slightly greater than that in the X axis. On the X path, the maximum stress is about 155 MPa, and the minimum stress is only 6.74Mpa. While in the Y axis the maximum stress is 155.17MPa, and the minimum stress is 6.74MPa. The maximum stress at the solder joint on the X path and on the Y path appears at same level.

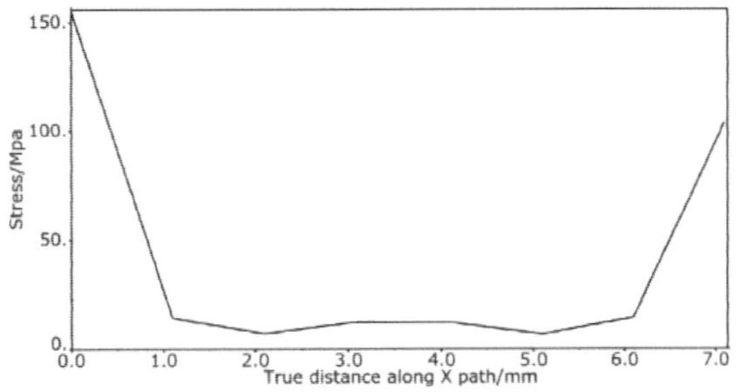

Figure 10. Distribution of Stress in X axis

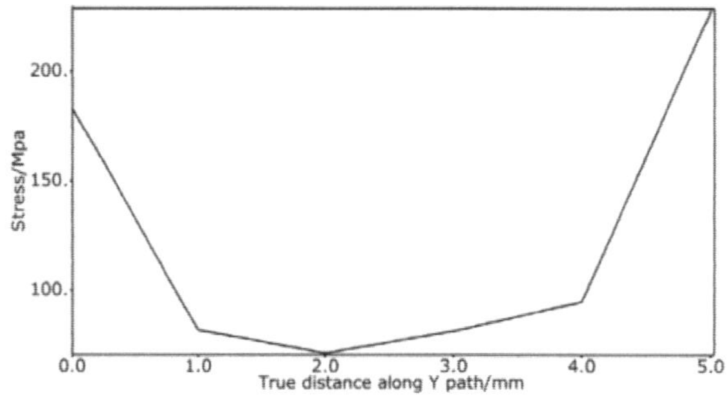

Figure 11. Distribution of the Stress in Y axis

5.2 Analysis of drop life under different parameters

5.2.1 Reliability analysis of solder joints under different solder joint materials. Materials not only play an important role in the vibration reliability of solder joints, but also play an important role in improving the reliability of drop impact reliability [7-8]. In this research, we study five different solder joint materials such as Pb90Sn10, Sn63Pb37, MIX, SAC387 and SAC305 respectively. The moduli, densities, and Poisson's ratios are given in Table 3. Under the same conditions, fatigue life can be obtained by changing the material of the solder joint.

Table 3. Fatigue life of solder joints

The material of solder joint	Maximum stress/MPa	Maximum peel stress/MPa	Fatigue life/ms
Pb90Sn10	714.48	264.54	21.34
Sn63Pb37	722.12	274.78	18.95
MIX	694.94	254.95	23.97
SAC387	676.26	246.89	26.52
SAC305	653.73	236.45	30.38

As shown in Table 3, the fatigue life of tin lead solder joints is worse than that of tin silver copper. With the rapid development of lead-free electronic products, SAC solder joints will gradually replace tin lead materials because of their superior properties.

5.2.2 Reliability analysis of solder joints under different impact accelerations. Select the working conditions A, F and B, the corresponding peak acceleration is 500G, 900G and 1500G, the corresponding pulse time is 1.0ms, 0.7ms, 0.5ms. Conduct transient dynamics analysis of solder joints under the condition of Sn63Pb73. The equivalent stress distribution of the solder joints is under three conditions, as shown in Figure 12. The solder joint reaches the maximum stress at 0.64ms under the pulse acceleration of 500G, the solder joint reaches the maximum stress at 0.49ms under the pulse acceleration of 900G, and the stress acceleration of the solder joint at 1500G is at 0.40ms. With the increase of pulse acceleration, the time of maximum stress of solder joint is smaller. The reason is that the greater the pulse acceleration, the greater the hardness of the impact surface, the harder the impact surface and the more likely the solder joint failure. In addition, in the same time, the solder joint in 500G acceleration experienced 2 sinusoidal stress changes. The solder joint in 900G acceleration experienced 2.5 sinusoidal stress changes. The solder joint in 1500G acceleration experienced 3.5 sinusoidal stress changes. In summary, the greater the pulse acceleration is, the faster the solder joint reaches the maximum stress, and the more the semi sinusoidal stress changes in the solder joint in the same amount of time. By analysing the stress distribution of the solder joint, the maximum stress and the maximum peeling stress of the critical solder joint can be obtained. The fatigue life of solder joints is obtained by the life prediction formula based on the Power criterion, under the working conditions A, F and B. Thus, the fatigue life of solder joints is 18.95, 5.65 and 2.07 respectively. As shown in Table 4, the maximum peeling stress and the maximum equivalent stress of the solder joint decrease with the increase of the pulse acceleration of the solder joint. And the fatigue life of solder joint decreases with the increasing of shock pulse amplitude.

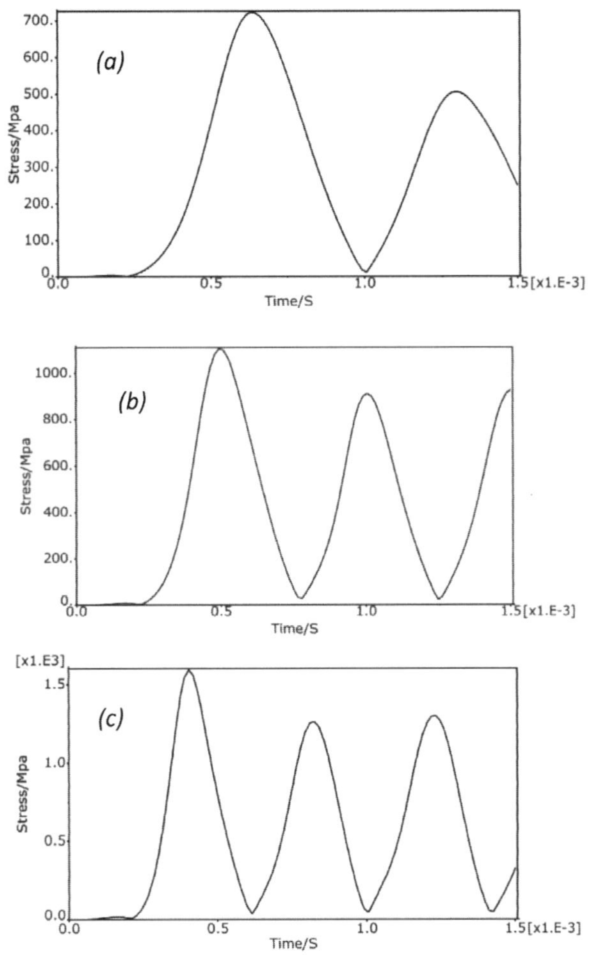

Figure 12. Stress distribution at different pulse acceleration: (a) at 500G, (b) at 900G and (c) at 1500G

Table 4. Fatigue life under different operating conditions

Condition	Maximum stress/MPa	Maximum peel stress/MPa	Fatigue life/ms
A (500G,1.0ms)	722.1	274.7	18.95
F (900G,0.7ms)	1105.3	403.4	5.65
B (1500G,0.5ms)	1595.9	554.3	2.07

IOP Conf. Series: Journal of Physics: Conf. Series **986** (2018) 012007 doi:10.1088/1742-6596/986/1/012007

6. Conclusion

In this paper, the drop process of the product is analyzed by means of explicit dynamic analysis. In addition, the application of Input-G method in drop test is studied, and this method can solve the problem of simulation of drop simulation. The fatigue life model of solder joints is established based on the Power criterion, and the failure mechanism of solder joints under working condition B is studied. Finally, the influence of different parameters on the fatigue life of solder joints is also investigated.

(1) In the same circumstances, the reliability of SnAgCu solder joints is better than that of tin-lead solder materials.

(2) In the same solder joint material, the maximum stress and the maximum peeling stress of the solder joint are measured respectively by changing the impact pulse acceleration. The results show that the fatigue life of the solder joint decreases as the amplitude of pulse impact acceleration increases.

Reference

[1] Yang X,Xiao G and Shu X.2013 Effect of solder joint shape on reliability of BGA packages under board level drop impact loads. J. Vib. Shock, 32 pp 104-107

[2] Jin C 2013 Failure analysis of board level electronic packaging *Mech. Electrl. Eng.* **30** pp565-9.

[3] Zhu G B2015 Reliability analysis of lead free solder joints at plate level under drop impact loading. *China High End SMT Acad. Conf.*

[4] Chen X, et al. **2014** Drop test and finite element analysis of a ball grid array package *Chin. Sci.Papers*, pp 961-4.

[5] Wen L, Fu X, Zhou J, et al. 2008Dynamic properties testing of solders and modeling of electronic packages subjected to drop impact. *Int. Conf. on Electron. Packag. Technol. High Density Packag.* pp 1-7.

[6] Fan P Y 2008 Study on fatigue crack behavior of BGA solder balls under plate level drop test. *Semicon. Technol.* **33** pp 585-588.

[7] Li C L and Li Q B. Improvement of BGA solder joint reliability with Sn-Ag-Cu solder paste doped with Mn. 2011 *Weld. Tech.* **40** pp 42-44.

[8] Yang J L, Lei Y P, Lin J and Xiao H 2013Influence of Silver Content on Fatigue Life and Failure Mechanism of Lead-free Solder Joints Under Drop Conditions. *J. Mater. Eng.* **12** pp 74-79.

Acknowledgments

This work was partially financially supported by the National Natural Science Foundation of China (NSFC Nos. 61571245 and 61474067) and Top-notch Academic Programs Project of Jiangsu Higher Education Institutions (TAPP: PPZY2015B135).

Study on Transmission Characteristics of TSV

P S Liu[1,2], G M Fan[1] and Y H Liu[1]

[1]Jiangsu Key Laboratory of ASIC Design, Nantong University, Nantong 226019, China

[2] School of Tongke Microelectronics, Nantong University, Nantong 226019, China

**E-mail: psliu@ntu.edu.cn*

Abstract. A three-dimensional model of Through-Silicon-Via (TSV) is established. The effects of TSV structural parameters and material parameters on signal transmission performance are studied. The influences of TSV height, radius, insulating layer thickness and material conductivity on the insertion loss (S21) are studied, and the saliency of influencing factors is examined by orthogonal design. The results show that TSV height and the substrate conductivity have the greatest effect on the transmission characteristics.

Keywords: TSV; Transmission characteristics; Saliency

1. Introduction

As the core technology of 3D packaging, the development of TSV technology still faces many challenges [1]. The vertically interconnected TSV structure improves signal transmission speed and reduces its power consumption [2]; the structure and material of TSV have great influence on its transmission performance [3]. The coaxial TSV has self-shielding function that can effectively improve EMI problems [4]; Air-Cavity TSV structure can not only improve the quality of signal transmission, but also have high stability [5]; Graphene, with its good electrical and thermodynamic properties, is likely to replace copper filled TSV in the future [6]. However, there is little research that studies on the influence of the TSV parameter on signal transmission characteristics. In this paper, the Significance ranking of TSV structural parameters and material conductivity on the insertion loss is examined by orthogonal analysis, which provides the optimized design basis for the research of signal integrity.

2. Geometric Structure and Material Parameters of TSV

The double TSV model is established by HFSS, as shown in Figure1. The model consists of silicon substrate, filler metal, insulating layer of silicon dioxide, and upper and lower oxide layers. Table 1 shows the structural parameters and the material properties of TSV.

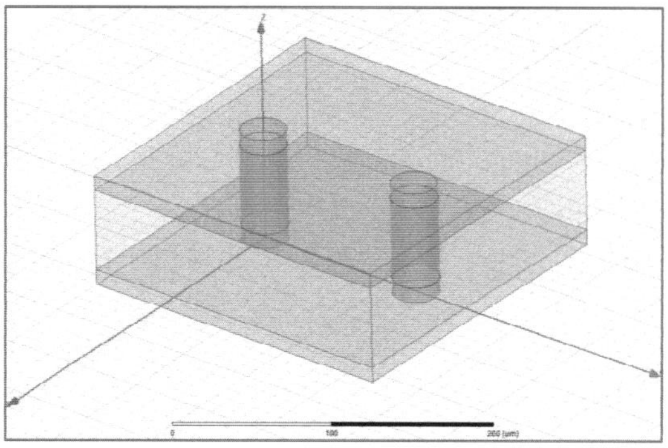

Figure 1. Double Channel TSV Model

Table 1. Structural parameters and material properties

Structural Parameter	Symbol	Value
TSV Height	H	70μm
TSV Radius	R	15μm
Insulation Thickness	t	0.4μm
TSV Spacing	P	120μm
Copper Resistivity	σ_{tsv}	$1.68\times10^{-8}\,\Omega.m$
Substrate Conductivity	σ_{si}	10s/m
Substrate Permittivity	ε_{si}	12

3. Influences of Structural Parameters on Transmission Characteristics

Frequency characteristics can analyze signal transmission performance and noise coupling, the effective working frequency of the frequency domain is set as 20GHz. To study the effect of different physical dimensions and material parameters on TSV transmission performance, the influence of TSV height, TSV radius, insulation thickness and material conductivity on signal insertion loss is analyzed in this paper.

3.1 The influence of TSV height on transmission characteristics

The change of TSV height changes the size of other components in the circuit. Keep TSV radius and the insulating layer thickness unchanged and increase TSV height from 70μm to 145μm and the step length is 25μm. The tendency of the insertion loss varies with the TSV height is shown in Figure 2. As shown in the diagram, the insertion loss increases as the height of the TSV increases, which means more signals couples to the substrate and the signal transmission performance reduce. Therefore, in the case of current process permits, TSV height can be short as much as possible to improve the transmission performance.

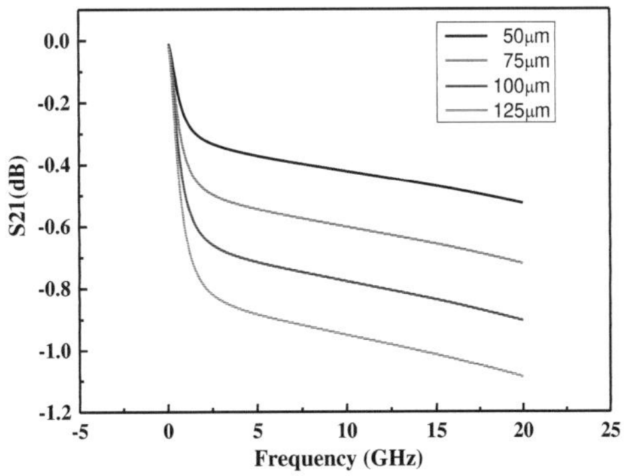

Figure 2. Insertion loss varies with TSV height

3.2 Influence of TSV radius on propagation characteristics

The radius of TSV has a great influence on thermal dissipation and it affects the density of TSV array in certain degree [7]. The TSV radius increases from 10μm to 25μm and the step length is 5μm. The tendency of the insertion loss varies with the TSV radius, as shown in Figure 3. To reduce the insertion loss, the TSV radius should be short as much as possible.

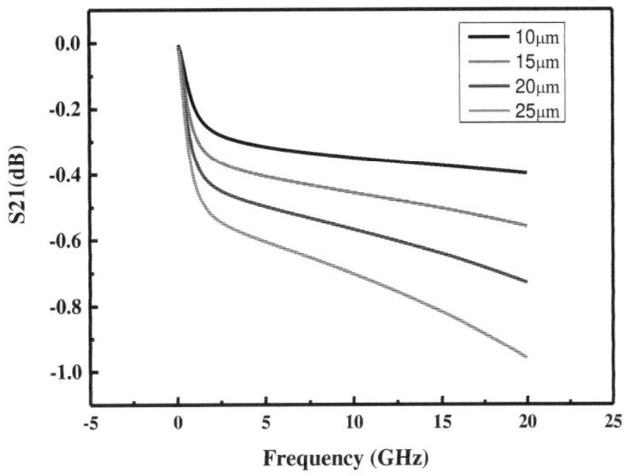

Figure 3. Insertion loss varies with the TSV radius

3.3 Influence of insulating layer thickness on transmission characteristics

As the insulating layer of silicon dioxide is relatively thin, parasitic capacitance is often produced and high frequency signal couples into the substrate [8], which will reduce signal intensity. Therefore, the study on the thickness of insulating layer is meaningful. The thickness of insulating layer increases from 0.2μm to 0.8μm and the step length is 0.2μm. The result shows that the tendency of the insertion loss varies with insulating layer thickness, as shown in Figure 4.

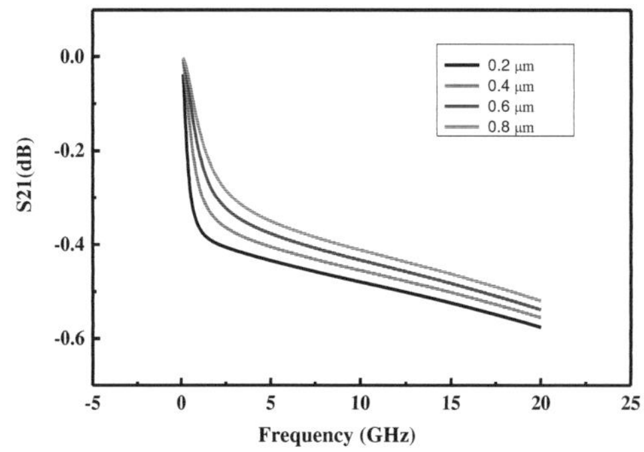

Figure 4. Insertion loss varies with the thickness of the insulating layer

3.4 Result analysis

In high frequency circuit, the electric field will pass through the dielectric layer between signal TSV and ground TSV, resulting in parasitic losses. The parasitic parameters of the silicon substrate and the insulating layer is expressed as [9]:

$$C_{si} = \frac{\pi\varepsilon_0\varepsilon_{r,si}h_{si}}{\ln[P_{tsv}/(2r_{tsv})+\sqrt{P_{tsv}/(2r_{tsv})^2-1}]} \quad (1)$$

$$G_{si} = (\frac{\sigma_{si}}{\varepsilon_{si}} + \omega\tan\delta_{c,si})C_{si} \quad (2)$$

$$C_{ox} = \frac{\pi\varepsilon_0\varepsilon_{r,ox}h_{ox}}{\cosh^{-1}(P_{tsv}/2r_{tsv})} \quad (3)$$

The C_{si} represents the parasitic capacitance of the silicon substrate, G_{si} represents the substrate conductance, and C_{ox} represents the parasitic capacitance of the insulate layer.

As the height (or radius) of the TSV increases, the parasitic capacitance of the substrate and the insulating layer increases but the impedance decreases. Hence, the signal couples to the substrate more easily. As the insulating layer thickness increases, the impedance between the silicon via and the substrate increases, because the signal is more difficult to couple to the substrate. Therefore, the quality of the signal transmission can be improved by increasing the thickness of the insulating layer and reducing the height and radius of TSV.

4. Orthogonal Analysis

Orthogonal design is a method that can test the influence of various factors on the experimental results without affecting the experimental results and can reduce the number of tests [10]. In this paper, orthogonal design and variance analysis are to investigate the influence of various parameters on insertion loss. Then, the optimized structures and materials are proposed.

4.1 Orthogonal design

4.1.1 Orthogonal design of structural parameters. The four levels of TSV height, radius and thickness of insulating layer are selected, as shown in Table 2. According to the principle of orthogonal design, the orthogonal design table of three factors and four levels is established in Table 3. Orthogonal test of 16 different combinations is arranged with orthogonal tables of $L_{16}(4^5)$. The insertion loss of each group is in the case of 10.05GHz, thus we can get the effect of each factor on the insertion loss.

Table 2. TSV structural parameter factor level

Level	Height/μm	Radius/μm	Insulation thickness/μm
1	70	10	0.2
2	95	15	0.4
3	120	20	0.6
4	145	25	0.8

The K_i in Table2 represents the sum of the test results for each factor at level i, and k_i represents the average result of each level. Range is the difference between the maximum and the minimum in k_i, which reflects the impact of various factors on the experimental results.

$$k_i = K_i / 3 \quad (4)$$

$$\text{Range} = \text{MAX}(K_i) - \text{MIN}(k_i) \quad (5)$$

The result of Table 3shows that the range of TSV height is max, followed by TSV radius and the range of the insulation thickness is the smallest.

Table 3 Orthogonal design of TSV structure factor

Test number	Height /(μm)	Radius /(μm)	Insulation thicknes /(μm)	Insertion Loss (S21)/(dB)
1	1	1	1	-0.314
2	1	2	2	-0.368
3	1	3	3	-0.422
4	1	4	4	-0.475
5	2	1	2	-0.452
6	2	2	3	-0.585
7	2	3	4	-0.650
8	2	4	1	-0.906
9	3	1	3	-0.562
10	3	2	4	-0.693
11	3	3	1	-0.978
12	3	4	2	-1.139
13	4	1	4	-0.655
14	4	2	1	-0.977
15	4	3	2	-1.150
16	4	4	3	-1.341
K1	-1.579	-1.983	-3.175	
K2	-2.593	-2.623	-3.109	
K3	-3.372	-3.200	-2.910	
K4	-4.123	-3.861	-2.473	
k1	-0.395	-0.496	-0.794	
k2	-0.648	-0.656	-0.657	
k3	-0.843	-0.800	-0.728	
k4	-1.031	-0.965	-0.618	
Range	-0.636	-0.470	-0.176	

4.1.2 Orthogonal design of electrical conductivity of materials. Orthogonal design is applied to TSV filling metal resistivity and substrate conductivity. The structural parameters of TSV refer to the parameters of the model in Table 1. Table 4 is a table of factor levels, and the materials are copper, silver and graphene [11]. The orthogonal design table of three materials shows in Table 5.

Table 4. Material parameter level

Level	Filler Resistivity /(Ω·m)	Substrate Conductivity / (s/m)
1	1.68×10^{-8}	16.8
2	1.59×10^{-8}	15.9
3	1.0×10^{-8}	10.0

The result of Table 5 shows that the range of the resistivity of filler is much less than that of the conductivity of the substrate. It means that the influence of substrate conductivity on the insertion loss is obviously greater than that of filler resistivity.

Table 5. Orthogonal design of material parameters

Test Number	Filler Resistivity (Ω·m)	Substrate Conductivity (s/m)	Insertion Loss (S21)/dB
1	1	1	-0.64700
2	1	2	-0.61113
3	1	3	-0.41096
4	2	2	-0.61108
5	2	3	-0.41091
6	2	1	-0.64695
7	3	3	-0.41053
8	3	1	-0.64653
9	3	2	-0.61066
K1	-1.66909	-1.94048	
K2	-1.66893	-1.83288	
K3	-1.66772	-1.23239	
Range	-0.00137	-0.70809	

4.2 Variance analysis

Variance analysis is a method of examining the significance of experimental factors on results based on experimental data. In the former analysis, It can be easily found that the TSV height has the greatest impact on the transmission characteristics, and the thickness of the insulating layer has minimal impact on the transmission characteristics. The effect of the substrate conductivity on the insertion loss is significantly greater than the resistivity of filler. Next, the variance analysis is to validate the former results.

4.2.1 Variance analysis of structural parameters of TSV. Table 6 shows the results of the variance analysis of the three structural parameters of TSV on the insertion loss, the F value in the table indicates the magnitude of the effect of each factor on the results, the critical values $F_{\alpha}(f_s, f_E)$ in the table can be inquired from the orthogonal distribution table. If $F>F_{0.01}(f_s, f_E)$, it is a high significant factor, if $F>F_{0.05}(f_s, f_E)$, it is a significant factor.

The result of Table 6 shows that the effect of TSV height on insertion loss is the most significant, followed by TSV radius and the thickness of the spacer layer has the smallest influence on the insertion loss, and the results are consistent with the analysis results of Table 3.

Table 6. Variance analysis results of TSV structure parameters

Source of variance	Sum of squares of Deviations (S_s)	Freedom (f_s)	Sum of squares of mean deviations	F	Critical	Saliency	Optimization Scheme
Height	0.889	3	0.296	64.505		**	H1
Radius	0.483	3	0.161	35.004	$F_{0.01(3,6)}$=9.78	**	D1
Insulation thickness	0.075	3	0.025	5.452	$F_{0.05(3,6)}$=4.76	*	T4

(The number of symbol "*" indicates the significance of the influence)

According to Table 7, the F value of the conductivity of the silicon substrate is much larger than that of the filler resistivity. The result is consistent with the analysis result of Table 5.

Table 7. Variance analysis results of TSV material parameters

Source of variance	Sum of squares of deviation	Freedom	F
A	3.738×10^{-7}	2	1.264×10^{3}
B	0.978×10^{-2}	2	3.282×10^{8}

Therefore, to reduce the value of insertion loss, we should reduce the height of TSV first, then decrease TSV radius, and finally increase the thickness of the insulating layer. In terms of the electrical conductivity, the method of reducing the substrate conductivity is more feasible and efficient.

4.3 Optimal design

Table 6 and Table 7 provide the optimized combinations of structural and materials for the transmission characteristics, in which TSV height is set to 70μm, TSV radius is 10μm, the insulating layer thickness is 0.8μm, the TSV filling material resistivity is $1\times10^{-8}\Omega$.m, and the conductivity of the silicon substrate is 10S/m. After analysing this TSV, the value of insertion loss varies with frequency, as shown in Figure 5. When the frequency is 10.05GHz, the insertion loss value is -0.274dB, and at this frequency, the insertion loss in the initial model is -0.368dB. It means that the insertion loss reduces by nearly 10%.

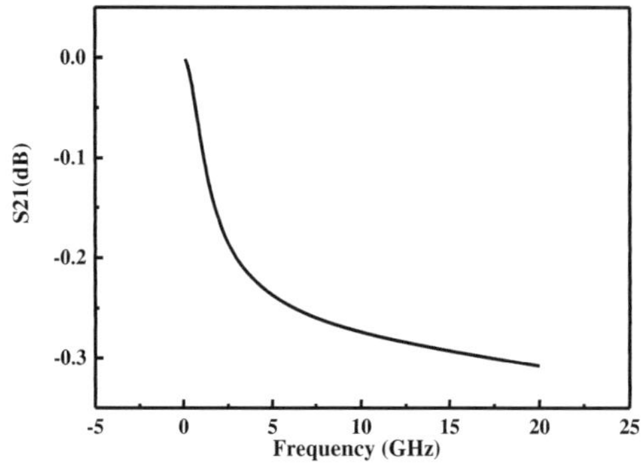

Figure 5. Insertion loss varies with frequency

5. Conclusion

The result of this study shows that the insertion loss increases when the TSV height and radius increase. When increasing the area of the parasitic capacitance between TSV and the substrate, the substrate capacitance and the isolation layer capacitance increase, and the impedance of the substrate and isolation layer decrease. Thus, the signal couples to the substrate more easily. Because the impedance between TSV and the substrate increases, the insertion loss increases with the increasing of isolation thickness. According to the results of the orthogonal analysis, TSV height has the greatest impact on the insertion loss value, followed by TSV radius, and the thickness of the spacer layer has the smallest influence on the insertion loss; the influence of substrate conductivity on insertion loss is far greater than the hole filling resistivity.

Reference:

[1] Liu P S, Huang J X 2012 Development and challenge of silicon through hole technology *Electron. Compon. Mater.* **31** pp 76-80

[2] Liang L, Miao M, Li ZS, Xu SF, Zhang YX and Zhang XQ 2011 3D modeling and electrical characteristics of through-silicon-via (TSV) in 3D integrated circuits *Int. Conf. on Electron. Packag. Technol. High Density Packag.* pp 471-5

[3] Par J S, Ryu C and Kim J 2007 Electrical characterization of through silicon via (TSV) depending on structural and material parameters based on 3D full wave simulation. *Proc. of Int. Conf. on Electron. Mater. Packag.* New York, USA: IEEE, pp1-6

[4] Srivastava N, Li H, Kreupl F and Banerjee K 2009 On the applicability of single-walled carbon nanotubes as VLSI interconnects *IEEE Trans. on Nanotech.* **8** pp 542-59

[5] Liu X X. 2015 Research on Parasitic Parameters and Equivalent Circuit of New Silicon vias. (XiAN：Xi'an Electronic and Science University).

[6] Van Noorden R. Moving towards a graphene world 2006 *Nature*, **442** pp 228-9

[7] Tan Y H 2013 Influence of size of silicon through hole and material on thermal stress *Semicon. Photoelectr.* **34** pp 255-8.

[8] Matthew N O Sadiku 2000 *Elements of Electromagnetics*, 3rd ed. (New York: Oxford Univ. Press)

[9] Kim J, Pak JS, Cho J, et al. 2011 High-Frequency Scalable Electrical Model and Analysis of a Through Silicon Via (TSV) *IEEE Trans. on Compon. Packag. Manuf. Tech.* **1** pp 181-95.

[10] Chen K 2004 *Experimental Design and Analysis* (Bei Jing：Tsinghua University Press) pp72-83.
[11] Du C H 2011 *Electronic Micro Joining Technology and Materials*. (Bei Jing：Machinery Industry Press) pp100-1.

Acknowledgments
This work was partially financially supported by the National Natural Science Foundation of China (NSFC Nos. 61571245 and 61474067) and Top-notch Academic Programs Project of Jiangsu Higher Education Institutions (TAPP: PPZY2015B135).

A flexible slip sensor using triboelectric nanogenerator approach

Xudong Wang[1], Jiaming Liang[2], Yuxiang Xiao[1], Yichuan Wu[2], Yang Deng[2], Xiaohao Wang[1,2], Min Zhang[1,*]

[1] Graduate School at Shenzhen, Tsinghua University, Shenzhen, Guangdong, China
[2] Tsinghua Berkeley Shenzhen Institute, Tsinghua University, Shenzhen, Guangdong, China

E-mail: zhang.min@sz.tsinghua.edu.cn (M. Zhang)

Abstract. With the rapid development of robotic technology, tactile sensors for robots have gained great attention from academic and industry researchers. Tactile sensors for slip detection are essential for human-like steady control in dexterous robot hand. In this paper, we propose and demonstrate a flexible slip sensor based on triboelectric nanogenerator with a seesaw structure. The sensor is composed of two porous PDMS layers separated by an inverted trapezoid structure with a height of 500 μm. In order to customize the sensitivity of the sensor, porous PDMS was fabricated by mixing PDMS with deionized water thoroughly and then removing water with heat. Laser-induced porous graphene and aluminium are served as the pair of contact materials. To detect slip from different directions, two sets of the electrode pair were used. Experimental results show a distinct difference between static state and the moment when a slip happens was detected. In addition, the output voltage of the sensors increased as the increase of slip velocity from 0.25 mm/s to 2.5 mm/s. The flexible slip sensor proposed here shows the potential applications in smart robotics and prosthesis.

1. Introduction

Human-like control of robotic systems draws great attention not only in industrial but also in people's daily life. In order to realize human-like steady control in dexterous robot hand, slip detection is essential. A slip sensor is a device used to detect slide or pre-slide between robotic hands and held objects. Their performance directly determines whether robots can successfully complete a soft grasping task. Slip sensors with different principles have been developed to meet the requirements, including piezoresistivity [1], piezoelectricity [2], capacitance [3], photoelastic [4], thermal [5] and so on. However, most of them are based on rigid substrates. In recent years, triboelectric nanogenerators have been widely studied owing to their unique physical properties [6], such as self-power, simple structure and so on. In this paper, we propose and demonstrate a flexible slip sensor based on triboelectric nanogenerators. The sensor is composed of two porous PDMS layers separated by an inverted trapezoid structure. The sensor can not only realize slip detection but also static normal force detection by piezoresistive effect through laser-induced graphene electrode. The sensor we fabricated is flexible, sensitive and repeatable which may have promising applications in robotics.

2. Design and simulation

The fundamental working principle of triboelectric Nano-generators(TENGs) is a combination of triboelectric and electrostatic induction [7]. Contact electrification generates immobilized charges on polymer materials' surface which shown in Figure 1a, whereas electrostatic induction is the main mechanism that converts mechanical energy to electricity. In this work, we adopted the single electrode contact-separation mode of TENGs. First, the metal and polymer electrode contact together which generates charge transfer on the contact surface shown in Figure 1a. During the separation of two electrodes, residual charges on the polymer surface will generate an external electric field which will induce the charges in bottom metal. When a load resistance is connected between the metal and ground, there will be current going through it.

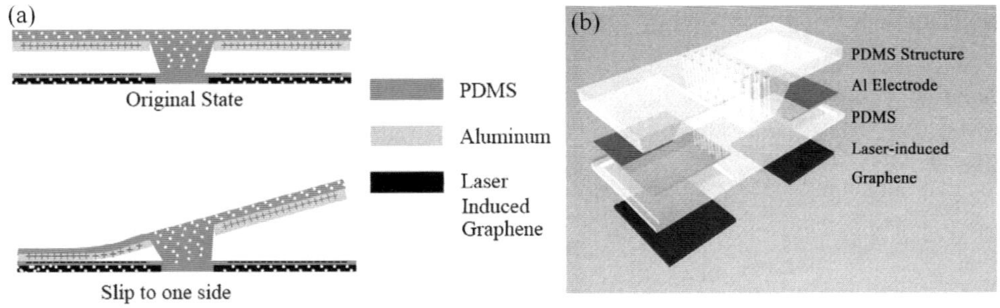

Figure 1. Schematic of the structure of the slip sensor: (a) Cross section view (b) 3D structure.

In our design, a seesaw structure of PDMS layer is adopted in the single unit in which each side of the seesaw serves as an isolated TENGs, as shown in Figure 1b. Two independent TENGs of single electrode mode in one unit was designed to detect the slip. The laser-induced graphene(LIG) is buried in the bottom PDMS layer which acts as the induction electrode. Laser-induced graphene is transferred from PI substrate into PDMS to keep its original porous structure [8]. The porous graphene structure has a large specific surface area which contributes to better induction charge capacity. In addition, the microcracking in porous laser-induced graphene results in the piezoresistive effect [9]. Therefore, the resistance of bottom electrode also has the potential to detect the static contact force.

The structure and static electricity simulation are conducted in COMSOL multiphysics. A spacer distance of 10μm is chosen. Zero contact pressure state of the structure is shown in Figure 2a. In order to clarify the phenomenon, we suppose that the sensor has been pre-charged to the original state. Figure 2b shows the static electric field on the bottom electrode when 8 nC/cm^2 [10] surface charge density is applied on PDMS. Average electric field potential of -42 volts and -40 volts is formed on each of the bottom electrodes. Then 100 mN contact pressure is applied on the left center of the unit. The contact force leads the deformation of PDMS structure, which generates a contact between the left electrodes. The static electric field becomes unbalanced due to the deformation, where the average electric field potential of left bottom increases to 25 volts and on the other side the average potential drops to -45 volts. The difference of the potential on the bottom electrodes will generate an output signal to indicate the specific contact of the electrodes. When a parallel movement happens on the sensors one side of the electrode will become contact while the other side will separate. Difference signal will generate from two electrodes.

IOP Conf. Series: Journal of Physics: Conf. Series **986** (2018) 012009 doi:10.1088/1742-6596/986/1/012009

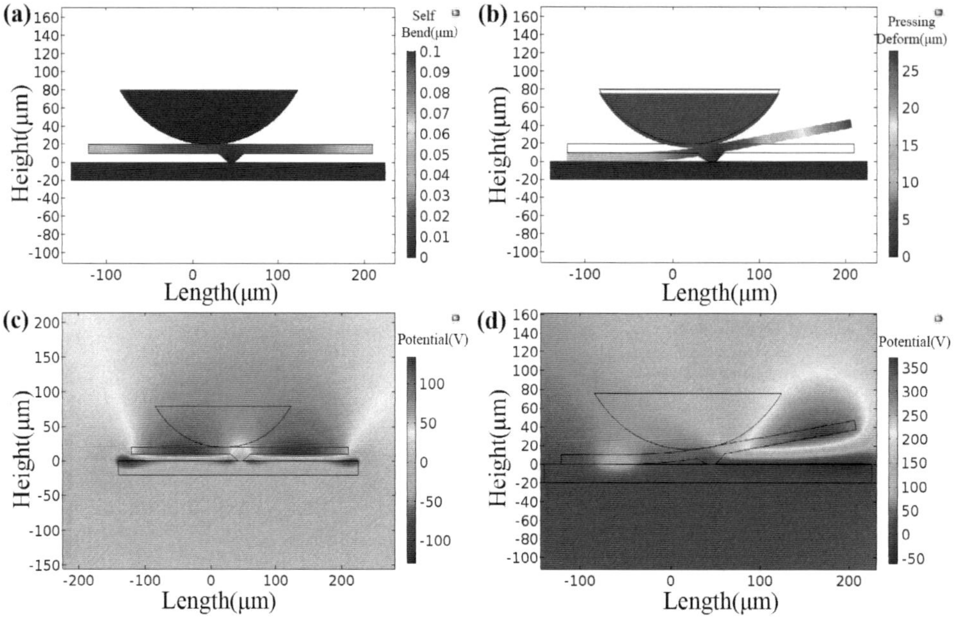

Figure 2. Sensor simulation in COMSOL: (a)Initial state (b)100 mN normal force applied on the sensor (c)Electric field of initial state (d) Electric field of sensor deformation.

3. Experimental

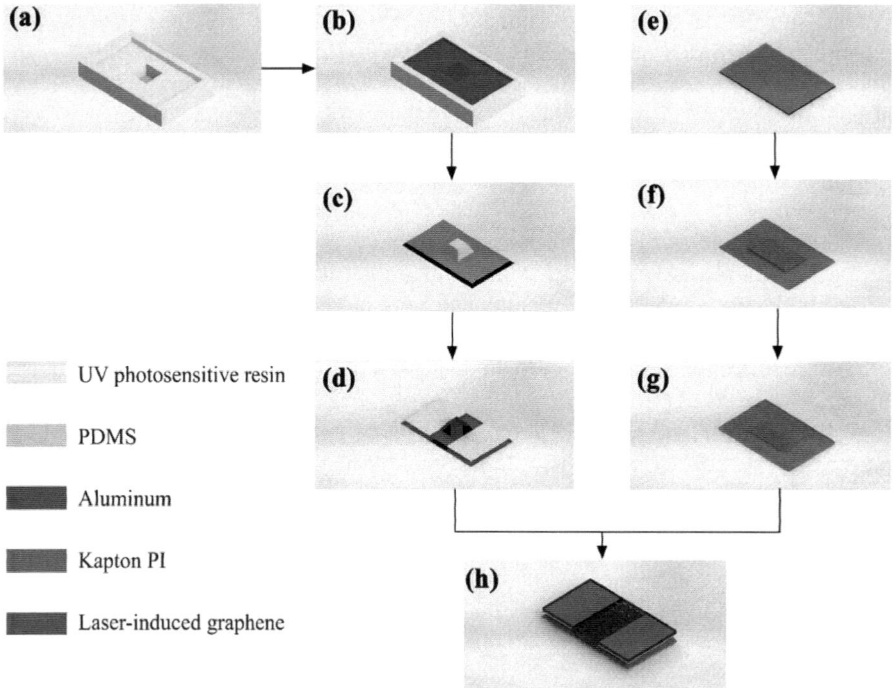

Figure 3. Fabrication process: (a) UV Curable structure by 3D printing. (b) Fill the mold with PDMS (c) PDMS after peel off (d) Deposit Aluminum (e) Clean Kapton PI substrate (f) Laser-induced graphene (g) Spin coating PDMS (h) Bonding two PDMS layer.

The fabrication process of the slip sensor is shown in Figure 3. A UV Curable Resin mold with pyramid structure is first fabricated by 3D printing. Then the mold is filled with a well stirred PDMS and deionized water mixture of 1:1 quality ratio (Figure 3b). After that, the mold is placed into a box at room temperature to stewing for 12 hours. Then the mold was put into an oven at 50 °C for 2 hours in order to evaporate the water inside the PDMS and leaves the holes. The SEM observation shows the average size of holes left in PDMS is about 12 μm, as shown in Figure 4. After evaporation, the PDMS is carefully demolded. A 200-nm-thick aluminum film is then deposited on the PDMS by magnetic sputtering method.

The bottom graphene electrodes of the sensor are generated by pulsed laser irradiation of polyimide （PI） films (Figure 3e)[8]. The LIG electrodes are directly writing on PI substrates through the laser. The sheet resistance of the fabricated LIG film is about 66.4 Ω/sq.

After the laser treatment, a thin film of PDMS is deposited on the electrodes surface by spin coating at 500 rpm. The PDMS was then cured at 80 °C for 2 hours. Then the PDMS layer is discreetly peeled off from the substrate together with LIG film. At last, the two layer of PDMS is bonded together with the assistant of oxygen plasma treatment.

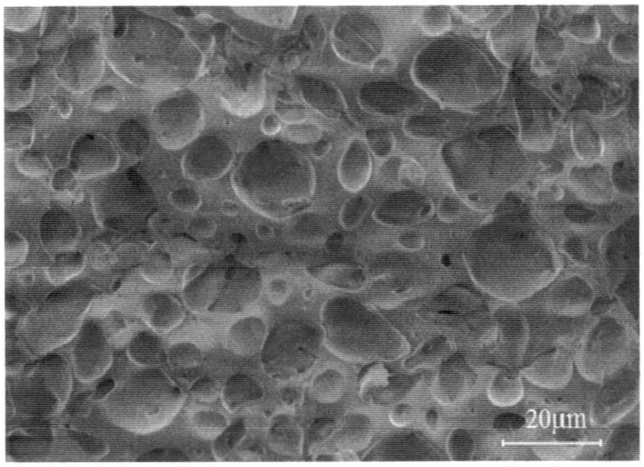

Figure 4. SEM image of the fabricated porous PDMS.

4. Result and discussion

As shown in Figure 5, the sensor test platform consists of THORLABS Z825B displacement platform, MARK-10 Series 7 digital force gauges, and KEYSIGHT 34470A digital multi-meter. In order to test a single sensing unit, a small normal force of 180mN is applied to the center of the unit at the beginning. Then a 300μm parallel movement of the pressure head at the different speed is generated from the translation stage. KEYSIGHT 34470A is set at the load of 10MΩ and the test wires are connected between the bottom electrode and ground separately.

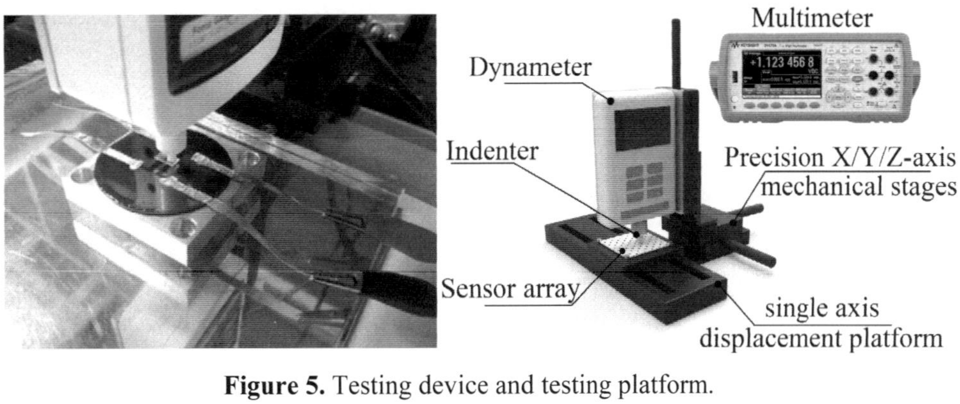

Figure 5. Testing device and testing platform.

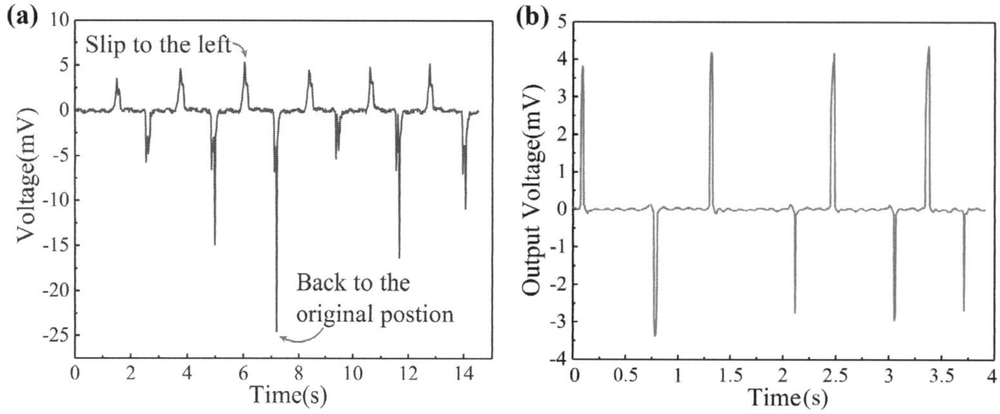

Figure 6. Slip test results of (a) signal output of a single electrode (b) the repeatability of the device.

The result of single electrode output with slip speed of 2mm/s is shown in Figure 6a. Distinct signals were observed when a slip happened on the device surface. Good repeatability of the device is also obtained, as shown in Figure 6b.

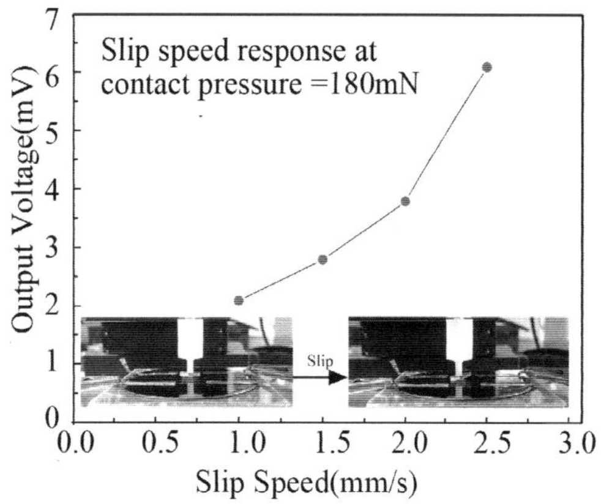

Figure 7. Output voltage of TENG structure at different slip speed.

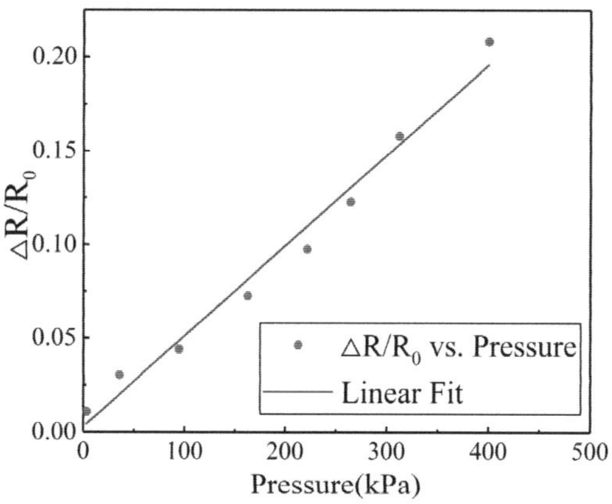

Figure 8. Relative resistance change ($\Delta R/R_0$) of the LIG structure at different pressure.

Figure 7 shows the output voltage of the slip sensor at different slip velocities varying from 0.25mm/s to 2.5mm/s. The peak output voltage of the sensor increased with the increase of the slip speed.

The static normal force response on LIG electrode was also tested by applying a normal force on one side of the unit. The test pressures begin with 3kPa to the end of 400kPa. The resistance of LIG changed linearly with pressure, as shown in Figure 8.

5. Conclusions

In this paper, we have demonstrated a flexible slip sensor based on single electrode mode triboelectric nano-generators. The sensor has a simple structure consisting of two layers of PDMS. Laser-induced graphene and aluminum are used as the electrode material pair. Experimental results show a distinct signal when a slip happens. In addition, the test of the piezoresistive effects of laser-induced graphene electrode indicates the slip sensor can be used to detect sliding and normal force simultaneously without changing the sensor structure. Further work will be conducted to minimize the size of the sensor to human fingerprints.

IOP Conf. Series: Journal of Physics: Conf. Series **986** (2018) 012009 doi:10.1088/1742-6596/986/1/012009

Acknowledgments
This work was supported by the National High Technology Research and Development Plan of China (2015AA043505) and the Shenzhen Fundamental Research Funds (No. JCYJ20150630170146829).

References

[1] Vatani M, Engeberg E D and Choi J-W 2013 Force and slip detection with direct-write compliant tactile sensors using multi-walled carbon nanotube/polymer composites *Sensors Actuators A. Phys.* **195** 90–7
[2] Cotton D P J, Chappell P H, Cranny A, White N M and Beeby S P 2007 A novel thick-film piezoelectric slip sensor for a prosthetic hand *IEEE Sens. J.* **7** 752–61
[3] Narendiran A and George B 2015 Capacitive tactile sensor with slip detection capabilities for robotic applications *2015 IEEE Int. Instrum. Meas. Technol. Conf. Proc.* **2015–July** 464–9
[4] Dubey V N and Crowder R M 2006 A dynamic tactile sensor on photoelastic effect *Sensors Actuators, A Phys.* **128** 217–24
[5] Francomano M T, Accoto D and Guglielmelli E 2012 Experimental characterization of a flexible thermal slip sensor *Sensors (Switzerland)* **12** 15267–80
[6] Wang Z L, Chen J and Lin L 2015 Progress in triboelectric nanogenerators as a new energy technology and self-powered sensors *Energy Environ. Sci.* **8** 2250–82
[7] Niu S and Wang Z L 2014 Theoretical systems of triboelectric nanogenerators *Nano Energy* **14** 161–92
[8] Lin J, Peng Z, Liu Y, Ruiz-Zepeda F, Ye R, Samuel E L G, Yacaman M J, Yakobson B I and Tour J M 2015 Laser-induced porous graphene films from commercial polymers *Nat. Commun.* **5** 1–8
[9] Wu Y, Beker L, Karakurt I, Cai W, Elwood J and Li X 2017 High resolution flexible strain sensors for biological signal measurements Tsinghua-Berkeley Shenzhen Institute , Tsinghua University , CHINA University of California , Berkeley , USA South China University of Technology , CHINA Southeast University , CHI 1144–7
[10] Peng J, Zhang H, Zheng Q, Clemons C M, Sabo R C, Gong S, Ma Z and Turng L-S 2017 A composite generator film impregnated with cellulose nanocrystals for enhanced triboelectric performance *Nanoscale* **9** 1428–33

Preparation and Luminescent Properties of the antibacterial materials of the La^{3+} Doped Sm^{3+}-Hydroxyapatite

Yuguang Lv [1,a,*], Qi Shi [1], Yuling Jin [2,b,*], Hengxin Ren [1], Yushan Qin [1], Bo Wang [1], Shanshan Song[1]
[1]College of Pharmacy, Jiamusi University, Jiamusi 154007, China
[2]First affiliated hospital of Jiamusi University, Jiamusi, Heilongjiang, 154002, China
[a] yuguanglv@163.com, [b] 975795903@qq.com
*Correspondence: yuguanglv@163.com; Tel.: +0454-86130808

Abstract: In this paper, the La^{3+}-doped Sm^{3+} hydroxyapatite (La/Sm/HAP) complexes were prepared by a precipitation method. The sample was defined by IR spectra, fluorescence spectra and X ray diffraction analysis et al. The structure of complexes were discussed. The emission wavelength of heat treatment of Sm^{3+} do not change, but will affect the intensity of the peak Sm^{3+} luminescence properties and the occupy hydroxyapatite in the lattice Ca(II)and Ca(I) loci with Sm^{3+} doped concentration and the proportion of the sintering temperature change and change: The nano hydroxyapatite complex of the La^{3+} doped samarium obtain the good fluorescence intensity, by La^{3+} doping content of Sm^{3+} were hydroxyapatite 6% (La^{3+}, Sm^{3+} mole ratio) device. The complex of La^{3+} doped samarium HAP have Stable chemical property, fluorescence property and excellent biological activity. The ligand HAP absorbs energy or captures an electron–hole pair and then transfers it to the lanthanide ions. The catalytic activity influence of the La^{3+}-doped Sm^{3+} hydroxyapatite was discussed, the La/Sm/HAP had excellent antibacterial property, which used as potential biological antibacterial material..

Keywords: Hydroxyapatite; Nano-lanthanum samarium complex; Luminescence; Biological antibacterial agent

1. Introduction

In nanotechnology, new materials in new applications is an interesting area of research. Preparation of newer material in Nano scale frequently show distinct characteristics on the luminescence, magnetism and structural property.Many rare earth complexes have been synthesized and used as the emitters in organic photoluminescence and electroluminescence devices. Lanthanide doped complexes were used in electrofluorescence, optical communication system, solid-state laser, microscopic imaging, catalytic activity and antibiosis. In previous references, the research focuses were concentrated on the photoreduction of Sm^{3+} ion to Sm^{2+} ion and increased hole burning efficiency of Sm^{2+} ion induced by fs laser. In recent decades, the up-conversion luminescences had been investigated in fluorescence material[1–4].As is well known, up to now, more and more conventional techniques, such as spin coating, nano self-assembly, colloidal deposit sediment, and solvent evaporation, have been explored to biological material and luminescent material self-assembly. Further, in order to meet the actual application, the chemical stability and material mechanical hardness should be high enough[5–7]. By the rise of Nanotechnology, hydroxylapatite study concentrate to the nano field. Nanotechnology focus on the preparing technology, which different sizes and shapes of nano hydroxylapatite[8–10]. Hydroxyapatite is the main inorganic components of the human and animal bones and teeth, itself does not have the obvious characteristics of light. Hydroxyapatite technology is mainly used in the following aspects: light emitting materials for lamp; Nanometer luminescent materials; Research in the field of chemical, medical attention[11–13]. In the paper, the La^{3+}-doped Sm^{3+} hydroxyapatite (La/Sm/HAP) complexes were prepared by a precipitation method. The sample was defined by IR spectra, fluorescence spectra and X ray diffraction analysis et al. The structure of complexes were discussed. The nano hydroxyapatite complex of the La^{3+} doped samarium obtain the good fluorescence intensity. The complex of La^{3+} doped samarium HAP have stable chemical property, fluorescence property and excellent biological activity. The catalytic activity influence of the La^{3+}-doped Sm^{3+} hydroxyapatite was discussed, the La/Sm/HAP had excellent antibacterial property, which used as potential biological antibacterial material.

2. Materials and Methods

Sm_2O_3, $Ca_{10}(PO_4)_6(OH)_2$ hydroxylapatite, $Sm(NO_3)_3{\cdot}6H_2O$, $La(NO_3)_3{\cdot}6H_2O$ and the other reagents are analytically pure.100ml 0.3mol/L concentration of the $(NH_4)_2HPO_4$ solution gradually drop 100ml 0.5mol/L $Ca(NO_3)_2$ aqueous solution, mixing and adjust system pH =10 by 1mol/L NaOH, continue to stirring 2 hour. In the same condition, using this way of preparation the lanthanum nitrate solution.Adjusting the solution to pH = 6.6 with 1mol/L HCl, different molar ratio of La^{3+} and Sm^{3+}account for 0%, 1%, 2%, 5%, 6%, 8%, 10%, 12%, 15%, 18%, 20%, 25% and 30% (La^{3+}+ Sm^{3+}+Ca^{2+}) gradually drop to the $La(NO_3)_3$ and $Sm(NO_3)_3$ solution, continue to stirring 2 hour. When Samples at 80°C aging 24 hour, products are complete centrifugal separation by deionized water, the Sm^{3+} samples by doped amount of apatite dry 24 hour at 80°C in dryer. Other different doping samples sintered 1 hour, respectively in 400, 450, 500, 600, 800, 1000 and 1200°C, getting to the nano La/Sm/HAP.

X-ray diffraction (XRD) was performed with Philips X Pert MPD diffractometer using Cu target as radiative source. FTIR spectra were recorded on a Nicolet-Nexus 670 infrared spectrophotometer by using the KBr. Hitachi F-4500 Fluorescence Spectrophotometer, and some other instruments.

3.Results and Discussion

3.1 X-ray diffraction and IR Spectrum analysis

In Figure 1, by the Scherrer formula, we calculated dimensions for 30.3 nm. La^{3+} and Sm^{3+} radius is smaller than Ca^{2+}, by La^{3+}, Sm^{3+} occupied location of Ca^{2+}, the nano hydroxyapatite sample add to the lattice, not only adsorb to surface. IR spectrum, 1042 and 971cm^{-1}, 606 and 568cm^{-1} is the phosphate characteristic band, 3346 and 641cm^{-1} is the hydroxyl absorption band, property band of the diverse component in HAP sample.

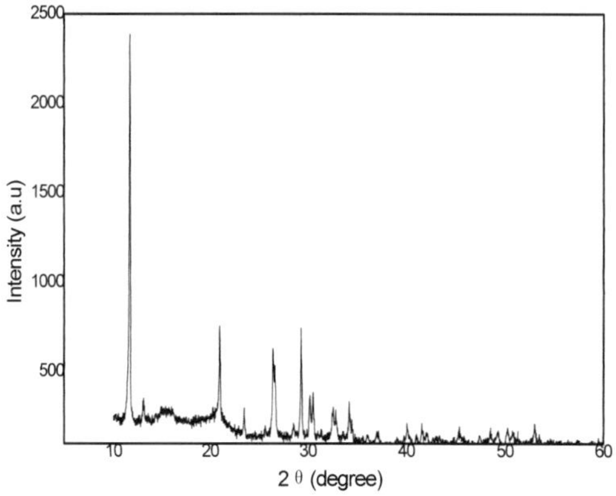

Figure 1. X-ray diffraction of 6% La^{3+} mix Sm $^{3+}$ HAP (La^{3+} and Sm^{3+} mole fraction ratio)

3.2 Luminescent properties

In Figure 2, Excited spectra are measured with λex=386nm. La^{3+}-doped Sm^{3+} HAP complexes emit key band, Sm^{3+} for ${}^4G_{5/2}\rightarrow{}^6H_{5/2}$, ${}^4G_{5/2}\rightarrow{}^6H_7$ and ${}^4G_{5/2}\rightarrow{}^6H_{9/2}$. The emission relative intensity at ${}^4G_{5/2}\rightarrow{}^6H_{9/2}$ is strongest. La/Sm/HAP emits the red fluorescence. It exhibits strong luminescence intensity[14].

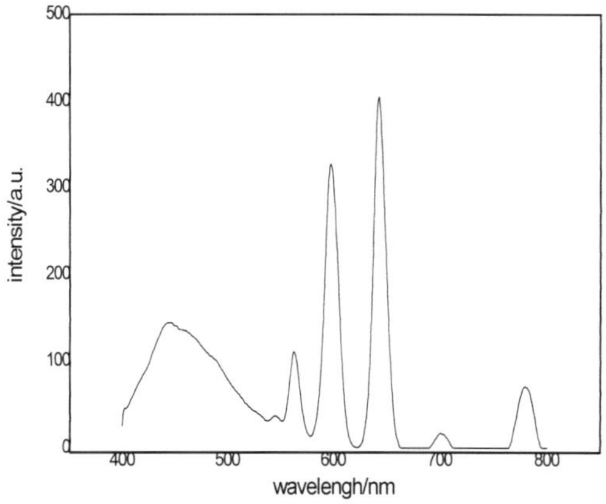

Figure 2. Fluorescent emission spectrum of 6% La^{3+} mix Sm^{3+} hydroxyapatite (La^{3+} and Sm^{3+} mole fraction ratio)

3.3 Influence of catalytic activity

Figure 3 shows the comparison of different photocatalysts active under visible light (a La/Sm/HAP, b Sm/HAP), the calcination temperature is 900°C, the same amount of sample. the 6% La^{3+} doped Sm^{3+} HAP, in the visible light (energy saving lamp 11W) under light activity patterns of the catalysts. Nanometer material La/Sm/HAP was higher than that of the two component Sm/HAP, this is because the absorption edge of La/Sm/HAP nanoparticles in the visible region is much larger than Sm/HAP, the catalytic activity of La/Sm/HAP nanoparticles under visible light is higher than Sm/HAP.

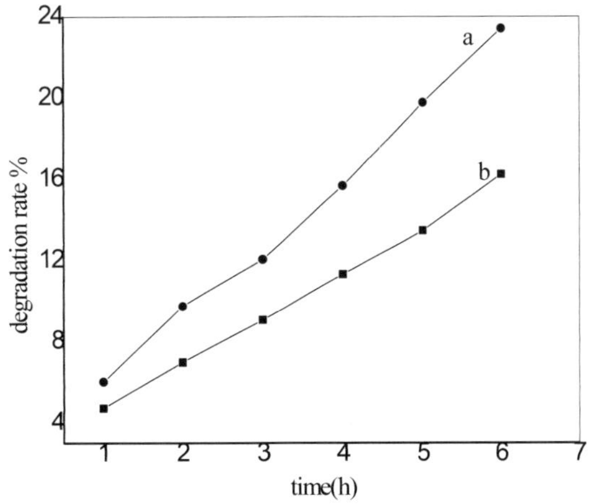

Figure 3. Comparison of different photocatalysts active under visible light

(a La/Sm/HAP b Sm-HAP)

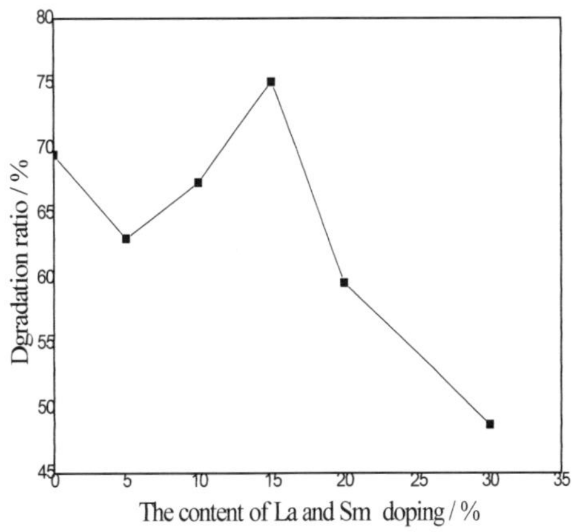

Figure 4. Effect of La/Sm/HAP P photocatalytic degradation of phenol by catalytic activity

Figure 4 shows lanthanum samarium doped amount on the hydroxyapatite figure of photocatalytic degradation of phenol by catalytic activity, comparison of lanthanum samarium addition on the catalytic performance of photocatalytic degradation of phenol. As it can be seen from the graph effect, with the increasing amount of lanthanum Samarium, the La-Sm-HAP activity of photocatalyst is decreased and then rose;To achieve the best in 15%, especially when the sintering temperature reaches 900°C. But with lanthanum samarium doped volume continues to increase and its catalytic activity is reduce, the catalytic activity of 15% La^{3+} doped Sm^{3+} HAP reached the maximum. With the increase of La and Sm content, hydroxyapatite ratio decreased, the catalytic active center is reduced, the catalytic activity decreased. The hydroxyapatite lanthanum samarium has good biological activity.

As everyone knows, rare earth elements have not filled 4d shell, a linear spectral of quite complex and optical properties of strange. The ligand HAP absorbs energy or captures an electron–hole pair and then transfers it to the lanthanide ions. HAP has the wide band gap spectrum, doping La and Sm ions produce a generation separated energy levels in the system, La^{3+} and Sm^{3+} acted as electron trapping center. When the effective excitation, La^{3+} and Sm^{3+} capture the part of the electronic.Through the energy transfer of La/Sm/HAP, it enhanced the antibacterial and the degradation of organic dyes properties of the inorganic nanocomposite material of La/Sm/HAP[15,16].

4.Conclusions

In this paper, the La^{3+}-doped Sm^{3+} hydroxyapatite (La/Sm/HAP) powder was prepared by a precipitation method. Samples were defined by IR spectra, fluorescence spectra and X ray diffraction analysis et al, the structure of complexes were discussed. The complex of La^{3+} doped samarium HAP have stable chemical property, fluorescence property and excellent biological activity. The catalytic activity influence of the La^{3+}-doped Sm^{3+} hydroxyapatite was discussed, the La/Sm/HAP had excellent antibacterial property, which used as potential biological antibacterial material.

5.Acknowledgments

This study was financially supported by National Science Foundation of China (No. 21346006), Department of scientific research project in Heilongjiang province (No. B2017015), Scientific research project of Heilongjiang province education department (No.12541783), National project training project of Jiamusi University (No.JMSUJCGP 2016-003), Principal innovation and entrepreneurship fund project of Jiamusi University (xzyf 2014-18); Scientific research project of Jiamusi University (JMSUJCMS2016-013), Support was also given by the Key Laboratory of Photochemistry, Institute of Chemistry, Chinese Academy of Sciences, and the Key Laboratory of Luminescence and Optical Information, Beijing Jiaotong University.

References

[1] Sen Qiu, Zeyuan Zhao, and Xiaoqi Sun. 2017 *J. Chem. Eng. Data.* **62** 469-476.
[2] Jiewei Rong, Wenwei Zhang and Junfeng Bai. 2016 *CrystEngComm.* **18** 7728-7736.
[3] Liu Wenbin, Adu and Lv Yuguang. 2015 *Applied Mechanics and Materials.* **716** 32-35.
[4] Jennifer Leduc, Yakup Gönüllü and Aida Raauf. 2017 *Semiconductors and Semimetal.,* **97** 185-215.
[5] M. Mortier, P. Goldner, C. Chateau and M.Genotelle. 2001 *J Alloy Compd.* **323** 245 -253.
[6] A. S. Gouveia-Neto, E. B. Da Costa, P. V. Dos Santos, L. A. Bueno and S. J. L. Ribeiro. 2003 *J Appl. Phys.* **94** 5678-5683.
[7] J. Kang, L. Zhuo, X. Lu, H. Liu, M. Zhang and H. Wu. 2004 *J. Inorg. Biochem.* **98** 79-83.
[8] A. Banerjee, A. Bandyopadhyay and S. Bose. 2007 *Mater. Sci. Eng. C.* **27** 729-735.
[9] B. Li, B. Guo, H. Fan and X. Zhang. 2008 *Appl. Surf. Sci.* **255** 357-362.
[10] L. H. Li, M. Y. Zhao, S. Ding and C. R. Zhou. 2011 *Express Polym. Lett.* **5** 545-549.
[11] R. Reyesr, M. Cremona, E. E. S. Teotonio, H. F. Brito and O. L. Malta. 2004 *Thin Solid Films.* **469** 59-65.
[12] B. Zhai, L. Yi, H.-S. Wang., B. Zhao, P. Cheng, D.-Z. Liao and S.-P. Yan. 2006 *Inorg. Chem.* **45** 8-13.
[13] C. Platas-Iglesias, C. Piguet, N. Andren and J.-C.G. Bunzli. 2001 *J. Chem. Soc Dalton Trans.* **18** 2655-2659.
[14] Y. Lv, C. Song, Y. Zhang, J. Sha, C. Liu and F. Zhang. 2010 *J. Alloys Compd.* **492** 259- 263.
[15] L. I. Yongjin, S. Zhiguo, L. I. Chen, Q. I. U. Jianbei, Y. Zhengwen, Y. I. N. Zhaoyi, Y. Yong, W. A. N. Ronghua and W. Qi. 2013 *J. Rare Earths.* **31** 400-405.
[16] E. Zelazowska, E. Rysiakiewicz-Pasek and M. Borczuch- Laczka. 2013 *Solid State Phenomena.* **200** 38-44.

IOP Conf. Series: Journal of Physics: Conf. Series **986** (2018) 012011 doi:10.1088/1742-6596/986/1/012011

A novel MEMS inertial switch with a reinforcing rib structure and electrostatic power assist to prolong the contact time

Jian Li [1,a], Yan Wang [1,b,*], Zhuoqing Yang [1,c], Guifu Ding [1,d], Xiaolin Zhao [1,e] and Hong Wang [1,f]

[1]Shanghai Jiao Tong University, Shanghai, 200240, CHINA

[a]lijian19930405@sjtu.edu.cn, [b]wyyw@sjtu.edu.cn (corresponding author), [c]yzhuoqing@sjtu.edu.cn, [d]gfding@sjtu.edu.cn, [e]xlzhao@sjtu.edu.cn, [f]wang90@sjtu.edu.cn

*Correspondence: wyyw@sjtu.edu.cn; Tel.: +86-021-34206687

Abstract. The MEMS inertial switch is widely used in various industries owing to its advantage of small size, high integration, low power consumption and low costs, especially in the timing of Internet of things, such as toys, handheld devices, accessories and vibration testing. This paper provided a novel inertial switch with a reinforcing rib structure and electrostatic power assist. The designed inertial switch can reduce the complexity of the post-processing circuit and broaden its application prospect. The continuous electrostatic force can extend the contact time of the designed inertia switch before the leakage of electricity ends. The moving electrode with a reinforcing rib structure can effectively restrain the bending of the lower surface of moving electrode caused by residual stress. The array-type fixed electrode can ensure stable contact between the electrodes when the device is sensitive to external shocks. The dynamic displacement-time curve can be simulated by the COMSOL finite element simulation software. The laminated plating process is used to produce the designed inertial switch and the drop hammer acceleration monitoring system is used to test the fabricated device. The results indicate that, compared with the traditional design, the bouncing phenomenon can be prevented and extend the contact time to 336μs.

1. Introduction

In the times of wireless sensor networks and the internet of things (IOT), the status of many things needs to be monitored in real time, such as storage and transportation of goods, toys and car collision [1-2]. In some remote or harsh areas, some devices require continuous vibration detection and transmit the status data to the central control system. In these practical applications, inertial threshold sensor plays a very important role and has a wide range of applications. The micro-accelerometer can accomplish the tasks of these applications well, but it often requires a constant supply of energy, which will lead to large power dissipation and limits its applications together with the difficulty of analog signal processing. Inertia switch is a passive device and outputs pulse signal in the practical applications, which make it better to meet the requirements of the above application environment, especially the vibration detection in the remote unattended harsh environment. However, the rigid contact collisions in traditional designs always lead to very short contact times and rebound phenomena, which makes the subsequent signal processing circuits very complex. The flexible contact electrodes have been designed to extend the

IOP Conf. Series: Journal of Physics: Conf. Series **986** (2018) 012011 doi:10.1088/1742-6596/986/1/012011

contact time, but the contact time is greatly limited (tens of microseconds) due to the characteristics of the flexible contact structure [3-4]. The designed inertial switch with a reinforcing rib structure and electrostatic power assist can greatly prolong the contact time (hundreds of times) and effectively restrain the bouncing phenomenon.

2. Design and Fabrication

2.1. Design

As shown in figure 1, a type of one-way-trigger inertial switch is proposed in this study. The designed structure consists of three parts: a moving electrode with a reinforcing rib structure, an array structure fixed electrode that is connected to each other through a layout connection, and a pulling electrode with a large surface area. The reinforcing rib structure of the upper surface of the moving electrode can effectively restrain the warping of the electrode produced by the residual stress of the electroplating. The good flatness of electrode surface ensures that the moving electrode and fixed electrode produce as much electrostatic attraction as possible through the parallel structure. The moving electrode is connected by four conjoined snake-like springs and is in a suspended state. The fixed electrode is composed of a contact array structure, which is different from the previous single contact structure [5]. The array structure extends the contact range of the fixed electrode and moving electrode, and can have a certain amount of redundancy under the accidental condition that the lack of flatness of the moving electrode to ensure the most stable contact of the two electrodes. A sufficient micro-gap is designed between the pulling electrode and the fixed electrode, which ensures that the two electrodes can be mutually insulated from each other. The fixed electrode is slightly higher than the pulling electrode, which can ensure the mutual insulation between the moving electrode and pulling electrode when the moving electrode is in contact with the fixed electrode. The method of electroplating growth can make the fixed electrode higher than the pulling electrode to maintain micron-level accuracy and plane consistency. A multi-hole cantilever beam is designed as a barrier-limiting structure above the moving electrode, which can effectively prevent the rebounding phenomena of the moving electrode in the non-sensitive direction in a practical application.

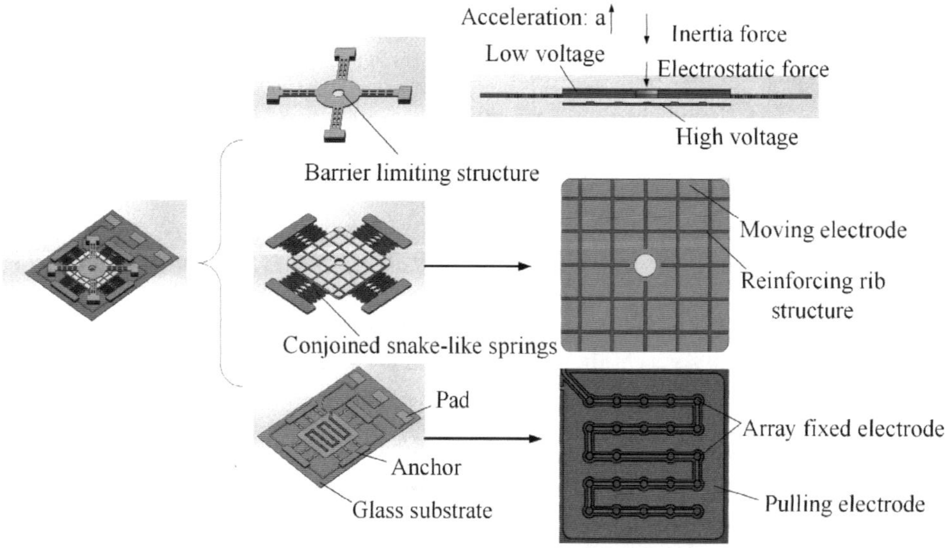

Figure 1. Schematic of the designed electrostatic force-assisted MEMS inertial switch.

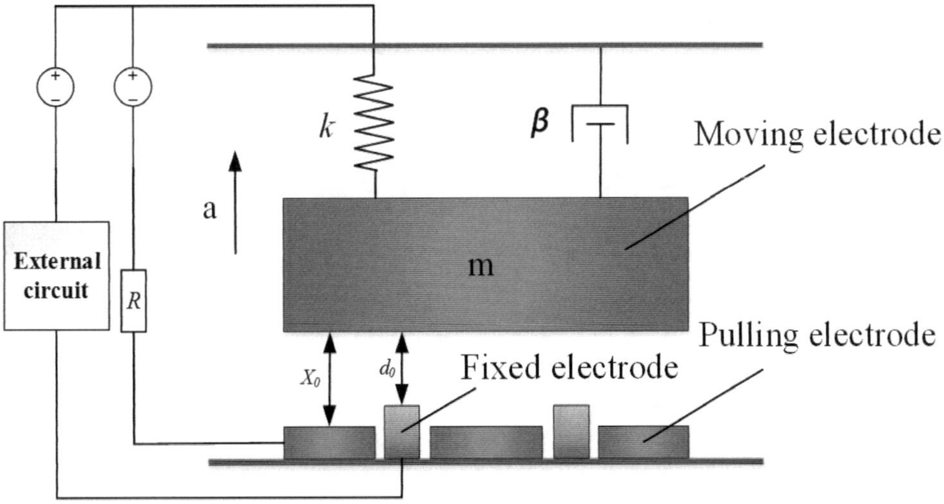

Figure 2. Model of the designed electrostatic force assisted inertia switch.

Figure 2 shows the model of the designed electrostatic force assisted inertia switch, which can be regarded as a spring–mass electrostatic-field-damping system. The equation of dynamic motion control can be described as follows:

$$m\frac{d^2x}{dt^2}+\beta\frac{dx}{dt}+kx+\frac{\varepsilon AV^2}{2(y-x)^2}+ma(t)=0 \quad (1)$$

where *m* is the mass of the moving electrode, *β* is the damping coefficient, *k* is the elastic coefficient, *ε* is the dielectric constant of air, *A* is the effective overlap area for the electrostatic force, *V* is the voltage between the moving and pulling electrodes, and *a(t)* is the time-varying external acceleration. The resistance of the damping system is mainly due to the air damping generated on the moving electrode in the movement process of the moving electrode.

Table 1. The structural dimensions of the designed device.

Structure	Structural dimensions	Size(μm)
Moving electrode	Rectangular side length L	1810
	Mass Thickness m	60
	Stiffener width a	30
	Stiffener thickness x	15
	Suspension gap d	10
Snake-like spring	Stretch across length l	1040
	Inner radius r2	25
	Thickness h	15
	Width b	25
Fixed electrode	Thickness h1	5
	Radius R	40

Table 1 summarizes the main structural dimensions of the designed device of the designed switch. In the design, the moving electrode thickness m is set to be variable, which can be changed through photolithography, lamination, electroplating, or other surface micromachining processes. In the device structure of the same shape and other parameters, the performance parameters of the device is changed by the preparation of the moving electrode with different thickness.

2.2. Fabrication

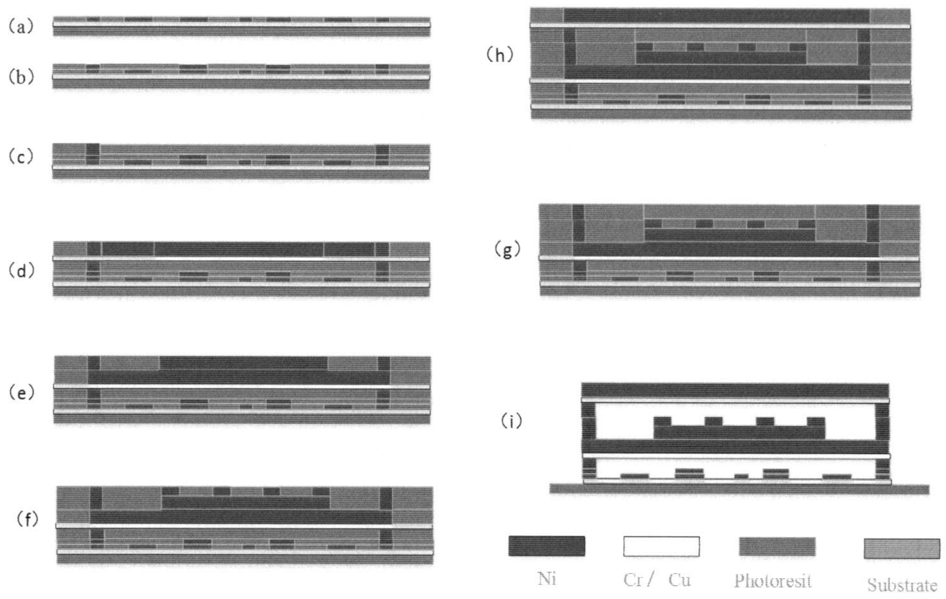

Figure 3. Main manufacturing process of micromechanical inertial: (a) Lead layer, (b) Support layer, (c) Suspended layer, (d) Spring layer, (e) Mass layer, (f) Reinforcing rib layer, (g) Second suspended layer, (h) Barrier layer and (i) Completed structure.

The designed MEMS inertial switch is fabricated by laminated plating process and nickel metal is selected as the structural material. The structure and morphology of the device are determined by lithography, development, and subsequent electroplating processes, and the thickness of each layer is determined by the thickness of the photoresist. The plating bath temperature is 45 °C, the pH value is about 4, and the plating rate is about 0.3μm/min. The composition of the nickel electrolyte is Ni $[NH_2SO_3]_2$(600 g/L), $NiCl_2 \cdot 6H_2O$ (10 g/L), H_3BO_3 (25 g/L). The main manufacturing process of the designed device is shown in figure 3 and is described as follows:

(a) Lead layer: select the quartz glass as an insulating substrate; sputter Cr/Cu on the substrate as the first seed layer; spin-coat a layer of positive photoresist and transfer graphics; electroplate the fixed electrode and pulling electrode;

(b) Support layer: electroplate the fixed electrode to make the fixed electrode higher than the suction electrode by 1–2μm;

(c) Suspended layer: electroplate the support column to support the spring–mass system and form the air gap between the electrodes;

(d) Spring layer: sputter the second Cr/Cu seed layer on the support layer and plate the conjoined snake-like spring and bottom mass block;

(e) Mass layer: construct the mass block out of a layer of electroplated nickel to make the mass block higher than the height of the spring;

(f) Reinforcing rib layer: construct the stiffener rib structure by electroplating the surface of the mass block to effectively prevent the uneven surface caused by residual stress;

(g) Second suspended layer: electroplate the supporting column to support the multi-hole barrier cantilever beam structure, and form an air gap between the barrier limiting structure and mass block;

(h) Barrier layer: sputter the third Cr/Cu seed layer and electroplate the multi-hole barrier cantilever beam structure;

(i) Release process: remove the photoresist with sodium hydroxide solution, remove the copper seed layer with ammonia and hydrogen peroxide solution, remove nickel seed layer with potassium permanganate solution acetone. The overall structure of the designed device is now obtained.

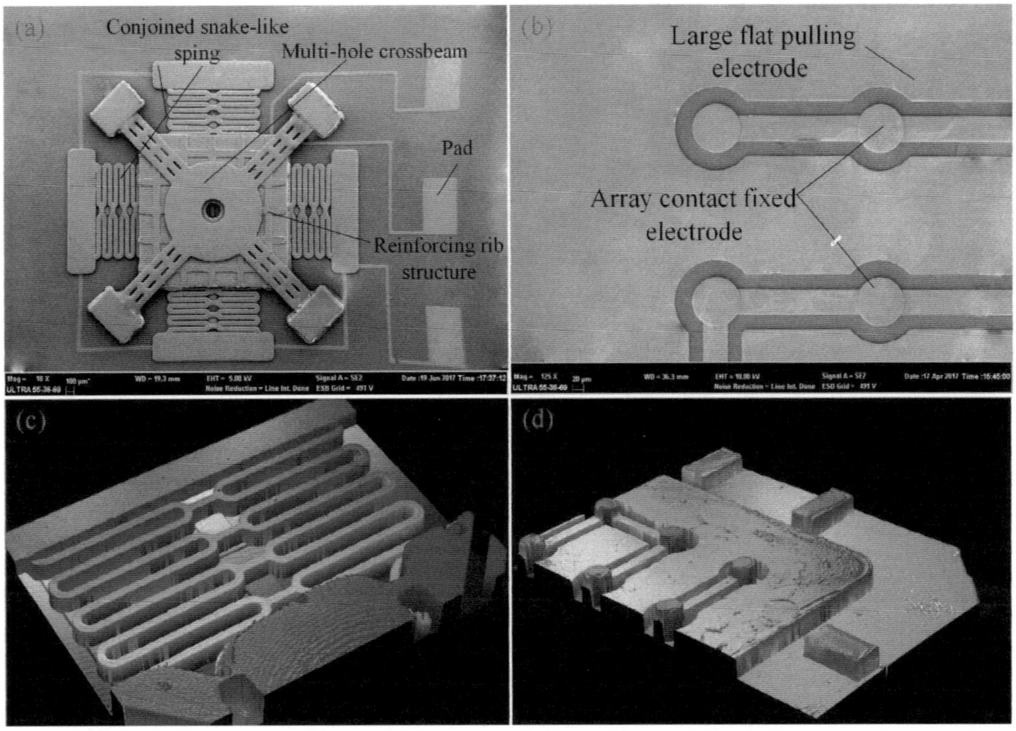

Figure 4. SEM images of (a) the fabricated device and (b) the designed large flat pulling electrode and array-type fixed electrode and VEECO images of (c) the spring-mass structure and (b) the pulling electrode and fixed electrode.

Figure 4 displays scanning electron micrograph (SEM) images and VEECO images of the overall structure of the released inertial switch with the electrostatic force assist enhanced contact effect. (a) shows the obtained overall structure of the designed device, which include Pad, Multi-hole crossbeam, Reinforcing rib structure and so on. (b) shows the top view of the array-type fixed electrode and the pulling electrode, which are insulated from each other by the micro-gap. (c) shows the VEECO three-dimensional oblique drawing of the conjoined snake-like spring and partial moving electrode. And (d) shows the pulling electrode and fixed electrode, and it clearly shows that the fixed electrode is slightly higher than the pulling electrode.

3. Simulation results and Experimental results

3.1. *Simulation results*

As shown in figure 5, a quarter of the finite element model is used as the simulation model because of the design of the MEMS inertial switch as a central symmetrical structure. This simulation model can effectively reduce the requirements for computer memory and other pieces of hardware, and greatly reduce the simulation time. The designed electrostatic force-assisted MEMS inertial switch includes the external acceleration of the half-sine wave, electrostatic force acting on the moving electrode, and elastic restoring force of the spring, which is a multi-physical field coupling problem. The COMSOL simulation software is used to simulate and analyze the dynamic response of the designed structure. The end of the spring, fixed electrode, and pulling electrode are set to a fixed constraint in the model, that is, the degree of freedom is 0. The displacement of the moving electrode can only exist in the Z-direction, and the displacement in the horizontal direction is set to 0. The free triangulation mesh and sweeping method are used to mesh the mesh. Nickel metal is selected as the structural material of the model because of its excellent mechanical properties. The parameters of the nickel metal are as follows: Young's modulus: 165GPa; density: 8.96g·cm^{-3}; Poisson's ratio: 0.31 [6].

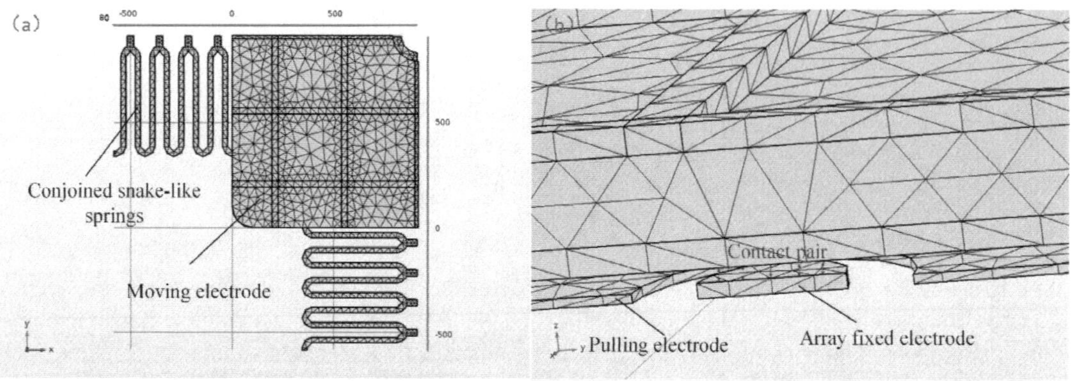

Figure 5. Schematic diagram of finite element simulation model.

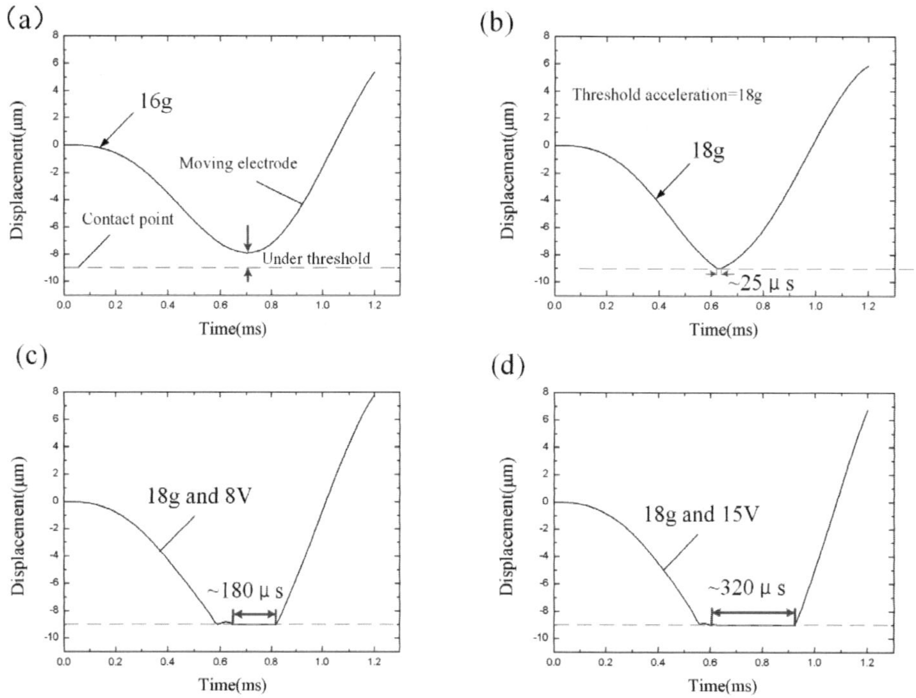

Figure 6. Dynamic Response curve of the designed device at the acceleration 16g, acceleration 18g, acceleration 18g and 8V, and acceleration 18g and 15V.

Figure 6 shows the dynamic response curve of the designed device in the case of the acceleration 16g, acceleration 18g, acceleration 18g and 8V, and acceleration 18g and 15V. It can be seen from the simulation that the threshold acceleration of the designed structure is 18g when there is no voltage applied. When the applied external acceleration amplitude is 18g, the simulation is also done in the condition of the applied pulling voltage is 8V and 15V, which shows that the contact time is 180μs in the case of an applied voltage of 8V, and the contact time is pro-longed to 320μs when the applied voltage is 15V.

3.2. Experimental results

The drop hammer system is used to characterize the designed micromechanical inertial switch, and Figure 12 shows the test principle. The acceleration of the drop hammer is detected by the standard accelerometer (ADXL-193, Analog Device Inc.) and the pulse signal generated by the MEMS inertial switch is detected by a multichannel oscilloscope (Agilent 6000 MSO6034A) when the fabricated device is triggered. The standard accelerometer and the inertial switch prepared are mounted on a drop hammer of a standard drop weight system. The fabricated device and standard accelerometer are connected to the external circuit. The inertial switch with a constant voltage source of 3V and current-limiting resistor resistance of 300Ω constitute the circuit. When the drop hammer falls from a certain height and collides with the substrate, a half-sine acceleration will occur on the device. Changing the height of the drop bar and the stiffness of the substrate can change the amplitude and pulse width of the resulting half-sine wave. When the applied external acceleration reaches or exceeds the inertia switch threshold, the inertia switch is turned on and a pulse signal is output.

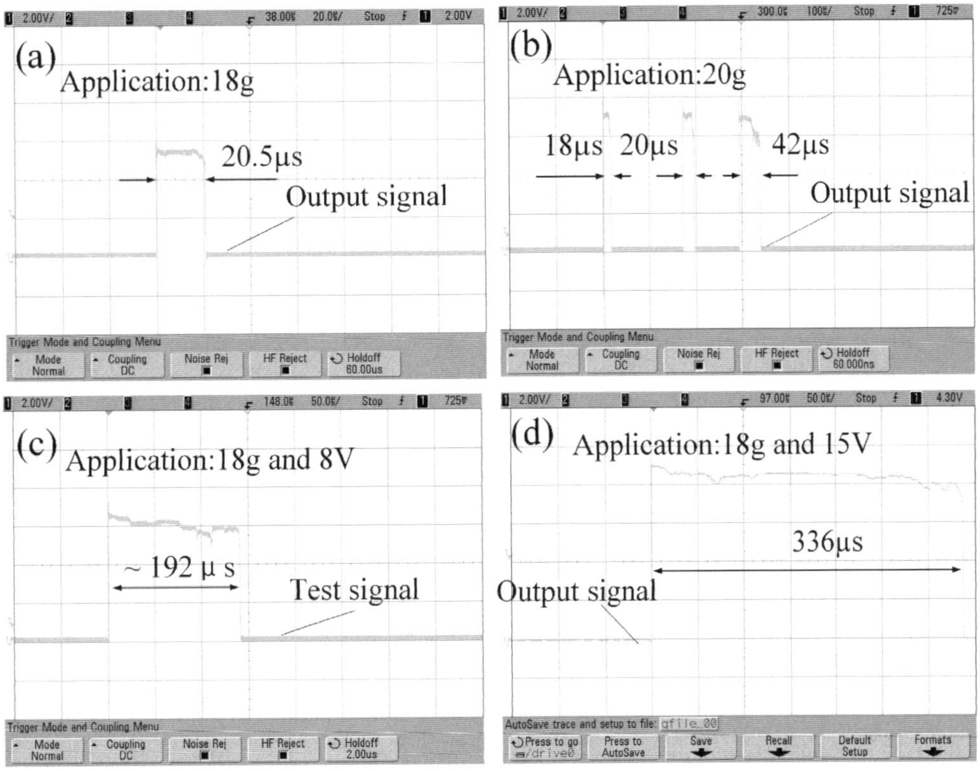

Figure 7. Dynamic response results of the fabricated MEMS inertial switch at the acceleration 18g, acceleration 20g, acceleration 18g and 8V, acceleration 18g and 15V.

Figure 7 shows the contrast test results under with and without the applied pulling voltage. Without the pulling voltage, the fabricated device switch on and the contact time is 20.5μs when the applied external acceleration is 18g, which is called the thresh-old acceleration of the device. The test result shows that the test result is slightly higher than the simulation result, which is due to the deviations from the microfabrication process. In the same condition that without voltage applied and when the applied acceleration is 20g, the device has a rebounding phenomenon and the minimum contact time is 18μs and the maximum contact time is 42μs. The contact time of the device is 192μs when the applied external acceleration is 18g and the applied pulling voltage of 8V. When the applied acceleration is equal to 18g and the applied voltage is 15V, the contact time of the device is 336μs and there is no rebounding phenomenon. When the applied pulling voltage continues to increase, the air film between the moving electrode and the pulling electrode may be broken down. The experimental results are in good agreement with the simulation results and verify that the designed MEMS inertial switch structure can effectively extend the contact time.

4. Conclusions

An inertial switch with an electrostatic force assist enhanced contact effect was proposed, simulated, and successfully fabricated by laminated plating process. The array-type fixed electrode and moving electrode with a reinforcing rib structure were designed and fabricated, which can ensure stable contact of the two electrodes and restrain the warping of the device caused by residual stress. The whole device is completed by laminated plating process. The test by the dropping hammer system indicates that the contact time of the fabricated inertial switch is about 336μs, which agrees well with the simulated contact time (320μs) in its sensing direction by the finite element COMSOL simulation software. It is demonstrated that the proposed electrostatic force-assisted MEMS inertial with the reinforcing structure is beneficial for extending the contact time.

5. Acknowledgments:

The authors would like to thank the support from National Key Laboratory of Micro/Nano Fabrication Technology, School of Electronic Information and Electrical Engineering, the Hi-Tech Research and Development Program of China (2015AA042701), the National Key Laboratory of Micro/Nano Fabrication Technology Foundation (9140C790403150C79332), and the Advanced Research Ministry of Education Joint Foundation (6141A02033403), SJTU SMC Young Scholar Program (16X100080044).

References

[1] Michaelis S, Timme H, Wycisk M, et al. Additive electroplating technology as a post-CMOS process for the production of MEMS acceleration-threshold switches for transportation applications[J]. Journal of Micromechanics & Microengineering, 2000, 10(2):120.

[2] Wycisk M, Tönnesen T, Binder J, et al. Low-cost post-CMOS integration of electroplated microstructures for inertial sensing[J]. Sensors & Actuators A Physical, 2000, 83(1–3):93-100.

[3] Yang Z, Ding G, Cai H, et al. A MEMS Inertia Switch With Bridge-Type Elastic Fixed Electrode for Long Duration Contact[J]. IEEE Transactions on Electron Devices, 2008, 55(9):2492-2497.

[4] Chen W, Wang Y, Wang Y, et al. A laterally-driven micromachined inertial switch with a compliant cantilever beam as the stationary electrode for prolonging contact time[J]. Journal of Micromechanics & Microengineering, 2014, 24(6):065020.

[5] Wang Y, Chen W, Yang Z, et al. An inertial micro-switch with compliant cantilever fixed electrode for prolonging contact time[C]// IEEE, International Conference on MICRO Electro Mechanical Systems. IEEE, 2013:600-603.

[6] Yang Z, Ding G, Chen W, et al. Design, simulation and characterization of an inertia micro-switch fabricated by non-silicon surface micromachining[J]. Application of Statistics & Management, 2007, 17(8):1598.

Thermocouple-based Temperature Sensing System for Chemical Cell Inside Micro UAV Device

Yanhui Han[1*], Yue Feng[1], Haozhe Lou[2], Xinzhao Zhang[1]

[1]National Key Laboratory of Electro-Mechanics Engineering and Control, School of Electro-Mechanics Engineering, Beijing Institute of Technology, Beijing, China, 100081. [2]No.101 Middle School, Beijing, China, 100091

Corresponding Author: Yanhui HAN, School of Electro-mechanics Engineering, Beijing Institute of Technology, Tel: 86-10-62772108, FAX: 86-10-62796707, email: hui920718@163.com

Abstract. Environmental temperature of UAV system is crucial for chemical cell component inside. Once the temperature of this chemical cell is over 259 ℃ and keeps more than 20 min, the high thermal accumulation would result in an explosion, which seriously damage the whole UAV system. Therefore, we develop a micro temperature sensing system for monitoring the temperature of chemical cell thermally influenced by UAV device deployed in a 300 ℃ temperature environment, which is quite useful for insensitive munitions and UAV safety enhancement technologies.

1. Introduction

In the process control of industrial production, temperature is an important parameter, thermocouple is one of the most widely used temperature sensors in engineering, its main characteristic is that the range of temperature measurement is wide, the performance is stable, the structure is simple and the dynamic response is well enough, it also easy to automatic control and centralized control and remotely transmit 4-20mA electrical signals. So, the thermocouple is important in temperature measurement. But because of the nonlinear relationship between the thermoelectric potential and the temperature of thermocouple, the complexity of display and processing is increase, and also with the development of industry and automation, the requirement of temperature accuracy is much higher. In modern industrial sites, thermocouples are commonly used to test high temperatures, and test results sent to the main control unit. Because the thermoelectric potential of the thermocouple is nonlinear with temperature, the thermocouple must be linearized to maintain the accuracy of the test.

The work environment is very important for all kinds of instruments or equipment to perform best and operate normally, One of the most important physical quantities in environmental factors is temperature. Thermocouple is one of the most widely used temperature sensors in Engineering. The utility model of thermocouple has the advantages of simple structure, convenient operation, well accuracy, small thermal inertia, good stability and reproducibility, wide temperature measuring range, etc. It is suitable for remote transmission, automatic recording and centralized control of signals, so it is important in temperature monitoring.

The main content of this work is using 89C52 single chip as the controller, K type thermocouple as the temperature sensor. The exclusive A/D convertor of thermocouple and the 5-industrial bit liquid crystal display HJ1602A formed the digital temperature monitoring system. Thermocouple acquisition temperature signal through signal conditioning, uses the A/D convertor to the microcontroller, and display the temperature in HJ1602A.

2. Design and Methods

Figure 1 shows thermal simulation of a chemical cell model in finite element analysis (FEA). The transient temperature of chemical cell is shown in figure 2, while the boundary temperature condition is set as 300 ℃ at the surface of UAV device. After 20 min elapsed, the temperature of chemical cell is heated up to 80 ℃, which is still quite lower than that of the threshold value.

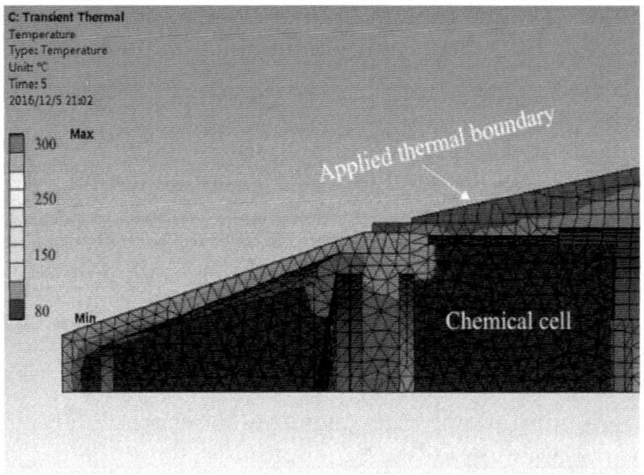

Figure 1. Thermal simulated UAV system using FEA.

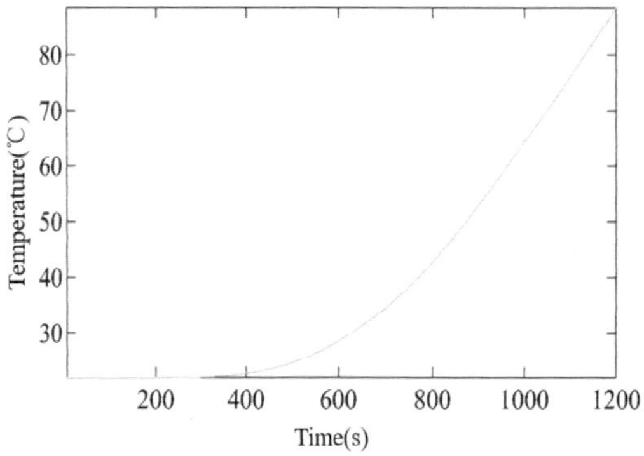

Figure 2. Varied temperature of the chemical cell

In the field of control, C52 Series MCU controller with better stability, faster and more accurate calculation accuracy is undoubtedly one of the goals that people pursue, the digital sensor and SCM can be organically linked, sensors need to collect the required information and make it digital. The microcontroller can process it directly so as to realize the interactive control between them. Secondly, because of the strong core control ability of the MCU, overlapping the auxiliary circuit, and then get the practical development system is necessary.

The working environment to achieve the best performance, whether the normal operation is very important for a variety of instruments or equipment, and a very important physical quantity of environmental factors is temperature, thermocouple temperature sensor engineering is one of the most widely used, it has the advantages of simple structure, convenient use, accuracy, small thermal inertia, stability and reproducibility good temperature, wide measuring range. It is also suitable for remote transmission, automatic recording and centralized control of signals, and plays an important role in temperature measurement. With the development of electronic information technology, new material and automation technology, sensor technology has also been changing with each passing day. SCM and automatic control system have been widely used in many fields.

We are no longer limited to analog signals collected from the environment, but to consider how to obtain the processing of digital signals, so the Wen Min devices, A/D converters and memory are integrated together to form a digital sensor. In the field of control, it has better stability, C52 Series MCU control faster and more accurate calculation accuracy is one of the goals that people pursue, it brings convenience is not negative, while the digital sensor and single chip to connect sensors to collect required information, and the digital, so SCM can be processed directly, so as to realize the interactive control. Secondly, because of the strong core control ability of the MCU, we can overlap the auxiliary circuit, and then get the practical development system. The temperature measurement system can adopt this mode. At present, the accuracy and convenience of measurement are getting higher and higher. If we start with micro controller technology, it will be a good solution to this problem.

The feature of this scheme is that the AT89S52 is a low-power, high-p erformance CMOS8 bit microcontroller with 8K in system programmable Flash memory. The AT89S52 is manufactured using Atmel's high-density nonvolatile memory technology, fully compatible with industrial 80C51 products, instructions, and pins. The on-chip Flash allows program memory to be programmable in the system and also suitable for conventional programming. On a single chip, the AT89S52 has a smart 8 bit CPU and programmable Flash in the system, enabling AT89S52 to provide high flexibility and ultra efficient solutions for many embedded control applications. MAX6675 is a high-precision integrated chip. It is small in size and does not need any other peripheral circuits. It greatly reduces the components and I/O connections in the circuit, thus simplifying the system structure.

Figure 3(a) demonstrates our proposed temperature monitoring system, including (b) STC89C52, (c) liquid crystal display and (d) power supply. The micro controller can process it directly and realize the interactive control between them. Because of the strong core control ability of the MCU, overlapping the auxiliary circuit, and then get the practical development system is necessary. The system we designed is using 89C52 single chip as the controller, K type thermocouple as the temperature sensor. The system structure diagram is shown in figure 3(a). The exclusive A/D convertor of thermocouple and the 5-bit industrial liquid crystal display HJ1602A formed the digital temperature monitoring system. This system takes the single chip microcomputer as the control core, samples and amplifies the thermocouple electromotive force through the high precision A/D converter, and a certain algorithm is used in the microcontroller to realize the linearization of the thermocouple and display the corresponding measurement data through the LCD screen.

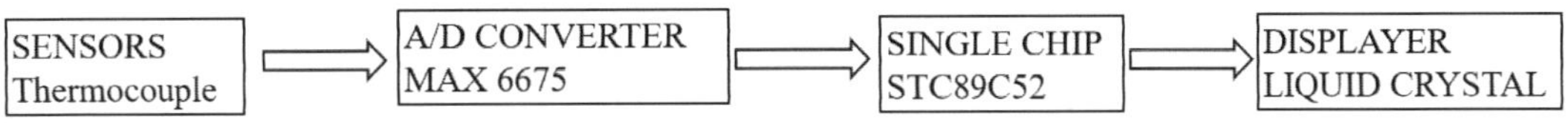

(a)

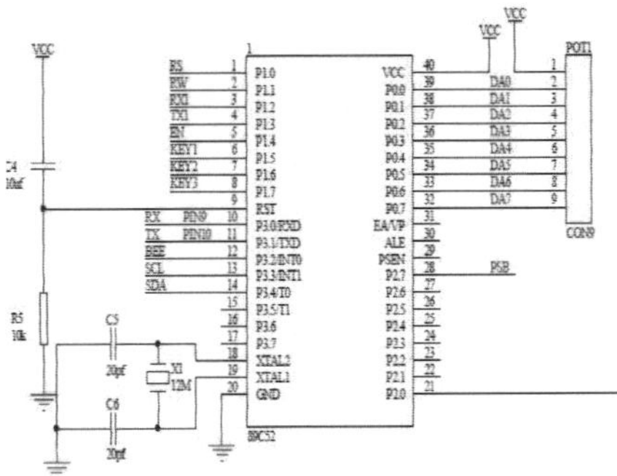

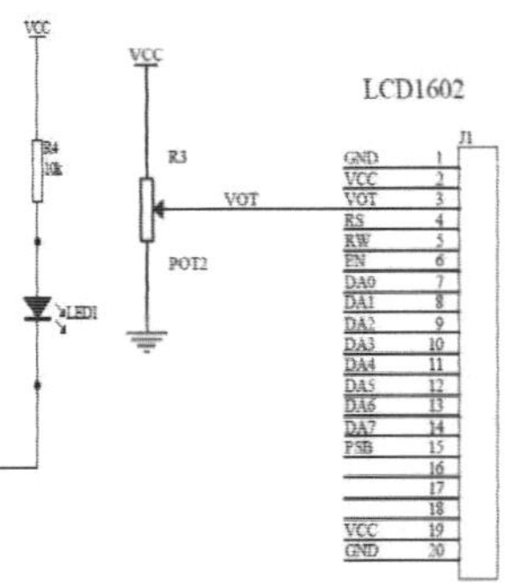

(b) (c)

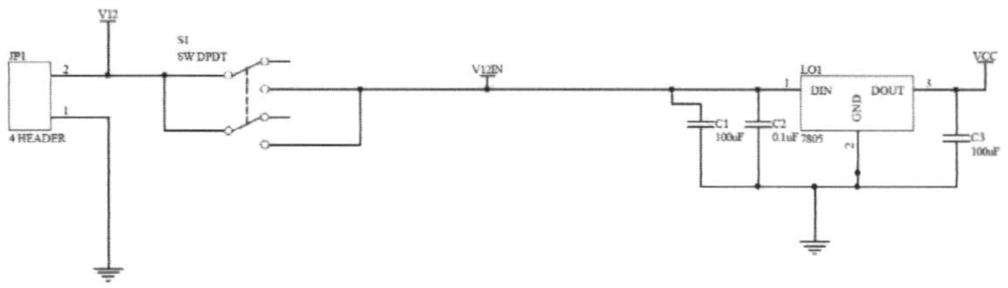

(d)

Figure 3. Schematic Interlinkage diagram (a) of temperature monitoring system including (b)STC89C52, (c) liquid crystal display and (d) power supply.

3. Experiment Results and Discussion

Our system uses AT89C52 as the microcontroller, uses MAX6675 as the temperature sensor, and temperature light crystal display. The whole system strives simple structure and well function. The use of temperature sensor MAX6675, thermocouple temperature measurement applications, complex linearization, cold end compensation, digital output and other problems focused on a chip to solve, the complex hardware and software design is simplified when the thermocouple temperature measurement scheme is applied to the embedded system field, therefore, the device is an ideal choice for the application of thermocouple temperature measurement in the field of embedded systems. With the rapid development of SCM technology, through the MCU control of the controlled object has become an important development direction in the future in the field of automatic control, current, voltage, temperature, pressure, flow, flow rate and switching capacity are mainly charged with common parameters. For example: in metallurgical industry, chemical industry, electric power engineering, paper industry, machinery and food processing and other fields, people need to all kinds of heating furnace, heat treatment furnace, furnace and boiler temperature detection and control. Using MCS-52 SCM to control the temperature, is not only easy to control, simple and flexible configuration advantages, and can greatly increase the temperature was charged with the technical indicators, which can greatly improve the quality and quantity of products. Therefore, the temperature control of single chip microcomputer is a problem that is often encountered in industrial production.

MAX6675 is a serial analog-to-digital converter K type thermocouple cold end compensation, linear correction, thermocouple break detection produced by American MAXIM company, it is the temperature resolution of 0.25 degrees, cold end compensation range is -20 to 80℃, the working voltage is 3 ~ 5.5V. This paper introduces the function characteristics, pin arrangement and operation sequence of MAX6675, and gives the interface circuit and programming design method of MAX6675 and 89C51. In the field of industrial temperature measurement and control, K thermocouple is widely used because of its good linearity, low cost and wide measuring range. But it often needs cold end compensation, and the circuit is more complex, debugging trouble. The K thermocouple serial analog to digital converter (MAX6675) produced by MAXIM can not only convert analog signals into 12-bit corresponding digital quantities, but also bring cold end compensation. The temperature resolving power is 0.25℃, which can satisfy most industrial applications. MAX6675 uses small size and good reliability SO-8 packaging.

The experimental setup is shown in figure 4. We employ a controllable hot plate as a source of thermal stimulation for UAV device and its chemical cell. The thermocouple is attached to the surface of chemical cell inside UAV device. The temperature data is shown in a display of temperature monitoring system. Figure 5 a tiny temperature difference between the set thermal value Ts (black line) and real one Tf (blue line) of a hot plate. Assuming the temperature of UAV surface is close to that of hot plate, the heat would transfer into the UAV body and thus chemical cell inside. A much temperature different between the chemical cell (Tc, red line) and UAV surface. When Tf reaches 300 ℃, Tc is about 90 ℃ on a stable state. These results can give us a conclusion that present design of temperature monitoring system can work well for sensing temperature of chemical cell inside UAV device, and also the chemical cell can hold a much lower temperature of 80 ℃ than that of thermal source of 300 ℃ within 20 min.

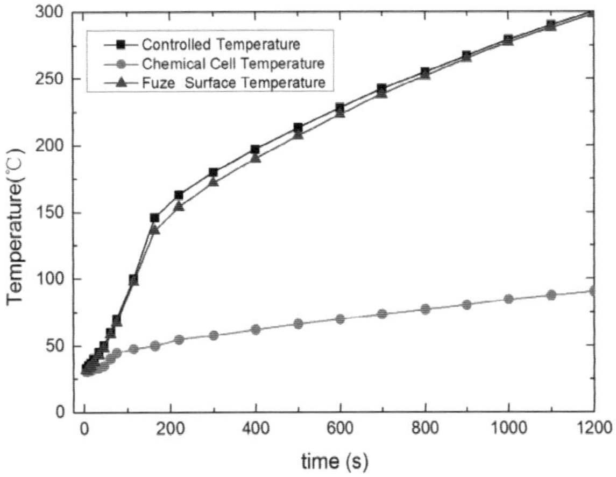

(a) (b)

Figure 4. Experiment system design and result diagram. (a) Experimental setup for heating fuse device and temperature monitoring, (b) Measured temperature of the chemical cell inside fuse device using a thermocouple

4. Conclusion

The main content of this work is using 89C52 single chip as the controller, K type thermocouple as the temperature sensor. The exclusive A/D convertor of thermocouple and the 5-industrial bit liquid crystal display HJ1602A formed the digital temperature monitoring system. Thermocouple acquisition temperature signal through signal conditioning, uses the A/D convertor to the microcontroller, and display the temperature in HJ1602A. This system takes the single chip microcomputer as the control core, samples and amplifies the thermocouple electromotive force through the high precision A/D converter, and a certain algorithm is used in the microcontroller to realize the linearization of the thermocouple and display the corresponding measurement data through the LCD screen.

Reference

[1] Z Ling, J Wang, K Zhu, Digital temperature sensor application in thermocouple compensation. Sensor Technology.2013. 22(6):45-46.
[2] Z Cui. Temperature Measurement and Metrology. (China Metrology Press,1997), pp 25-50.
[3] J Liu. Modern Test Technology and System Integration. (Electric Industry Press, 2004). pp 47-53.
[4] Fenlei J, Will H A. Thin - film thermocouples and strain gauge technologies for engine application [J]. Sensors and Actuators, 1988, 65: 187–193.
[5] Cattani M, Saivadori M C, Vaz A R , Thermoelectric powerin very thin film thermoeouples Quantum size effects[J]. Journal of Applied Physics, 2006, 100: 114905–114909.
[6] Lepieovsky J, Smith F A, Zhang L J. Thin-film thermocouples for turbine hot-cascade testing[R]. ISABE 99–7241.
[7] J.F. Li, W.S. Liu, L.D. Zhao, M. Zhou, "High-performance nanostructured thermoelectric materials", Npg Asia Mater, vol. 2, pp. 152-158, 2010.
[8] A.I. Soldatov, A.A. Soldatov, P.V. Sorokin, O.A. Kozhemyak, S.I. Bortalevich, "Control system for device thermotest", 2016 International Siberian Conference on Control and Communications SIBCON 2016 - Proceedings, pp. 7491869.

IOP Conf. Series: Journal of Physics: Conf. Series **986** (2018) 012013 doi:10.1088/1742-6596/986/1/012013

Metal organic frameworks enhanced graphene oxide electrode for humidity sensor

Wen Zhang[1], Siyu Meng[1], Hui Wang[2] and Yongning He[3]

[1] School of Science, Xi'an Jiaotong University, Xi'an, Shaanxi 710049, China
[2] MOE Key Laboratory of Thermo-Fluid Science and Engineering, School of Energy and Power Engineering, Xi'an Jiaotong University, Xi'an, Shaanxi 710049, China
[3] School of Electric and Electronic, Xi'an Jiaotong University, Shaanxi 710049, China

Corresponding author's e-mail address: zhangwen@mail.xjtu.edu.cn

Abstract. Copper benzene-1,3,5-tricarboxylate (Cu-BTC), a typical metal organic framework, is deposited on the graphene oxide (GO) film to prepare a resistance humidity sensor (Cu-BTC/GO) for improving humidity sensing. The characteristics of Cu-BTC, GO and Cu-BTC/GO were measured by scanning electron microscopy (SEM), X-ray diffraction (XRD), nitrogen isotherm adsorption and electrochemical impedance spectroscopy (EIS). The humidity sensing properties of the Cu-BTC/GO were investigated in detail. The obtained Cu-BTC/GO demonstrates good sensitivity and repeatability over 11%-85% relative humidity (RH) measurements. The Cu-BTC/GO coated device shows high normalized response (S) value (6200%), which is much higher than that of pure GO coated device. Sensing mechanism of Cu-BTC/GO is discussed based on different RH and the results indicate that moderate amounts of Cu-BTC deposition can enhance sensing abilities of GO. High specific surface area and interfacial conductivity are crucial factors to fabricate humidity sensors with high performance.

1. Introduction

Humidity sensors have attracted considerable attention because of their wide applications in many fields, such as industry, agriculture, medicine, meteorology, environment monitoring, home appliance and so on. Recently, some researches also find that relative humidity (RH) level can affect photochemical reaction speed of O_3 and NO_2 in the air [1]. Therefore, humidity is an important factor for the air quality evaluation. Various transduction techniques are used to develop humidity sensors, such as capacitance [2], resistance [3], surface acoustic wave (SAW) [4], quartz crystal microbalance [5] and optical fiber [6]. Among these, resistive humidity sensors are gaining much more attention because they have the features of small form factor, easy to fabricate and able to be integrated with circuits [3].

During the past decade, various materials have been adapted to the sensing layers of humidity sensors (*e.g.*, graphenes [3], carbon nanotubes [7], metal oxides [8], polymers [9], metal-organic frameworks (MOFs) [10] and composites [11]), which are expected to improve the performance by virtue of their intrinsic properties, porous morphology, and large surface-to-volume ratio. Among these sensing materials, graphene, a two-dimensional monolayer of sp^2-bonded carbon atoms exhibiting exceptional mechanical thermal and electrical properties, holds great potential for ultra sensitive detection. Graphene oxide (GO) is a graphene derivative covered by oxygen-containing functional groups. The hydroxyl and the epoxy groups decorated on the basal plane and carboxyl groups at the edges according to the Klinowski model of GO [12]. These oxygen-containing functional groups are highly hydrophilic and reactive, therefore, GO-based materials are considered to be good candidates for fabricating efficient humidity sensors. A. De Luca et al. used GO to produce a sensor for the humidity monitoring in indoor environment, which shows a strong response in humidity due to high proton conduction and fairly high specific area [3]. A flexible humidity sensor made by D. Zhang et al. [13] exhibits excellent sensing behaviour to humidity, fast response to recover time and good repeatability by using reduced GO/polymer nanocomposite film as the sensing materials. Although there have a few successful cases, the application of graphene-based electronic sensors is still in its infancy.

In a humidity sensor, high surface-to-volume ratio of the sensitive material is the most important part to get good signals. GO has high surface area, but it's easy to aggregate and lost its advantages. MOFs are kinds of porous crystal materials which have high porosity because of their unique structures. Furthermore, MOFs can be modified easily through changing their surface functional groups [14-16]. Among thousands of MOFs, copper benzene-1,3,5-tricarboxylate noted as Cu-BTC ($[Cu_3(BTC)_2(H_2O)_3]$) coating is a promising candidate because of its high porosity and sensitivity to water. The framework of Cu-BTC composed of dimeric cupric tetra-carboxylate units [16]. Each metal completes its pseudo octahedral coordination sphere with an axial aqua ligand opposite to the Cu-Cu vector [17]. Therefore, Cu-BTC has high porosity and huge gas storage potential due to its unique structure. Li's group [16] proves that the composite of Cu-BTC@GO can remain its structure and porosity and its Brunauer-Emmett-Teller (BET) surface area is 1205 m^2/g. Küsgens [18] et al. demonstrate that the water stability of Cu-BTC is sufficient for molecular recognition if the framework is not directly exposed to liquid water. Experiments proves that even though Cu-BTC is unstable in liquid water, it is a promising material for applications in removal of trace water or molecular sensing.

Herein, a novel type of resistance humidity sensor with high conductivity and large porosity was prepared by depositing GO and Cu-BTC layer on Ag inter-digital electrodes. The Cu-BTC/GO humidity sensor shows significant enhancement of sensing performance with the humidity ranging from 11% to 85% RH, indicating that MOFs have great potential application in humidity sensor. In addition, the sensing mechanism has also been discussed.

2. Materials and methods

All the chemicals were bought from Sigma-Aldrich and used without further purification. The Cu-BTC powder was synthesized according to a previous method [19]. Copper nitrate tri-hydrate ($Cu(NO_3)_2{\cdot}3H_2O$, 1 mmol) and 1,3,5-benzenetricarboxylic acid (H_3BTC, 0.67 mmol) were added in an autoclave with DMF (12 mL), ethanol (12 mL) and DI water (8 mL). The mixture was stirred for 15 min and the autoclave was placed in an oven at 80 °C for 24 h. The product was isolated by filtration and rinsed with 10 mL of ethanol for three times, and then immersed in methanol for 3 days, during which the activation solvent was decanted and replaced freshly everyday. After that, the sample was dried at 200 °C in vacuum to yield Cu-BTC. The final powder was dispersed in ethanol to form 50 mg/mL Cu-BTC ethanol solution for further use.

GO was prepared by a modified Hemmer method [20]. Firstly, flake graphite powder (1 g) and concentrated sulfuric acid (23 mL) were added into a 500 mL of glass bottle, which was stirred at room temperature for 24 h. Secondly, the mixture was heated up to 40 °C, $NaNO_3$ (0.1 g) and $KMnO_4$ (3 g) were added with stirring for 30 min. Then 5 mL of deionized water was added slowly and stirred for 5 min for two times. Next, 40 mL of water was added and stirred for 15 min. Subsequently, 140 mL of water and 10 mL of H_2O_2 (30%) was added with continuous stirring for 5 min. Finally, the bright yellow reaction system was centrifugal separated at 8000 rpm, then the precipitation was ultrasonic washed with HCl (5%). The collected supernatant was washed with ethanol for several times. The obtained product was mixed to form 1 mg/mL GO ethanol solution for further use.

The humidity sensor was fabricated on a glass substrate through microfabrication technology, including spin coating of photoresist, exposure, development, and lift-off technique. Then, Ag interdigitated electrode (IDE, ~120 nm thick) was fabricated on the glass substrate. The width and gap of IDE finger was both 0.5 mm. After that, the sensing film was formed by dropping the GO ethanol solution (0.1 mg/cm^2) and the Cu-BTC ethanol solution (1 mg/cm^2) successively on the interdigital part to cover it entirely. At last, the sensing layers were dried at 50 °C for 6 h.

The crystal structure for Cu-BTC is investigated by X-ray powder diffraction (XRD) using a LabX XRD-6000 with Cu K radiation (1.54 Å), operating at 40 kV and 35 mA. The morphologies of GO and Cu-BTC were examined using field emission scanning electron microscopy (SEM, JEOL S4800, Japan) operating at an accelerating voltage of 15 kV. N_2 adsorption isotherms of GO and Cu-BTC were carried out at -196 °C and 1atm. The electrochemical impedance spectroscopy (EIS) were also measured by CHI660D using three electrode system. The testing frequency range is 1 Hz to 100000 Hz, and the electrolyte is 0.01 M $K_3Fe(CN)_6$ in 0.1 M KCl.

The humidity sensing measurement was carried out by exposing the prepared sensor to different relative humidity levels, and the humidity environment was provided by different saturated aqueous solutions in closed vessels. In this work, five saturated aqueous solutions of LiCl, $MgCl_2$, $Mg(NO_3)_2$, NaCl and KCl were adopted at room temperature to yield approximately 11%, 33%, 54%, 75% and 85% RH levels, respectively. The resistance of the Cu-BTC/GO film was measured by Agilent B2902A. The Fig. of merit used for the evaluation of sensor performance is the normalized response(S), determined by S= $(R_{RH}-R_{11})/R_{11}\times 100\%$, where R_{RH} and R_{11} are the electrical resistance of the sensor in the given RH and 11% RH, respectively [11].

3. Results and discussion

3.1. Material characterization

Figure 1 shows the X-ray pattern of pure GO, Cu-BTC and Cu-BTC/GO. The sample of Cu-BTC is powder, while the other two samples are prepared by dropping the materials dispersed in ethanol on substrates, so the intensity of XRD for Cu-BTC is much stronger than those of the other two samples. For comparison purposes, the intensity of XRD for Cu-BTC has been deduced to 1/100. It can be found that the powder sample shows the typical patterns of Cu-BTC [17, 24]. The appearance of sharp peaks in XRD patterns is a testimony of good degree of crystallinity in synthesized products. The main peak of GO usually located in the range of 8 to 11° [20]. According to Bragg's law, the major peak of GO in this work is at 2θ = 8.59°, which reveals that there has a distance between the carbon layers of 10.3 Å. The peak indicating GO becomes 2θ = 10.17° (8.8 Å) due to the deposition of Cu-BTC on GO. Other X-ray diffraction patterns of the Cu-BTC/GO contain both of the characteristic peaks of Cu-BTC and GO.

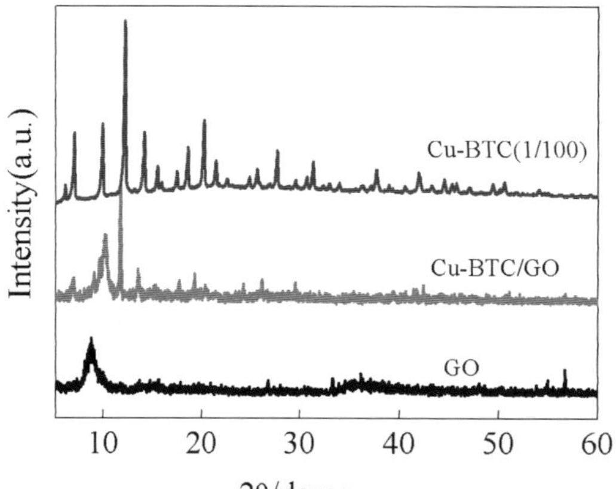

Figure 1. X-ray diffraction for Cu-BTC, Cu-BTC/GO and GO.

The morphological structures and size distribution histograms of Cu-BTC, GO, and Cu-BTC/GO are displayed in Figure 2. It can be observed that the typical shape of Cu-BTC is octahedron and the size of Cu-BTC is about 2-10 μm. Figure 2(b) shows that GO has several thin layers folded together. After dropping GO and Cu-BTC ethanol solutions on the Ag IDE, the surface of GO is covered by Cu-BTC crystal practically (figure 2c and d). The average diameter of the Cu-BTC/GO is 3-10 μm, and MOFs and GO maintain their original structures.

Figure 2. The SEM images of Cu-BTC(a), GO(b) and Cu-BTC/GO with different magnification(c and d).

The nitrogen adsorption isotherm of Cu-BTC and GO were investigated. (Figure 3). As expected, GO has no much nitrogen adsorption. [19] According to the standard Brunauer Emmett Teller (BET) method, the specific surface area of Cu-BTC is calculated to be 1272 m^2/g. It is hard to directly detect the N_2 adsorption isotherm for the Cu-BTC/GO on the Ag IDE, we assume that the BET surface area of Cu-BTC/GO might be similar with that of pure Cu-BTC. Because there is only physical attachment and the proportion of the two raw materials is 10:1 (Cu-BTC:GO=10:1).

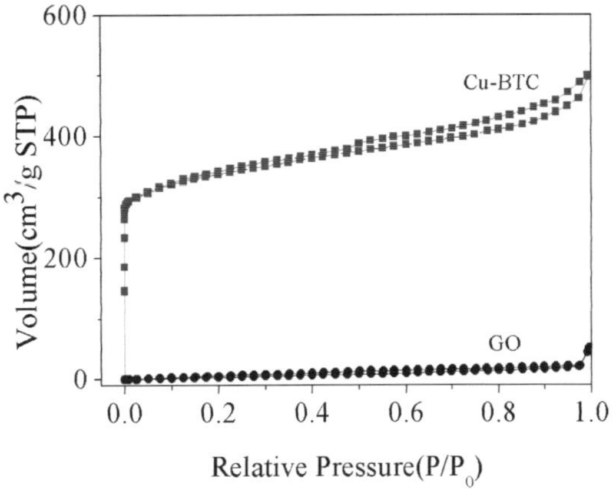

Figure 3. The N_2 adsorption isotherms of Cu-BTC and GO.

The electrochemical impedance spectroscopy (EIS) of Cu-BTC, GO and Cu-BTC/GO were measured and presented in Figure 4. It is clear that the impedance of Cu-BTC is much larger than those of GO and Cu-BTC/GO. The semicircle is invisible. The resistant of the testing materials are extremely high, but they are unimpeded for ion and electron in the electrolyte. The straight line might stem from the ionic and/or electrolytic conductivity [23]. The impedance of Cu-BTC/GO and GO are almost the same, suggesting that coating Cu-BTC layer on the GO could not change the behaviour of conductivity of GO.

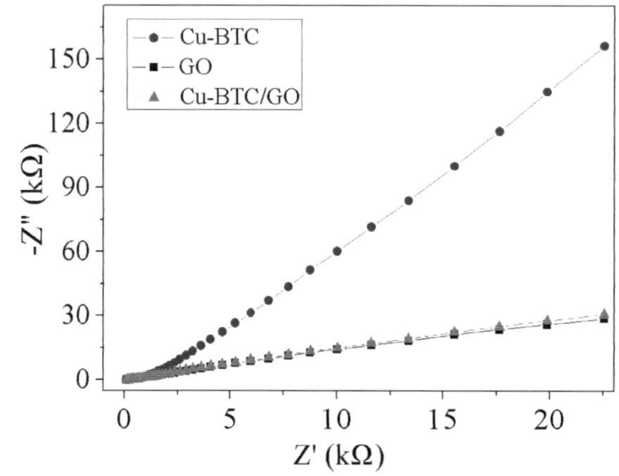

Figure 4. The electrochemical impedance spectroscopy (EIS) plots of GO, Cu-BTC and Cu-BTC/GO.

3.2. Response property of the sensors

The loading amount of the sensing materials can greatly affect the humidity sensing signal. Excessive amount of GO might reduce the adhesive force to the substrate and lead to the layer exfoliate from the substrate, while too much Cu-BTC may enhance the resistance and result no effective signal. According to the experimental results, the optimized amount ratio for GO: Cu-BTC =0.1:1 mg/cm^2 is chosen in this paper.

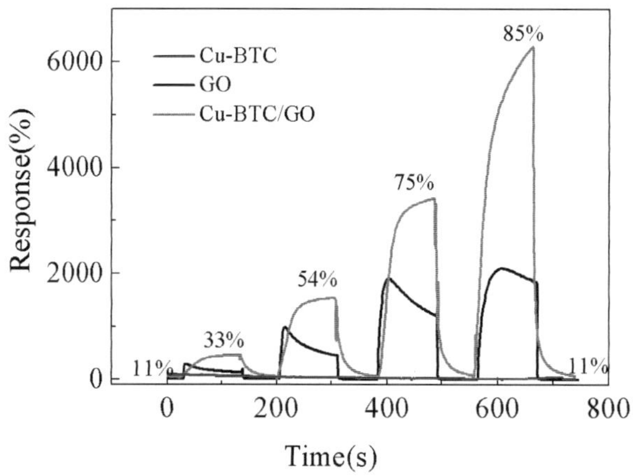

Figure 5. The humidity sensitive performance of GO, Cu-BTC and Cu-BTC/GO.

Figure 5 shows the response of the sensors exposing to different relative humidity levels. For comparison, Cu-BTC, GO and Cu-BTC/GO are deposited on the electrodes separately. Each exposure/recovery cycle was carried out by an exposure interval of 100 s followed by a recovery interval of 100 s at 11% RH. The Cu-BTC device has almost no peak and shows very weak response compared with GO and Cu-BTC/GO. The sensitive peak significantly increases with the rising of RH in the range of 11% -85% for both GO and Cu-BTC/GO. The sensitive peak of the Cu-BTC/GO coated device can reach up to 6200%, which is much higher than that of GO device (2100%).

We know that GO has good electrical conductivity, but pure Cu-BTC shows very low electrical conductivity (10^{-6} S/m) [14]. It could combine the merits of Cu-BTC and GO when depositing appropriate amount of Cu-BTC/GO. Cu-BTC not only has high surface-to-volume ratios, but also possesses a large density of Cu-Cu unsaturated sites [21], which make Cu-BTC facilitate water molecule to be captured from external environment. Meanwhile, hydrophilic functional groups decorated on the basal planes and the edges of GO can also absorb water molecules. Based on this, the humidity sensing mechanism for the Cu-BTC/GO can be explained (Figure 6).

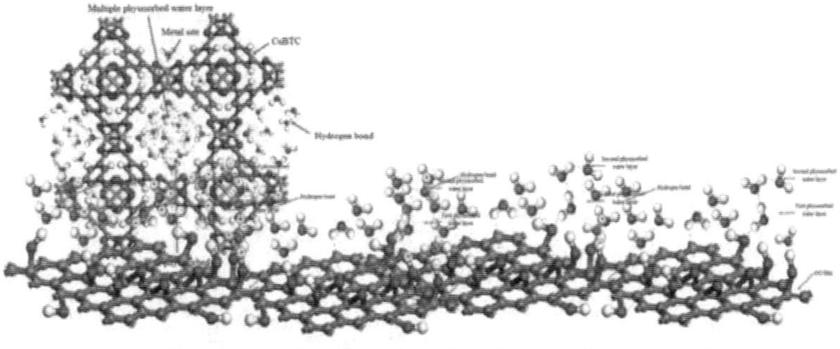

Figure 6. The mechanism for Cu-BTC improving the moisture sensitivity of GO.

At low RH, only a few water molecules are absorbed on the sensing materials. Water molecules can be quickly absorbed on the hydrophilic groups of the exposed GO. Although the ionize protons are restricted by discontinuous mobile layers, they contribute to the leak conduction due to the proton hopping between the adjacent water molecules [23]. Meanwhile, water molecules can also be absorbed on Cu-Cu unsaturated sites of Cu-BTC, which is a slower process associating with diffusion. The accumulated water molecules in the MOFs pore can provide a proton conducting channel to gain higher response current. Therefore, the sensing ability of water vapour can be ascribed to the coordination between the oxygen groups in GO and Cu^{2+} mental centre in Cu-BTC.

As the RH increases, the water molecules are absorbed in the pores of framework and the surface of the GO by the multilayer physical adsorption. At the second physical absorption layer, water molecules are absorbed through hydrogen bonds. Then the hydroxyl groups on the surface will be combined with water molecules, and protons can be transferred from hydroxyl group to water molecules to come into hydronium ions (H_3O^+). The hydronium ions are the main charge carriers and conductivity is generated by a Grotthuss chain reaction [25], which can cause an increase of the leak conduction. Thereafter, the water molecules become mobile and progressively more identical to those in the bulk liquid.

At higher RH, the physical absorbed water layers gradually exhibit liquid-like behaviour, which is beneficial for the hydrolysis of the functional groups (carboxyl, sulphonic and/or hydroxyl) on GO sheets. These ions will contribute to the ionic conductivity [10, 23]and the physical absorbed water can penetrate into the interlayer of GO films [3]. In addition, there is another ionic conductivity channel formed by absorbing a large amount of water molecules in Cu-BTC porous crystals, which greatly increases the response signals.

The repeatability of Cu-BTC/GO humidity sensor was measured for five exposure/recovery cycles repeatedly for 54%, 75% and 85% RH (Figure 7). The ratio of maximum deviation to full-scale measurement is under the same conditions, and the repeatable error is less than 1.3%. The results demonstrated that the adsorption of water molecules in the Cu-BTC and on the GO are both physical adsorption, and the Cu-BTC has enough stability for enhancing humidity sensing.

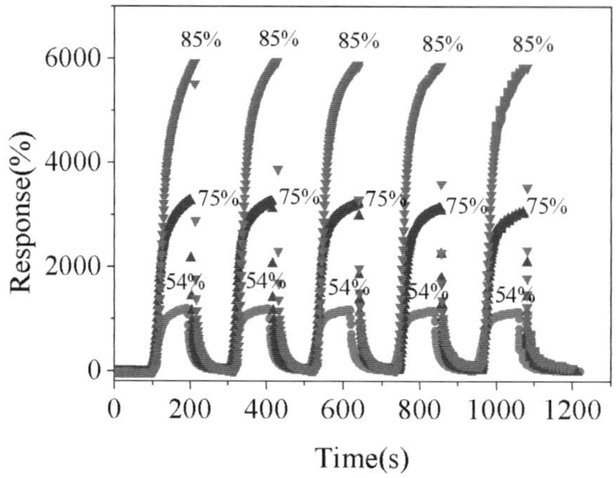

Figure 7. The humidity sensitive repeatability of Cu-BTC/GO sensor at different humidity.

4. Conclusions

In this work, a Cu-BTC/GO resistance humidity sensor has been fabricated by depositing GO and Cu-BTC crystal on silver interdigitated electrodes. The Cu-BTC/GO coated device shows high normalized response (S) value (6200%) and good repeatability for humidity sensing. This can be explained by rich oxygen groups on the GO and high density of Cu^{2+} mental centre in the Cu-BTC. The satisfactory capability of Cu-BTC/GO suggests the potential practical application of preparing better humidity sensor or other gas sensors using MOF.

References

[1] Hu G, Xu Y, Jia L. Effects of relative humidity on the characterization of a photochemical smog chamber. 2011 *Journal of Environmental Sciences* **23** 2013.

[2] Guo R, Tang W, Shen C, Wang X. High sensitivity and fast response graphene oxide capacitive humidity sensor with computer-aided design. 2016 *Computational Materials Science* **111** 289.

[3] De Luca A, Santra S, Ghosh R, Ali S Z, Gardner J W, Guhac P K, Udrea F. Temperature-modulated graphene oxide resistive humidity sensor for indoor air quality monitoring. 2016 *Nanoscale* **8** 4565.

[4] Yao Y, Xue Y. Influence of the oxygen content on the humidity sensing properties of functionalized graphene films based on bulk acoustic wave humidity sensors. 2016 *Sensors and Actuators B:Chemical* **222** 755.

[5] Yao Y, Xue Y. Impedance analysis of quartz crystal microbalance humidity sensors based on nanodiamond/graphene oxide nanocomposite film. 2015 *Sensors and Actuators B: Chemical* **211** 52.

[6] Ohira S I, Miki Y, Matsuzaki T, Nakamura N, Sato Y, Hirose Y, Toda K, A fiber optic sensor with a metal organic framework as a sensing material for trace levels of water in industrial gases. 2015 *Analytica Chimica Acta* **886** 188.

[7] Zhao Z G, Liu X W, Chen W P, Li T. Carbon nanotubes humidity sensor based on high testing frequencies. 2011 *Sensors and Actuators A: Physical* **168** 10.

[8] Narimani K, Nayeri F D, Kolahdouz M, Ebrahimi P. Fabrication, modeling and simulation of high sensitivity capacitive humidity sensors based on ZnO nanorods. 2016 *Sensors and Actuators B: Chemical* **224** 338.

[9] Zhuang Z, Li Y F, Qi D, Zhao C J, Na H. Novel polymeric humidity sensors based on sulfonated poly (ether ether ketone)s: Influence of sulfonation degree on sensing properties. 2017 *Sensors and Actuators B: Chemical* **242** 801.

[10] Zhang J J, Sun L, Chen C, Liu M, Dong W, Guo W B, Ruan S P. High performance humidity

sensor based on metal organic framework MIL-101(Cr) nanoparticles. 2017 *Journal of Alloys and Compounds* **695** 520.

[11] Zhang D Z, Tong J, Xia B. Humidity-sensing properties of chemically reduced graphene oxide/polymer nanocomposite film sensor based on layer-by-layer nano self-assembly. 2014 *Sensors and Actuators B: Chemical* **197** 66.

[12] Lerf A, He H Y, Forster M, Klinowski J. Structure of Graphite Oxide Revisited. 1998 *The Journal of Physical Chemistry B* **102** 4477.

[13] Zhang D Z, Tong J, Xia B, Xue Q Z. Ultrahigh performance humidity sensor based on layer-by-layer self-assembly of graphene oxide/polyelectrolyte nanocomposite film. 2014 *Sensors and Actuators B: Chemical* **203** 263.

[14] Talin A A, Centrone A, Ford A C, Foster M E, Stavila V, Haney P, Kinney R A, Szalai V, Gabaly F E, Yoon H P, Léonard F, Allendorf M D. Tunable electrical conductivity in metal-organic framework thin-film devices. 2014 *Science* **343** 66.

[15] Zhang Z J, Zhao Y G, Gong Q H, Li Z, Li J. MOFs for CO_2 capture and separation from flue gas mixtures: the effect of multifunctional sites on their adsorption capacity and selectivity. 2013 *Chem. Commun* **49** 653.

[16] Zlotea C, Phanon D, Mazaj M, Heurtaux D, Guillerm V, Serre C, Horcajada P, Devic T, Magnier E, Cuevas F, Férey G, Llewellyn P L, Latroche M. Effect of NH_2 and CF_3 functionalization on the hydrogen sorption properties of MOFs. 2011 *Dalton Transactions* **40** 4879.

[17] Israr F, Israr F, Kim D K, Kim Y, Oh S J, Ng K C, Chun W. Synthesis of porous Cu-BTC with ultrasonic treatment: Effects of ultrasonic power and solvent condition. 2016 *Ultrasonics Sonochemistry* **29** 186.

[18] Küsgens P, Rose M, Senkovska I, Fröde H, Henschel A, Siegle S, Kaskel S. Characterization of metal-organic frameworks by water adsorption. 2009 *Microporous and Mesoporous Materials* **120** 325.

[19] Liu S, Sun L X, Xu F, Zhang J, Jiao C L, Li F, Li Z B, Wang S, Wang Z Q, Jiang X, Zhou H Y, Yang L N, Schick C. Nanosized Cu-MOFs induced by graphene oxide and enhanced gas storage capacity. 2013 *Energy & Environmental Science* **6** 818.

[20] Marcano D C, Kosynkin D V, Berlin J M, Sinitskii A, Sun Z Z, Slesarev A, Alemany L B, Lu W, Tour J M. Improved Synthesis of Graphene Oxide. 2010 *ACS Nano* **4** 4806.

[21] Hulvey Z, Lawler K V, Qiao Z W, Zhou J, Fairen-Jimenez D, Snurr R Q, Ushakov S V, NavrotskyA, Brown C M, Forster P M. Noble Gas Adsorption in Copper Trimesate, HKUST-1: An Experimental and Computational Study. 2013 *The Journal of Physical Chemistry C* **117** 20116.

[22] Seredych M, Petit C, Tamashausky A V, Bandosz T J. Role of graphite precursor in the performance of graphite oxides as ammonia adsorbents. 2009 *Carbon* **47** 445.

[23] Bi H C, Yin K B, Xie X, Ji J, Wan S, Sun L T, Terrones M, Dresselhaus M S. Ultrahigh humidity sensitivity of graphene oxide. 2013 *Scientific Reports* **3** 1.

[24] Li Y J, Miao J P, Sun X J, Xiao J, Li Y W, Wang H H, Xia Q B, Li. Z. Mechanochemical synthesis of Cu-BTC@GO with enhanced water stability and toluene adsorption capacity. 2016 *Chemical Engineering Journal* **298** 191.

[25] Chui S S Y, Lo S M F, Charmant J P H, Orpen A G, Williams I D. A Chemically Functionalizable Nanoporous Material $[Cu_3(TMA)_2(H_2O)_3]n$. 1999 *Science* **283** 1148.

[26] Agmon N. The Grotthuss mechanism. 1995 *Chemical Physics Letters* **244** 456.

Analysis of Poiseuille Flow Property in Two-Dimensional Micro Channels of Microfluidic Pneumatic Micro-Valve

Shaohua Yang[1], Wei Long[2]* and Yajun Chen [3]

1 Faculty of Electromechanical Engineering, Kunming University of Science and Technology, Kunming ,650093 , China ,18725020778@163.com
2 Faculty of Electromechanical Engineering, Kunming University of Science and Technology, Kunming ,650093 , China ,827161564@qq.com
3 Faculty of Electromechanical Engineering, Kunming University of Science and Technology, Kunming ,650093 , China ,yajun918@foxmail.com

Abstract: In this paper, the control mechanism and mathematical description of the microfluidic flow in the microfluidic process of the PDMS membrane type pneumatic micro-valve were studied. The velocity and pressure variation law of the velocity field inside micro valve was analyzed by numerical simulation method. The influence of the two kinds of inlet drive modes on the working effect and the pressure flow characteristics of the pneumatic micro-valve was studied. The structure of the elastic solid valve diaphragm under the dual action of the airway and the liquid channel was analyzed. Deformation and stress distribution. The results show that the gas flow in the gas flow channel under the diaphragm by the vacuum part of the role of the formation of a suction gas vortex, pressure-driven mode was easier under the diaphragm to produce a strong gas vortex, resulting in internal and external pressure to promote diaphragm cut-off liquid channel ; In the pressure pneumatic mode, the stress at both ends of the diaphragm was smaller, the membrane was not easy to tear failure.

Keywords: Micro - valve; Micro - nano flow; Gas - liquid coupling; Burnett equation

1. Introduction

Microfluidics is a technique that mainly controls the fluid in the microscale space [1]. The basic characteristics of microfluidic chips and the biggest advantage is the flexible combination of a variety of cell technology in a small controllable platform, Micro-pipelines, micro-pumps, micro-valves, micro-reservoirs, microelectrodes, micro-detection devices, etc. are integrated on a specific material substrate through microfabrication techniques [2,3]. The flow in the microfluidic chip is mainly of various channel flows, and its flow characteristics are generally in the range of 0.1-1 mm, which belongs to the field of hydrodynamic study [4]. However, as an integrated system, the flow in a microfluidic chip consists of a fluid drive, transmission, detection and control unit. The aerodynamic micro-valve is a typical microfluidic control unit. Due to the different control objects and control functions, the aerodynamic micro-valve has a variety of internal flow media and complicated nature. Driven by micro-electromechanical systems and microfluidic chip technologies, microfluidics is studied around the practicality of continuous assumptions, the manipulation of fluid motion, and local nanoscale flow. In the early 1990s, Pfahler et al. [5] found that the coefficient of flow resistance did not match the theoretical predictions through the pressure flow test in micro-channels, which attracted people's concern about the applicability of fluid continuity equations at microscale [6]. Since 2000, attention has been focused on the design of microfluidic chips. Since the transport of samples between the various units depends on the flow of liquids, how to achieve flow control at the micro-scale becomes the key to microfluidic chip technology. In 2004, Stone et al. [7] published an article in the Yearbook of Fluid Mechanics, pointing out that the design of microchip needs to consider the influence of various physical factors and it is necessary to carry out further research on microscale fluid dynamics. Darhuber et al. [8] analyzed the physical mechanism of normal and tangential stress-driven surface fluids for discrete or continuous micro-control of small-scale systems and pointed out the contact-related boundary phenomena (such as separation pressure, line tension and slip Shift, etc.) still need to be theoretically considered. In recent years, with the development from the microfluidic system to nanoscale, the micro-nano flow theory and nanotechnology are gradually introduced. In 2008, Han's research group [9] reviewed the flow phenomena in nanofluids. In 2010, Chemical Society Review published a special issue titled "From Microfluidic Applications to Nano-fluidics" [10], providing an overview of the latest international developments in micro and nano flow control.

Many scholars have studied Poiseuille flow in the Wiener Passage in the past. Harley et al. [11] studied the characteristics of sub-sonic gas flow in the Wiener channel using experimental and analytical methods and found that velocity slip conditions must be introduced at the boundary Arkilic [12] used perturbation method to solve the N-S equation describing the gas flow in the micro-channels. It was found that the calculated results and the experimental values were only in good agreement when the Knudsen number was small. Beskok et al. [13] studied the Poiseuille flow characteristics using the N-S equation and the high-order slip boundary conditions, and also performed numerical simulations with the DSMC method. It was found that the two methods agree well with the Knudsen number. When Knudsen number, the calculation results of N-S equation gradually deviate from the results obtained by DSMC. Experiments show that when the characteristic size of the flow field is on the order of micrometers or submicrometers, the velocity and temperature jumps of the fluid on the solid wall, as well as the thermal peristalsis, the electrokinetic effect, the viscous heating, the anomalous diffusion and the chemical effects, The internal play a leading role, leading to abnormal micro-scale flow phenomenon [14,15].

In recent years, some scholars began to use the Burnett equation to simulate the gas flow in the micro-nanochannels. Agarwal et al. [16] studied the planar Poiseuille flow using the augmented Burnett equation and found that solving the Burnett equation relative to the N-S equation yields better agreement with the results obtained using the DSMC method. Fang [17] also uses the augmented Burnett equation to simulate the flow and heat transfer problems in microchannels and gives the results. However, one of the biggest obstacles to using the Burnett equation is the numerical stability problem [18, 19]. In the past, the Burnett equation was used to solve the Poiseuille flow in two-dimensional flat plate only when it was convergent [16]

However, more and more gas flows in the micro-nano-electromechanical systems are in the area of slip flow (ie, [19, 20]), so it is necessary to further explore finding suitable mathematical methods in a wider range. Road flow description.

In this paper, the typical micro-nano-scale pneumatic valve flow for the object, The flow characteristics and mechanical laws of two-dimensional Poiseuille inside the passage were analyzed. The flow characteristics of the rectangular cross-section gas passage and the controlled arcuate liquid passage inside the control passage were analyzed, and the double role of elastic solid valve diaphragm in airway and liquid passage Under the structural deformation and stress distribution. The control mechanism and mathematical description of different media channels in the microfluidic process and the overcurrent characteristics of aerodynamic micro-valve under the gas-liquid coupling and the related influencing factors are clarified.2. Materials and Methods

Materials and Methods should be described with sufficient details to allow others to replicate and build on published results. Please note that publication of your manuscript implicates that you must make all materials, data, computer code, and protocols associated with the publication available to readers. Please disclose at the submission stage any restrictions on the availability of materials or information.

IOP Conf. Series: Journal of Physics: Conf. Series **986** (2018) 012014 doi:10.1088/1742-6596/986/1/012014

New methods and protocols should be described in detail while well-established methods can be briefly described and appropriately cited.

Research manuscripts reporting large datasets that are deposited in a publicly available database should specify where the data have been deposited and provide the relevant accession numbers. If the accession numbers have not yet been obtained at the time of submission, please state that they will be provided during review. They must be provided prior to publication.

Interventionary studies involving animals or humans, and other studies require ethical approval must list the authority that provided approval and the corresponding ethical approval code.

2. Working principle and control model

In this paper, two-channel thin-film gas-actuated micro-valve is taken as an example to study the flow and heat transfer rules in micron-sized gas channels, as well as the deformation and deformation of elastic solidvalve membrane during airway expansion and micro-valve closure.

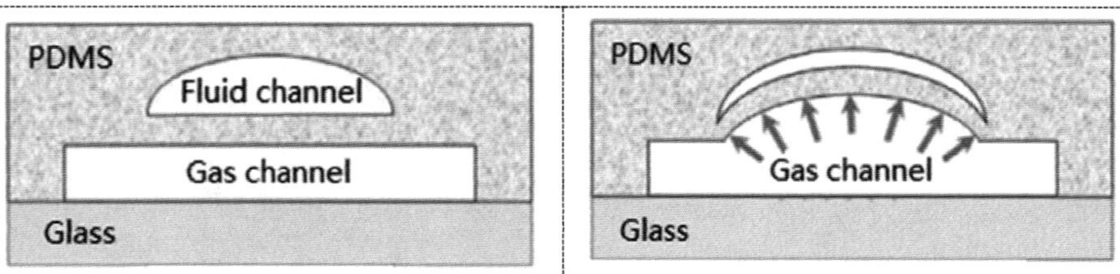

a) open b) closed

Figure 1. Operating principle of pneumatic valve

As shown in Figure 1, an active microfluidic aerodynamic micro-valve consisting of glass and poly-dimethylsiloxane (PDMS) was investigated. The micro-valve is a multi-layer structure with a gas control flow path, a liquid controlled flow path, and a middle flexible PDMS valve membrane. As shown inFig.1. (a), the gas control channel and the semi-arc (cross-section high) liquid flow channel with rectangular section () are separated by a PDMS membrane with a thickness of 1 mm and are co-bonded to a glass substrate. When the lower air flow channel leads to the pressure gas, the gas film is deformed under the internal pressure and the upper liquid channel is squeezed to shrink the liquid flow passage cross section (as shown in Fig 1. b). When the deformation amount reaches a certain level , Liquid channel is cut off, micro-valve to achieve closure.

3. Control equations

3.1. Control of airway

For the gaseous media in slip flow at the micron scale, this paper uses the unconditionally stable augmented Burnett equation and the boundary conditions to describe the working mechanism mathematically. The two-dimensional augmented Burnett equation in Cartesian coordinates can be written as:

$$\frac{\partial \rho}{\partial t}+\frac{\partial \rho u}{\partial x}+\frac{\partial \rho v}{\partial y}=0 \tag{1}$$

$$\frac{\partial \rho u}{\partial t}+\frac{\partial \rho u^2}{\partial x}+\frac{\partial \rho uv}{\partial y}=-\frac{\partial p}{\partial x}-\frac{\partial \sigma_{11}}{\partial x}-\frac{\partial \sigma_{12}}{\partial y} \tag{2}$$

$$\frac{\partial \rho v}{\partial t}+\frac{\partial \rho uv}{\partial x}+\frac{\partial \rho v^2}{\partial y}=-\frac{\partial p}{\partial y}-\frac{\partial \sigma_{21}}{\partial x}-\frac{\partial \sigma_{22}}{\partial y} \tag{3}$$

$$\frac{\partial e_t}{\partial t}+\frac{\partial e_t u}{\partial x}+\frac{\partial e_t v}{\partial y}=-\frac{\partial pu}{\partial x}-\frac{\partial pv}{\partial y}-\frac{\partial \sigma_{21}}{\partial x}-\frac{\partial \sigma_{11}u+\sigma_{12}v+q_1}{\partial x}-\frac{\partial \sigma_{21}u+\sigma_{22}v+q_2}{\partial y} \quad (4)$$

among them:

$$\sigma_{ij}=\sigma_{ij}^{(0)}+\sigma_{ij}^{(1)}+\sigma_{ij}^{(2)}+\sigma_{ij}^{(a)}$$

$$q_i=q_i^{(0)}+q_i^{(1)}+q_i^{(2)}+q_i^{(a)}$$

Due to the effect of levitation, there are tangential momentum fluxes of gas molecules in the isothermal wall motion, in which the approximate molecule comes from the near-wall region [20], and the other half reflects back from the wall. When the inlet temperature and the wall temperature are different, there is a tangential temperature gradient on the wall surface. The existence of such a temperature jump causes the gas temperature to be less than the wall temperature, especially near the entrance, and a larger tangential temperature gradient results in a greater thermal creep effect. As a result of hot creep, the body is forced to flow from the lower temperature side to the higher temperature side, exacerbating the speed slip at the boundary. Therefore, in this paper, for the gas flow in the micro-channel of the valve, the velocity slip boundary condition considering thermal creep effect of the boundary is adopted:

$$u_s=\frac{1}{2}\left[(2-\sigma_v)u_\lambda+\sigma_v u_w\right]+\frac{3}{4}\frac{\mu}{\rho T}\frac{\partial T}{\partial x}\bigg|_w \quad (5)$$

Of which: u_λ is the inlet gas tangential velocity; u_w is the wall tangential velocity; σ_v is the proportion of diffuse reflection gas molecules, the other part of the gas molecules mirror reflection law .

The corresponding wall temperature of gas changes can be expressed as:

$$T_s=\frac{\frac{2-\sigma_T}{\mathrm{Pr}}\frac{2\gamma}{\gamma+1}T_\lambda+\sigma_T T_w}{\sigma_T+\frac{2-\sigma_T}{\mathrm{Pr}}\frac{2\gamma}{\gamma+1}} \quad (6)$$

Where: σ_T is the temperature adaptation factor; Pr is Prandtl number; γ is the specific heat ratio; T_λ is the temperature near the wall of a molecule free path; T_w is the wall temperature.

In order to increase the stability of the calculation, the relaxation method can be introduced respectively in the above speed slip boundary and Burnett term:

$$u_s=u_s^{old}+R_f\left(u_s^{new}-u_s^{old}\right) \quad (7)$$

$$B=B^{old}+R_f\left(B^{new}-B^{old}\right) \quad (8)$$

Where: B represents the Burnett term and super Burnett term in the equation, old represents the value of the previous iteration, and new represents the newly calculated value.

3.2. *Elastic membrane*

For the PDMS diaphragm that separates the airway from the liquid passage, the nonlinear solid equations of the elastomer are used to describe the forces acting on the structure during deformation. The deformation force of the film structure is impulse pressure, liquid pressure and viscous force, where the gas pressure is a constant related to time, and the resultant force of liquid pressure and viscous force can be expressed as

$$F=\vec{n}\cdot\left\{-pI+\left[\eta\left(\nabla\vec{V}+\nabla\vec{V}^T\right)\right]\right\} \quad (9)$$

Where: $\vec{n}$ is the normal vector of the boundary; $\vec{V}=(u,v)$ is the velocity vector of the two-dimensional flow, u、v is the flow velocity in the x、y direction; p is the pressure of the liquid; η is the viscosity coefficient; I is the identity matrix.

In this paper, we mainly analyze the flow characteristics and energy conversion of the micro-scale Poiseuille flow in the gas control channel. The finite volume method is used to solve the system's Burnett equation. The second-order windward Roe flux difference splitting scheme is adopted for the non-sticky item. The second-order central difference scheme is used for the stress tensor and heat flux. The Gauss-Equations, momentum equations and energy equations, Finally, the control airway flow velocity and pressure distribution, controlled flow passage through the flow rate changes and shut-off pressure, as well as the diaphragm stress and deformation and other indicators of parameters, with the micro-valve structure parameters and driving changes in the law.

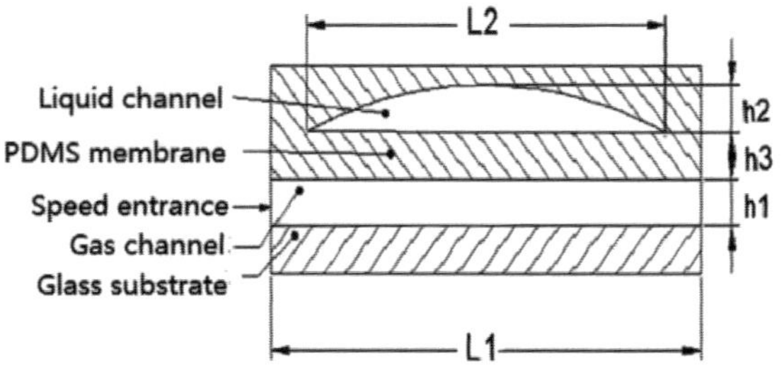

Figure 2. Schematic structural parameters

Table 1. Parameter Planning Table

Name Symbol Value (μm)	Name Symbol Value (μm)	Name Symbol Value (μm)
Airway length L1 500 ~ 600	Airway length L1 500 ~ 600	Airway length L1 500 ~ 600
Airway height h150	Airway height h150	Airway height h150
Diaphragm thickness h3 50	Diaphragm thickness h3 50	Diaphragm thickness h3 50
Liquid height h2 50	Liquid height h2 50	Liquid height h2 50
Liquid channel width L2 500 ~ 1100	Liquid channel width L2 500 ~ 1100	Liquid channel width L2 500 ~ 1100

4. Analysis of influencing factors

4.1. Airway width L1 of the impact

When the length of the gas channel is 800μm, the influence of velocity and pressure on the flow characteristics of the aerodynamic micro-valve is studied.

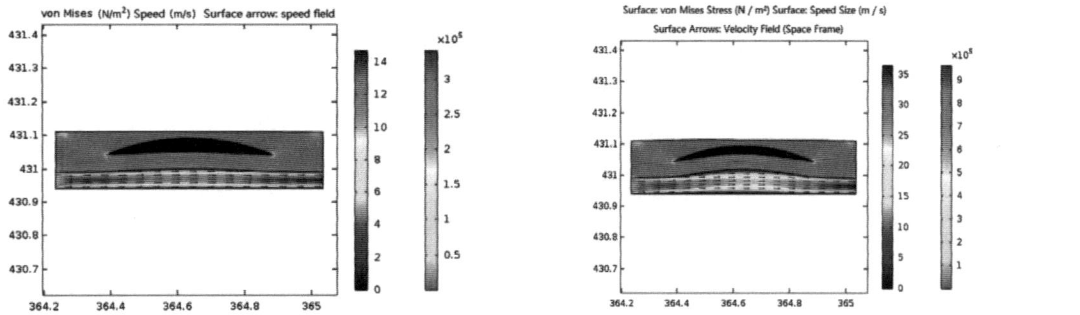

IOP Conf. Series: Journal of Physics: Conf. Series **986** (2018) 012014 doi:10.1088/1742-6596/986/1/012014

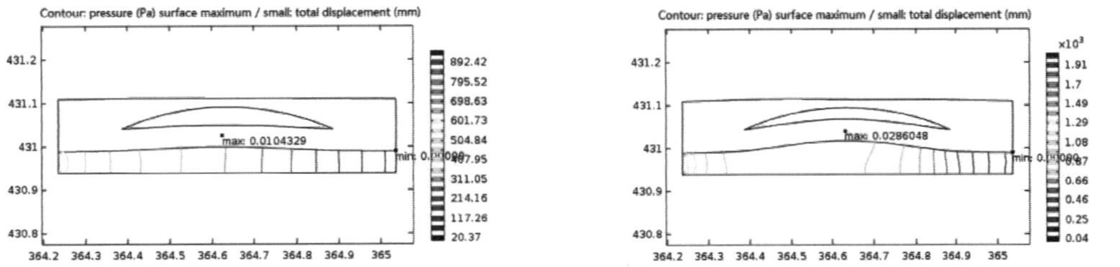

a) 20m / s speed drive b) 30m / s speed drive

Figure 3. velocity-driven airway flow field distribution

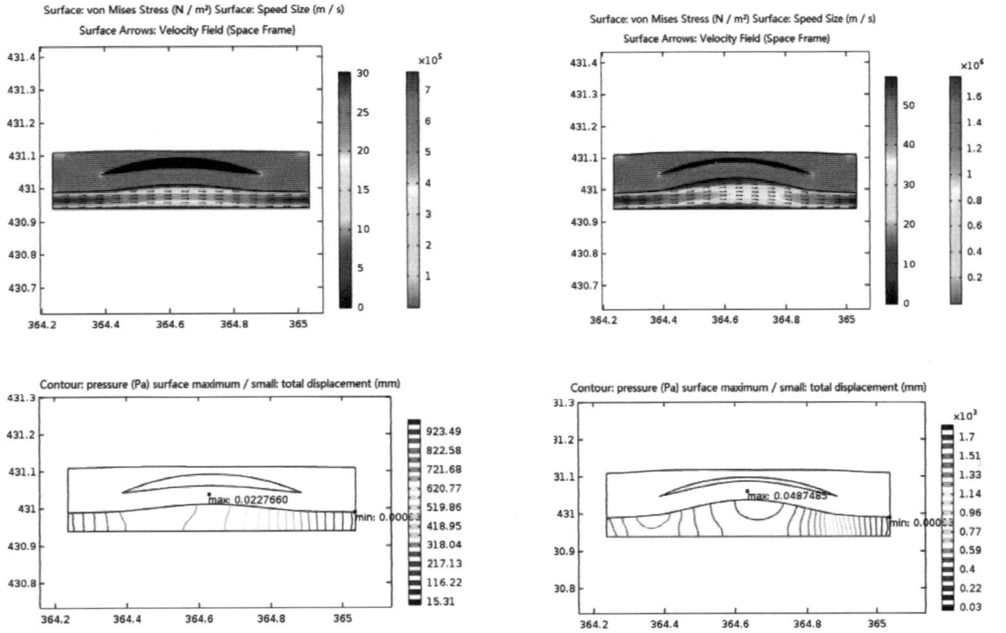

a) 1000 Pa b) 1500 Pa

Figure 4. pressure-driven flow field distribution in the airway

Comparing Fig. 3 and Fig. 4, we can see that regardless of whether the control airway is driven by speed inlet or pressure inlet, the pressure distribution in the airway flow field can be changed to warp the valve diaphragm to gradually close the flow passage cross section of the liquid passage, Different driving methods of control effect is not the same. As shown in Figure 3: As the inlet velocity increases from 50% to 30 m / s from 20 m / s, the velocity of the flow field inside the airway decreases gradually, and the velocity in the region right below the diaphragm decreases to the minimum. Pressure reached the maximum, pushing the valve diaphragm upward warping increased from 10.4 to 28.6; As shown in Fig. 4, as the inlet pressure increases from 1000 Pa to 1500 Pa, the warp effect of the valve diaphragm becomes more pronounced with increasing inlet pressure, and the amount of upward warp of the diaphragm increases from 22.8 to 48.7. Compared with the speed-driven mode, the pressure-driven closure of the liquid channel is better; The gas pressure under the diaphragm forms an isostatic ring in the warpage region, which increases the driving force of the diaphragm deformation. At this moment, most of the kinetic energy and pressure of the gas can be transformed into the elastic potential energy of the warpage of the supporting valve diaphragm.

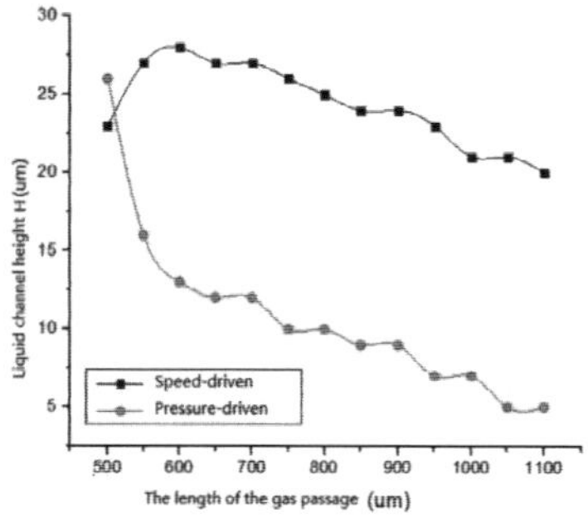

Figure 5. actuation of different airway length of the micro-valve closing characteristics

Further analysis of different driving modes on the pneumatic micro-valve cut-off characteristics, As shown in Fig. 5, when the gas passage length L1> 550μm, the pressure-driven effect is obviously better than the speed-driven effect, and the flow rate decreases slowly as the gas length increases. When L1 <1100μm, the micro-valve closing effect is not obvious under the speed-driven condition, and the flow rate is greater than 20μm. Under the pressure-driven condition, the closing effect of the micro-valve is obvious with the increase of the gas passage length. When the gas passage length L1> 600μm, the flow rate is less than 13μm, the closure effect is more obvious.

4.2. The influence of L2 / h2 ratio of liquid channel

Fixed controlled liquid channel height of 50μm, in order to ensure the micro-valve closure effect, so that the airway width L1= L2 +100μm. When L2 / h2 = (10 ~ 20), the influence of velocity and pressure on the flow characteristics of aerodynamic micro-valve were studied.

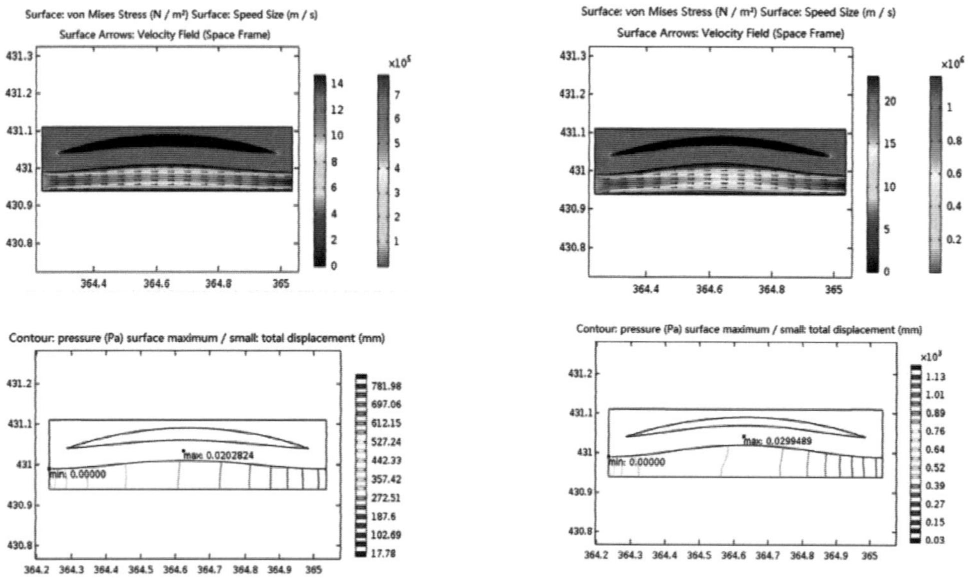

a) 10m / s speed drive b) 15m / s speed drive

Figure 6. velocity-driven airway flow field distribution

IOP Conf. Series: Journal of Physics: Conf. Series **986** (2018) 012014 doi:10.1088/1742-6596/986/1/012014

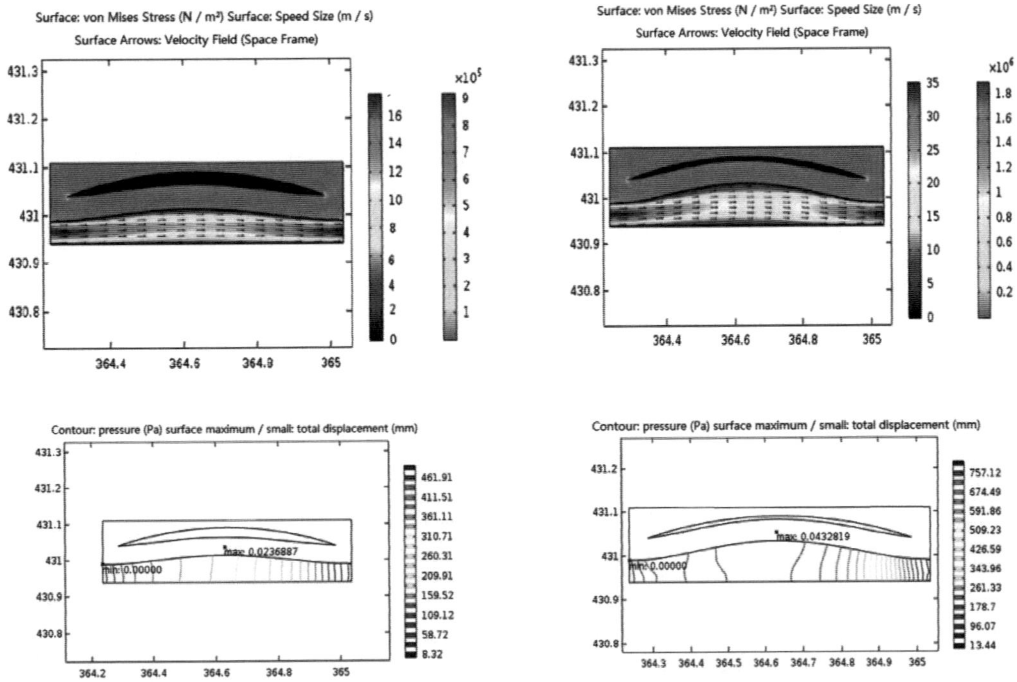

a) 500 Pa b) 750 Pa

Figure 7. pressure-driven airway flow field distribution

It can be seen from Fig. 6 and Fig. 7 that regardless of whether the control air passage is driven at a speed inlet or a pressure inlet, the pressure distribution in the air flow field can be changed to warp the valve diaphragm to change the flow cross section of the liquid passage. As shown in Fig. 4: As the inlet pressure increases from 500 Pa to 50O Pa, The degree of bending of the valve diaphragm gradually increases with the increase of inlet pressure, and the degree of warpage of the diaphragm changes significantly with the increase of inlet pressure. The amount of upward warpage of the diaphragm increased from 23.7 to 43.3. This shows that: compared to speed-driven approach, the pressure-driven closure of the liquid channel better At this time, the pressure in the gas channel under the diaphragm becomes obvious asymmetric distribution, and most of the kinetic energy of the gas is transformed into the elastic potential energy of the warpage of the diaphragm supporting the valve, and the valve diaphragm is upwardly closed to close the liquid passage upwardly.

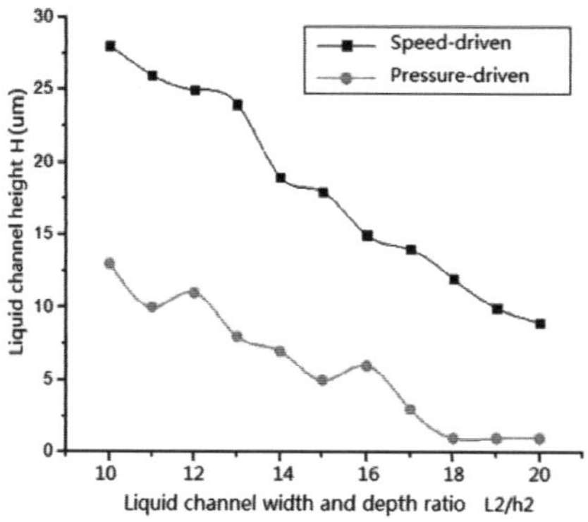

Figure 8. drive mode for different airway length of the micro-valve closing characteristics

Further analysis of different driving modes under the aspect ratio of the valve effect of micro-valve as shown in Figure 8 shows: the pressure-driven effect is significantly better than the speed-driven effect. As the liquid channel aspect ratio increases, the flow rate decreases rapidly. When L2 / h2> 18, the micro-valve closed more than 80% under the condition of speed driving, and the passage height of the liquid passage was less than 10μm. The micro-valve closing effect was excellent under the pressure driven condition and the flow passage height of the liquid passage was less than 1μm. The reason why 100% did not reach 100% at this moment was mainly caused by the use of two-dimensional model. The follow-up experimental research shows that when L2 / h2> 18, the pressure-driven membrane micro- The closure of the closure, and the efficiency of repeated operations up to 30 times. Thus, the film-type aerodynamic micro-valve, the use of inlet pressure control, appropriately increase the aspect ratio of the controlled liquid channel, you can effectively and accurately achieve the switch control and flow control.

5. Valve diaphragm deformation and airway flow field relationship

Because the PDMS material used in the diaphragm is a mixture of organosiloxanes with chain structures of different degree of polymerization, it has the structure of high elasticity and also has a certain breath-ability. Scanning electron microscopy was used to observe the cross-sectional morphology of the PDMS film. It was found that the PDMS homogeneous coating structure, the temperature gradient in the micro-channel, and the local pressure change on the hydrophobic surface all affect the permeability of the valve membrane and finally the flow inside the micro-valve Field characteristics.

5.1. Speed

In this paper, when the inlet pressure is 5000 Pa, the gas flow characteristics in the micro-channels are analyzed. As shown in Fig. 9, as the valve membrane warps upwards, while changing the cross-sectional shape of the flow passage, a partial vacuum is generated inside the air passage close to the membrane so that the air flow is partially entrained by the vacuum when it is moved below the membrane Form a counterclockwise gas swirl. Further studies showed that with the increase of inlet pressure, the gas vortex intensity under the diaphragm gradually increased, the range of action gradually expanded, the gas slip velocity near the diaphragm increased, the Reynolds number inside the airway was con-trolled, and the degree of turbulence was enhanced.

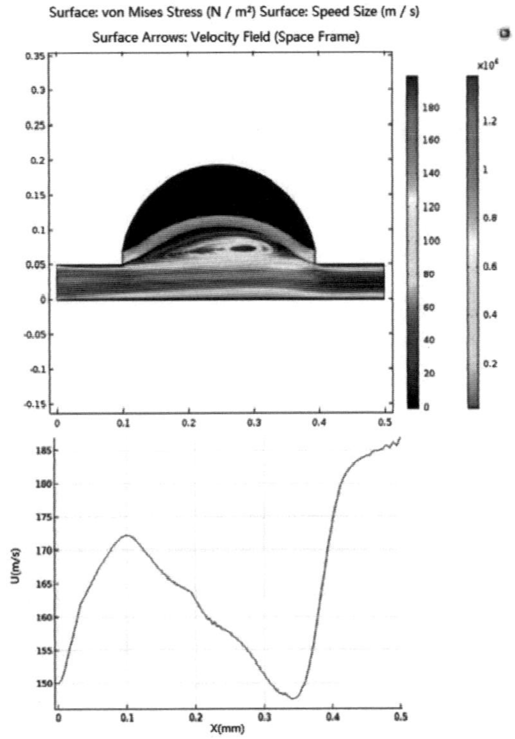

Figure.9. Speed streamline along gas channel in micro valve

As shown in Fig. 9, as the valve membrane warps upwards, while changing the cross-sectional shape of the flow passage, a partial vacuum is generated inside the air passage close to the membrane so that the air flow is partially entrained by the vacuum when it is moved below the membrane Form a counterclockwise gas swirl. Further studies showed that with the increase of inlet pressure, the gas vortex intensity under the diaphragm gradually increased, the range of action gradually expanded, the gas slip velocity near the diaphragm increased, the Reynolds number inside the airway was controlled, and the degree of turbulence was enhanced. Due to the existence of the gas vortex below the diaphragm, the gas flow path has obvious speed difference. The gas flow velocity is the lowest at the center of the gas vortex. The outlet velocity from the gas swirl center to the gas flow path gradually increases and reaches the maximum value.

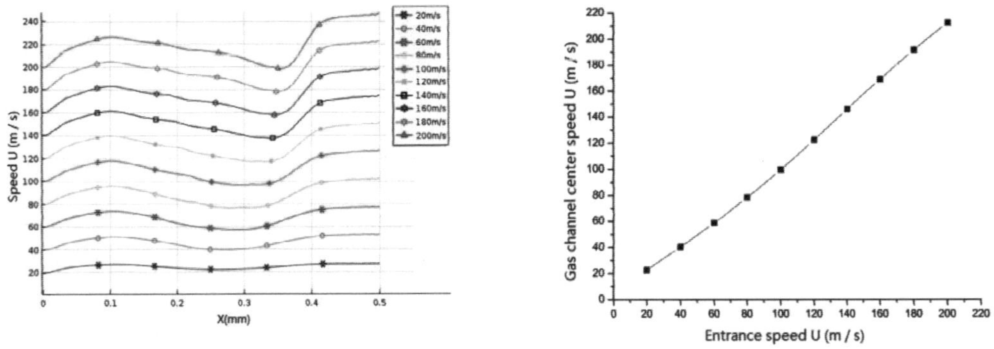

a) **b)**

Figure.10. Velocity characteristics in gas channel with different inlet pressure

Further studies show that as the inlet velocity increases, the vortex range of the gas gradually increases and occupies the entire cavity; the position of the gas vortex center gradually moves away from the gas inlet and the formation speed of the gas vortex is faster. As can be seen from Fig. 10 a), the gas flow rate at the center just below the diaphragm in the gas flow path increases with inlet velocity as the inlet gas velocity increases. As shown in Fig. 10 b), it can be seen that with the increase of inlet velocity, the velocity of air in the airway decreases obviously. And as the inlet velocity increases, the velocity decreases more. The flow velocity near the inlet of the gas channel descends more slowly and the velocity near the outlet increases rapidly.

5.2. Pressure

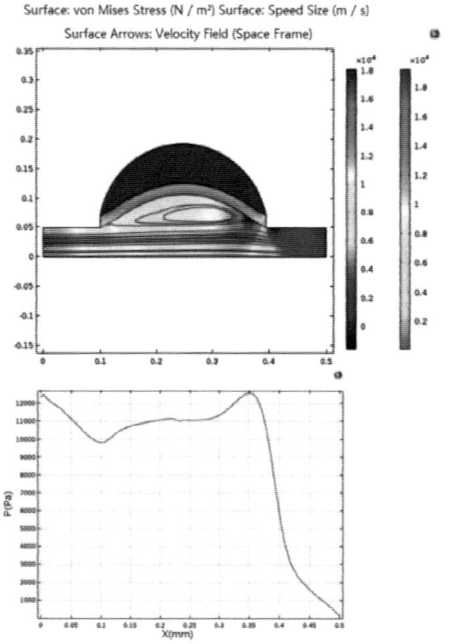

Figure 11. Pressure distribution along gas channel when inlet pressure is 1.26 kPa

As can be seen in Fig. 11, the pressure in the gas passage decreases first and then increases and then decreases slightly. The main reason for the pressure change is caused by the pressure difference between inside and outside of the gas vortex. The pressure below is asymmetrically distributed and the pressure near the airway exit is greater.

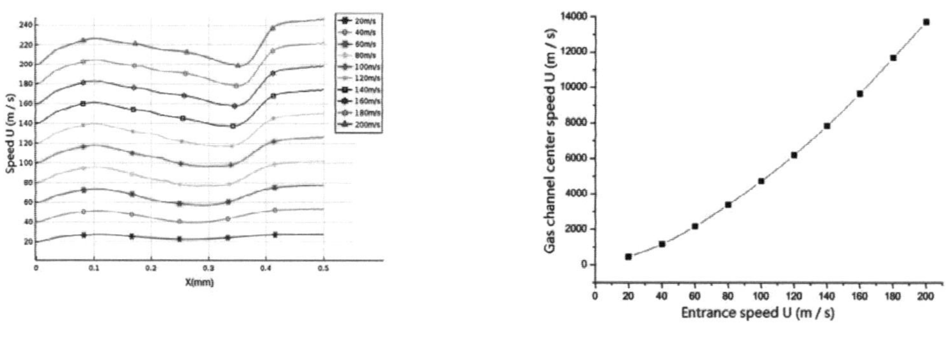

a) b)

Figure 12. Pressure characteristics in gas channel with different inlet pressure

As can be seen from Fig 12. a), as the inlet air velocity increases, the pressure in the air passage increases gradually. As Fig 12. (b) shows, the pressure fluctuation in the air passage becomes more pronounced as the inlet pressure increases.

5.3Diaphragm force analysis

In this paper, the deformation and stress analysis of the elastic diaphragm in the middle of the micro-valve are studied respectively under the two different driving modes of air inlet and velocity inlet.

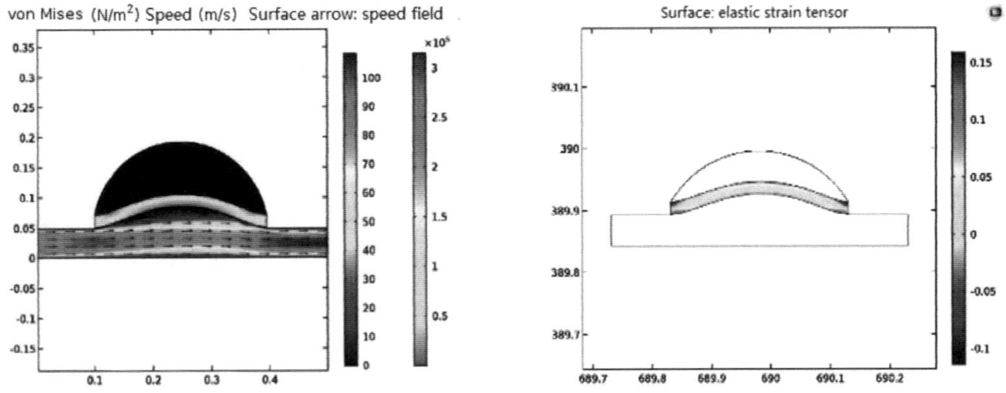

a) Stress tensor b) Strain tensor

Figure 13. Stress and strain of valve diaphragm with inlet velocity 100 m/s

As can be seen from Fig 13, under the inlet speed, the central area of the diaphragm is pushed up-wards to squeeze the fluid passage. The two ends of the diaphragm are respectively subjected to tensile stress near the airway side and the compressive stress near the side of the liquid channel. Straight direc-tion to form a shear effect.

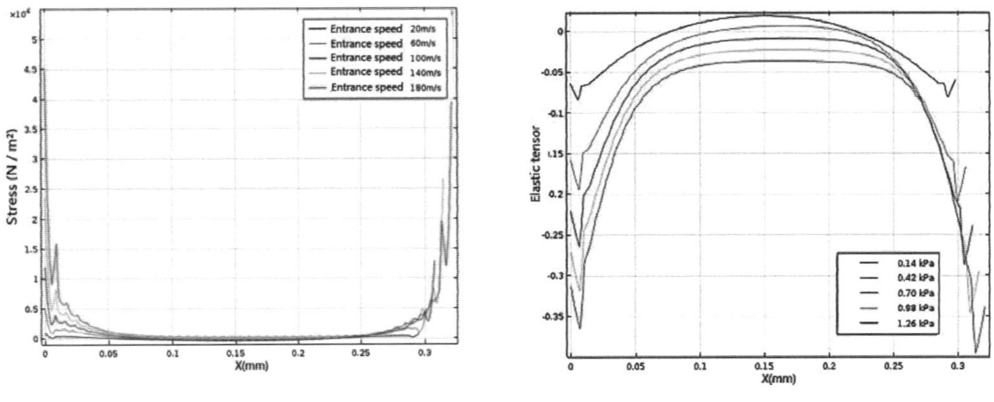

a) Stress tensor b) Strain tensor

Figure 14. Mechanical characteristic of valve diaphragm with different inlet pressure

From Fig 14 (a), it can be seen that the stress acting on the two ends of the diaphragm gradually increases with the increase of inlet velocity, and the greater the inlet driving velocity, the greater the stress gradient at the edge. The strain tensor of Fig. 14 (b) further adds that the smaller the inlet velocity is, the smaller the overall strain of the valve in the vertical direction is. However, as the inlet velocity increases, the steady area of the valve center increases, However, the strain at both ends changed drastically, an increase of 7 to 10 times over the central area. This shows that the increase of inlet speed will aggravate the stress concentration on the edge of the diaphragm, which will easily cause the fatigue of the diaphragm of the valve.

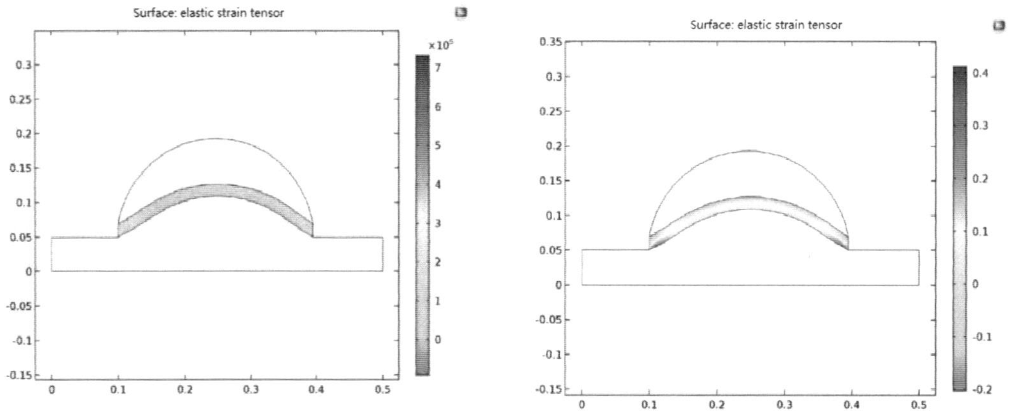

a)Stress tensor b) Strain tensor

Figure 15. Stress and strain of valve diaphragm with inlet pressure 0.7 kPa

It can be seen from Fig. 15 that under the inlet pressure, the central region of the diaphragm is pushed upwards to squeeze the fluid channel. The two ends of the diaphragm are respectively subjected to tensile stress near the airway, and the diaphragm near the fluid channel is under compressive stress. The whole piece of the tensile stress, the diaphragm near the central part of the gas channel by the local compressive stress.

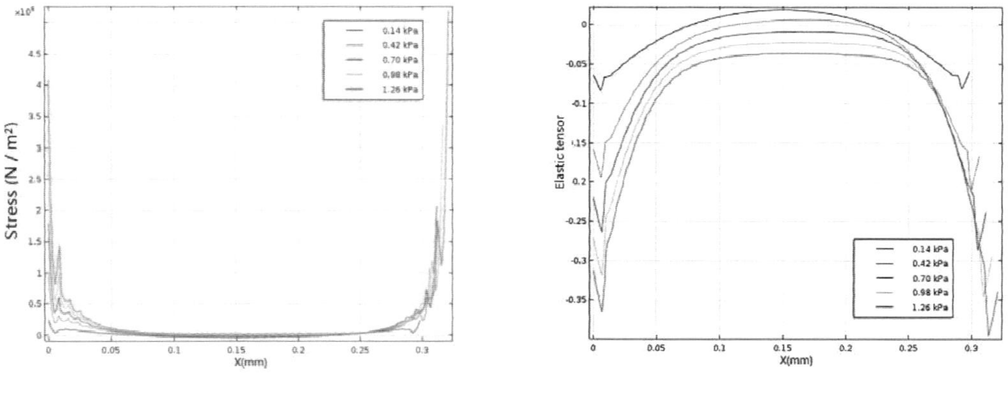

a)Stress b) Strain

Figure 16. Mechanical characteristic of valve diaphragm with different inlet pressure

The pneumatic diaphragm valve at 0~200m / s inlet pressure on the diaphragm analysis. As shown in Fig .16, during the deformation of the planar film, stress on both ends of the film is relatively large, which is the main reason that the two ends of the film fail to rupture. As the inlet velocity increases, the stress at both ends of the planar film increases sharply, while the stress in the rest increases slightly. In terms of strain, increasing the driving pressure and increasing the starting speed have basically the same changes on the strain of the valve diaphragm, which are all stable in the central region, and the strain at both ends of the diaphragm increases with the inlet pressure. This can be understood as both a velocity inlet and a pressure inlet, in fact, the source of the initial energy, the enhancement of the energy of the driving source, the increase of turbulence in the change of the internal flow in the airway, The energy transfer and exchange become more complex, the effect of the diaphragm is also more and more effective, and this effect due to the structural characteristics of the diaphragm determined, the diaphragm will play a role in the destruction of the impact of micro Valve life. Therefore, to improve the effectiveness of micro-valve control at the same time, we must fully consider the pressure resistance of the diaphragm to ensure control effectiveness.

6. Experimental program and effect

In order to further verify the above analysis is true and reliable, this paper designs and designs the test device as shown in Figure 17 to validate the effectiveness of the pneumatic micro-valve driven by different speed inlet and pressure inlet respectively.

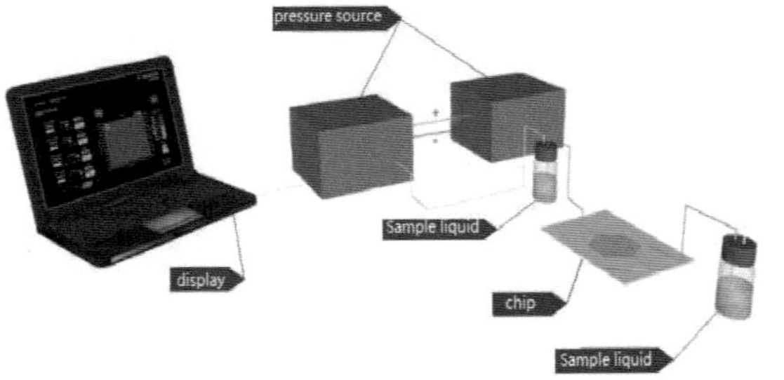

a) Test principle

b) Test device

Figure 17. Test rig

According to Figure 17 a) Test Schematic Figure 17. b) set up the experimental test device. The controller adopts Wen-Chou chip company WH-PMPP-12-type constant pressure pump. Liquid channel outlet flow measurement OMEGAdpg120-type flow meter. This paper tests the different inlet velocity (500,1000,1500,2000) $\mu L / min$, the flow channel with the control of the gas channel can achieve a clear flow control effect, as shown in Figure 18.

a) Micro valve open

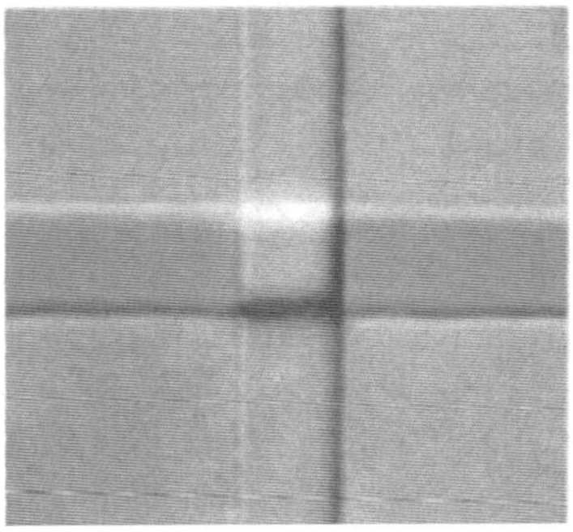

b) Micro valve closed

Figure 18. Operating performance of pneumatic micro valve

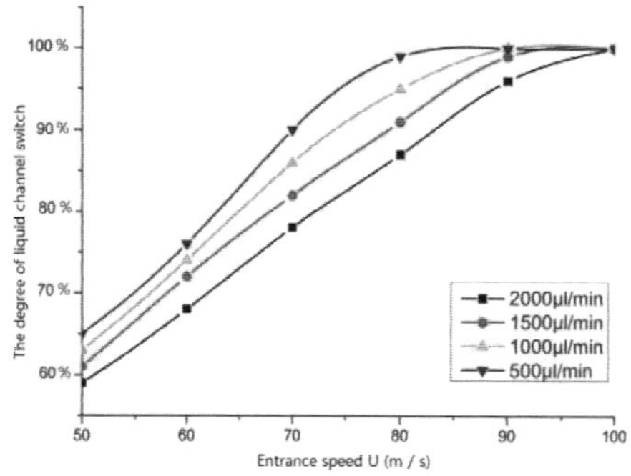

Figure 19. Control effect of vecolity inlet

As shown in Fig 19. At the same air supply rate, the degree of liquid channel closure gradually decreases as the liquid flow rate increases.At the same liquid flow rate, with the increase of gas supply rate, the degree of liquid channel closure gradually increases, and the increasing speed increases first and then decreases.

In this paper, we test the pressure flow characteristics and shutdown response characteristics of micro-valves with different diaphragm area.

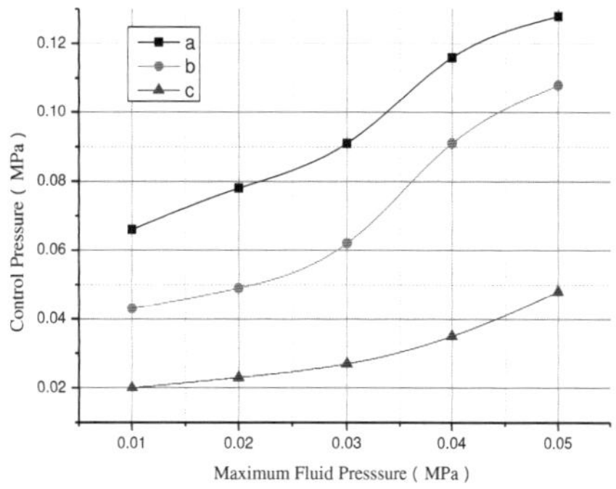

a) Pressure-flow characteristics of diaphragms of different cross-sectional area under pressure（a 400μm×400μm、b 600μm ×600μm、c 800μm ×800μm）

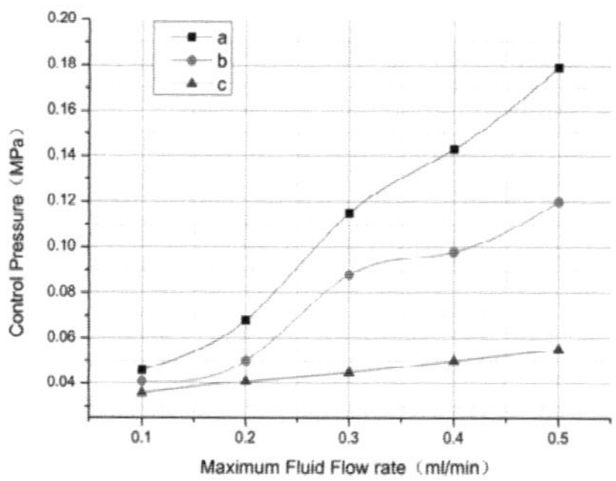

b) Flow-driven pressure-flow characteristics of diaphragms of different cross-sectional areas

（a 400μm×400μm、b 600μm ×600μm、c 800μm ×800μm）

Figure 20. Pressure flow characteristics of diaphragms with different cross sections under different driving modes

As shown in Fig 20, a wide width with a large actuation area can withstand a relatively large hydraulic pressure at the same control pressure. This case can be explained for the following reason. When the control pressure is applied to deform the elastic film to cut off the flow path, a part of the applied control pressure is used to counteract the stress caused by film deformation and the remaining part is used to resist the passage in the liquid passage Hydraulic. Due to the same channel depth, the deformation rate of the film (the linear ratio in the direction of deformation before and after the deformation) is larger for a narrower liquid passage, and in this case, a larger partial pressure is required to counteract the film deformation Stress, and can be used to resist the partial pressure of the hydraulic pressure correspondingly reduced, making the micro-valve of the critical pressure value becomes smaller.

7. Conclusions

In this paper, passive gas brake flow control micro-valve to complete the following research work:

(1) The gas flow problems for micro and nanoscale channel, analyse the internal air flow characteristics of the micro-valve, is obtained based on the given two-dimensional augmented Burnett equations, a clear boundary conditions considering creep effect under micro-nano scale, A relaxation method is introduced to increase the computational stability. The elastic solid nonlinear equation is used to describe the stress in the process of structural deformation, and the formulas for calculating the pressure and the viscous force of the liquid subjected to the deformation of the film structure are given.

(2) The velocity and pressure variation of the velocity inlet to the micro-valve internal flow field are analyzed numerically. The results show that with the increase of the supply pressure, the horizontal velocity in the gas flow path below the valve diaphragm decreases, and the kinetic energy of the gas partially transforms to the rotational kinetic energy of the gas at the cavity. With the increase of inlet velocity, the position of swirling vortex center in airway gradually moves away from the gas inlet, and obvious pressure drop phenomenon occurs near the outlet of gas film, At this time, the energy of the gas is mainly represented by the rotational kinetic energy of the gas at the cavity.

(3) Combining experiments, this paper analyzes the working effect and pressure flow characteristics of flat-diaphragm pneumatic micro-valves under different driving modes. The results show that the pressure-driven method is more likely to produce a stronger gas vortex below the diaphragm, resulting in the pressure difference between the inside and outside of the diaphragm to push the diaphragm to cut off the liquid passage, thus having a better control effect. Under the pressure-driven mode, the stress and strain on both ends of the diaphragm is small, and the diaphragm is not easily damaged.6. Patents

8. References

[1] Thor Sen T, Maerkl SJ, and Quake SR 2002 Microfluidic large-scale integration. Science. 298 (5593) pp 580-4
[2] Mc Donald JC and White Sides GM 2002 Poly as a material for fabricating microfluidic devices. Accounts of chemical research. 35(7) pp 491-499.
[3] Qu XF. Preparation and application of microfluidic chips. Shanghai: Jiao Tong University. 2006
[4] Li ZH, Wu JK, and Hu GQ 1991 Fluid Flow in Microfluidic Chip. Beijing: Science Press. 2012
[5] Pfahler J, Harley J, and Bau H, et al. Gas and liquid flow in small channels. ASME DSC. 32 pp 49-60
[6] Ho CM and Tai YC 1998 Micro-electro-mechanical-systems (MEMS) and fluid flow. Annual Review of Fluid Mechanics. 30 pp 579-612
[7] Stone HA, Stroock AD, and Ajdari A 2004 Engineering flows in small devices: Microfluidics toward a lab-on-a-chip. Annual Review of Fluid Mechanics. pp 36:381-411
[8] Dar Huber AA and Troian SM 2005 Principles of microfluidic actuation by modulation of surface stresses. Annual Review of Fluid Mechanics. 37 pp 425-455
[9] Schoch R, Han J, and Renaud P. Transport phenomena in nanofluidics. Review of Modern Physics. 80 pp 839-883
[10] Van der Berg A, Craighead H, and Yang P 2010 microfluidic application to nanofluidic phenomena issue reviewing the latest advances in microfluidic and nanofluidic. Chemical Society Reviews. 39(3) pp 948-956
[11] Harley JC, Huang Y, and Bau HH 2004 Gas flow in micro-channels. Journal of Fluid Mechanics. 36 pp 381-411
[12] Arkilic EB, Schmidt MA, and Breuer KS 1997 Gaseous slip flow in long micro channels. Journal of Microelectromechanical Systems. 6(2) pp 167-178
[13] Beskok A. Simulations and models for gas flows in micro geometries. PHD thesis. Princeton University. 1996
[14] Wang CT and Sun JM 2007 Unsteady Analysis of Micro valves With No Moving Parts. Journal of Mechanics. 23(1) pp 9-14
[15] AK Au, Lai H, Utela BR and Folch A 2011 Micro valves and Micro pumps for BioMEMS. Micro machines. 2(2) pp 179-220
[16] Agarwal R K, Yun K Y, and Balakrishnan R 2001 Beyond Navier-Stokes: Burnett equations for flows in the continuum-transition regime. Physics of Fluids. 13(10) pp 3061-3085
[17] Fang YC. Parallel simulation of micro flows by DSMC and Burnett equations. PHD Thesis. Western Michigan University. 2003
[18] Zhong XL, Mac Cormack RW, and Chapman DR 1993 Stabilization of the Burnett equations and application to hypersonic flows. AIAA Journal. 31(6) pp 1036-1043
[19] Bao FB and Lin JZ 1991 Linear stability analysis for various forms of one-dimensional Burnett equations. International Journal of Nonlinear Sciences and Numerical Simulation. 6(3) pp 295-303
[20] Ning FW, Wei L, and Yan L 2016 Analysis of gas flow and pressure characteristics between flat plates under micro-scale. China Mechanical Engineering.14 pp 1862-1865

IOP Conf. Series: Journal of Physics: Conf. Series **986** (2018) 012015 doi:10.1088/1742-6596/986/1/012015

Method of Preparation AZP4330 PR Pattern with Edge Slope 40°

Jie Wu[1], Hongyuan Zhao [1], Yuanwei Yu [1]and Jian Zhu [1]

[1]Nanjing Electronic Devices Institute, Nanjing zhongshan road 524, China

ckwujie@foxmail.com
Abstract. When the edge which is under the multi-film is more steep or angular, the stress in the multilayer film near the edge is concentrated, this situation will greatly reduce the reliability of electronic components. And sometimes, we need some special structure such as a slope with a specific angle in the MEMS, so that the metal line can take the signal to the output pad through the slope instead of deep step. To cover these problems, the lithography method of preparing the structure with edge slope is studied. In this paper, based on the Kirchhoff scalar diffraction theory we try to change the contact exposure gap and the post-baking time at the specific temperature to find out the effect about the edge angle of the photoresist. After test by SEM, the results were presented by using AZP4330 photoresist, we can get the PR Pattern with edge slope 40 ° of the process and the specific process parameters.

1. Introduction

In the IC process, there always be some structure under the multi-film, but the stress of the film much higher near the step of the structure rather than other smooth place. This situation will greatly reduce the reliability of electronic components. This phenomenon is from the lithography and etch process, we usually make the process like figure 1 The conventional process (left), after normal lithography from process (a) to (b), the edge of the PR pattern is steep, and then ,after etch process, the pattern is almost copy onto the substrate, the edge can't be changed.

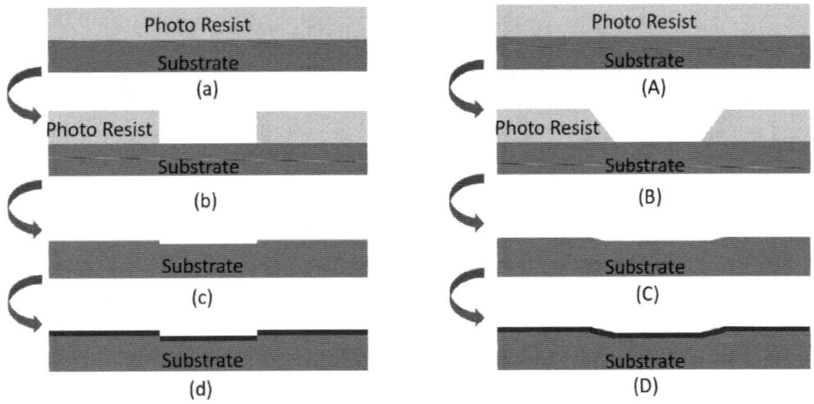

Figure 1. The conventional process (left) and the special process(right)

And at the same time, in the MEMS, we also need some special structure such as a slope with a specific angle[1,2]. so that the metal line can take the signal to the output pad through the slope instead of deep step or we need some specific structure to make the light refraction and this requirement is more extreme because the structure always more than tens or even hundreds microns deep. Conventional processes are far from meeting this need. New processes in this area are rarely published. we need to try to develop a new method to solve this problem by lithography such as the process in figure 1 The special process(right).

2. Theories and Methods

In the lithography process, the first step is coating photoresist, we usually adopt Spin coating at a high rotating speed to throw the photoresist from the center of the wafer to outside so that the centrifugal force and viscous force make the photoresist forming a layer of thin film on the flat silicon substrate. The second step is soft bake, it can make the organic solvent of the photoresist volatile, and then the photoresist thin film become hard enough so that it can contact reticle with not deformed. The third step is exposure, in the ultraviolet radiation, the solubility of the photoresist which is exposed in the developer is changed, after developing, we can get the graphics the same with the reticle. The last step is hard bake, it can make the rest of the organic solvent of the photoresist volatile, and then the photoresist graphic thin film is cured.

In the above process steps, the most important factors affecting the edge morphology of the photoresist are exposure mode, contact exposure gap, exposure dose, hard-bake film temperature and time[3-5].

When we use the proximity exposure mode, the principle model as shown in figure 2.Based on the Kirchhoff scalar diffraction theory, the light dose of the pattern edge is progressively increased, so that we can get the slope by developing.

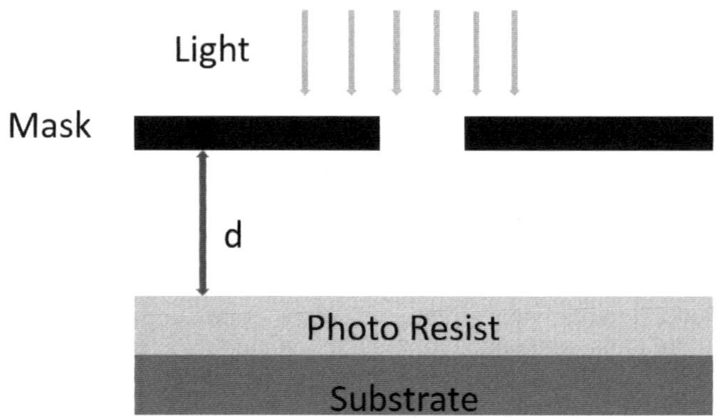

Figure 2. Proximity exposure mode

In the MEMS dry etching process, we usually use thick photoresist as the etching masking layer, the AZ4330 PR is widely used in this process, in the main sizing speed of 3000 r / min, the AZ4330 PR film thickness can be controlled at 4um or so, and its critical dimension can reach to a few microns, by using of the SUSS MA6 contact lithography exposure, set the exposure dose at 400mj / cm2, in the proximity mode, by setting the exposure gap between the wafer and the mask, to make the parallel ultraviolet light diffraction, the edge of the photoresist pattern is irradiated by partially diffracted ultraviolet light, while the other regions are normally exposed or masked, so that the edges of the photoresist pattern are subject to gradient changes in the light, in the subsequent development, a slope-like edge will be formed, but at this time the edge of the slope is still far from our needs, because the surface roughness is very large. By adjusting the temperature and time of the bake film, the edges of the photoresist are softened and self-leveling to form a smooth slope.

Based on the above principles, we designed an experiment, only change the exposure gap, to test the result of different and then, change the hard bake temperature and time, to test the effect of temperature. First, we set the exposure gap at 10 microns, 30 microns, 50 microns ,and keep the other steps the same, to get 2 sets of samples , and then, take 1 set of the samples to process with hard bake at 120 ℃.

After the multi-round test, we obtained a set of photomask mask pattern samples, since the photoresist was relatively thin on the silicon substrate and was easily damaged in the physical cutting, so it is hard to get a clear photo of the profile in the intuitive test, we try to manually split the samples with external force, to make sure that the split direction is the same to the silicon wafer crystal direction, in the early lithography we must make sure that the observed photolithography profile parallel to the silicon crystal direction, so that we can get a neat profile, after observed by Hitachi 8230 scanning electron microscopy, we can get a clear profile photo, and by the test of the photoresist profile slope, we can get a certain angle.

3. Results and Discussion

After the multi-round test, we get the test photo by SEM , after software measurement, we found that when the exposure gap at 10 microns, 30 microns, 50 microns, the slope edge angle is about 44° , 42 ° , 40° , as figure 3, figure 4, figure5 respectively.

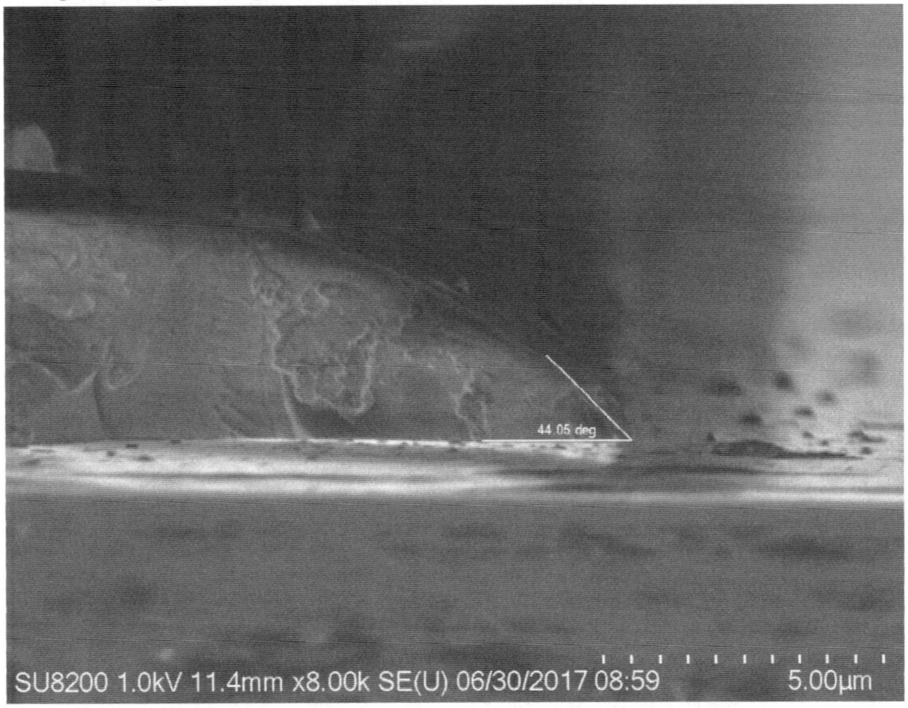

Figure 3. The PR slope at Contact exposure gap 10um

When the baking temperature is set to 120 ℃, and the baking time is set to 4 minutes, 8 minutes, 12 minutes, The slope angle will gradually decrease, the surface will gradually become smooth, but the overall angle change is not very obvious, as figure 6. From the above test, we can draw a set of empirical values, by changing the exposure gap, post-bake time and other parameters, you can get different slope of the photoresist graphics, so as to provide more options for future process.

IOP Conf. Series: Journal of Physics: Conf. Series **986** (2018) 012015 doi:10.1088/1742-6596/986/1/012015

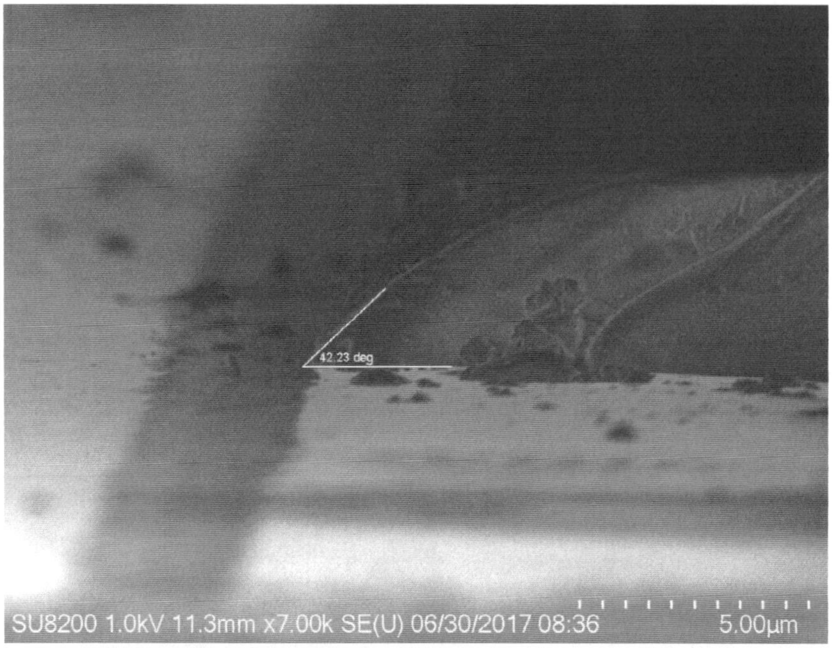

Figure 4. The PR slope at Contact exposure gap 30um

Figure 5. The PR slope at Contact exposure gap 50um

Figure 6. The PR slope at Contact exposure gap 50um and hard-bake

Through the above SEM test picture data, We can draw out the table 1, it can show us directly that the effect of the contact exposure gap and the hard bake. When the exposure gap is 0, we can get a steep steps which is conventional process requirements, when we set the exposure gap to 10 microns, the steps become to a slope with an angle 44° , when the exposure gap is set to 30 microns and 50 microns, the slope angle become sharper and sharper correspondingly.

Table 1. Test results about the gap changing.

Exposure gap	Slope angle	After hard bake
10 microns	44.05°	43.98°
30 microns	42.23°	42.18°
50 microns	40.68°	40.60°

4. Conclusions

In this paper, by changing the exposure gap and the post-bake time under the contact lithography condition, the edge of the photoresist pattern is ramped at a certain angle, and the slope PR is provided for the subsequent dry etching process, which is more favorable for obtaining the edge Structure, to ease the multi-layer film cover when the edge of the stress concentration, to improve the reliability of the device.

References

[1] B. Morgan, C . Waits , J . Krizmanic, R .Ghodssi: Development of a deep silicon phase fesnel lens using gray-scale technology and deep reactive ion etching, J .Microelectromech . Sys.(JMEMS) 13 (2004) 113-120.

[2] P . Yao , G . J . Schneider, D .Prather, E .Wetzel, D .O' Brein : Fabrication of three-dimensional photonic crystals with multilayer photolithography, Opt. Express 13 (2005) 2370-2376.

[3] Yu Hang :Method for Improving the Steepness of Phone Resist , Journal of Changchun teachers College 33(2014)28-32.
[4] Lai Wu-xing ,Xuan Jian-ping : Comparison of Characteristics of Photo lithography Process on Micromachining ,Semiconductor Technology 29 (2004) 22-25.
[5] Liu Quan ,Wan Hua, Wu Jian-hong :Research on Acquisition of Low Gradient Photoresist Gratings , Acta Photonica Sinica 37(2008)1401-1405.

Research of a smart cutting tool based on MEMS strain gauge

Y Zhao, Y L Zhao, Y W Shao, T J Hu, Q Zhang and X H Ge

The State Key Laboratory for Manufacturing Systems Engineering, Xi'an Jiaotong University, Xi'an, Shaan Xi, China

Corresponding author's e-mail address: zhaoyulong@xjtu.edu.cn

Abstract. Cutting force is an important factor that affects machining accuracy, cutting vibration and tool wear. Machining condition monitoring by cutting force measurement is a key technology for intelligent manufacture. Current cutting force sensors exist problems of large volume, complex structure and poor compatibility in practical application, for these problems, a smart cutting tool is proposed in this paper for cutting force measurement. Commercial MEMS (Micro-Electro-Mechanical System) strain gauges with high sensitivity and small size are adopted as transducing element of the smart tool, and a structure optimized cutting tool is fabricated for MEMS strain gauge bonding. Static calibration results show that the developed smart cutting tool is able to measure cutting forces in both X and Y directions, and the cross-interference error is within 3%. Its general accuracy is 3.35% and 3.27% in X and Y directions, and sensitivity is 0.1 mV/N, which is very suitable for measuring small cutting forces in high speed and precision machining. The smart cutting tool is portable and reliable for practical application in CNC machine tool.

1. Introduction

Cutting force is generated between workpiece and cutting tool during machining process, it is closely associated with cutting heat, cutting vibration and tool wear. Cutting force is a key factor that affects machining smooth, tool life and energy consumption [1]. Any tiny variation of cutting status can be reflected by cutting force. Thus, measuring cutting force is useful for monitoring machining status, improving accuracy and forecasting failure [2], which is of great importance for intelligent machining. Cutting force dynamometer is an effective tool for cutting force measurement.

So far, many kinds of cutting force dynamometer based on different principles have been developed [3,4]. Among them, strain gauge and piezoelectric cutting force dynamometers are widely used because of high accuracy and good reliability. For example, Kistler Company (Switzerland) is a famous piezoelectric sensor manufacturer [5], its piezoelectric cutting force dynamometer is qualified with large rigidity (2×10^9N/m) and low cross-interference error ($\leqslant$2%), which is a main supplier of cutting force dynamometer around the world. YDC series piezoelectric cutting force dynamometers developed by Dalian University of Technology possesses satisfying resolution (1N) and high natural frequency ($\geqslant$ 4kHz), it is very suitable for dynamic cutting force measurement [6,7]. Strain gauge cutting force dynamometer developed by TeLC Company (Germany) is able to measure and analyze cutting forces [13], its DKM type cutting force dynamometer can match with different types of cutting insert for variable application occasions. Xi'an jiaotong University has proposed a strain gauge cutting force dynamometer based on octagonal ring structure, the developed dynamometer owns high accuracy ($\leqslant$ 0.84%) and high natural frequency ($\geqslant$1122Hz)[8.9].

However, strain gauge and piezoelectric cutting force dynamometer usually own comprehensive structure and heavy weight, which is incompatible with CNC lathe for installation. In recent 5 years, researchers have focused on smart cutting tool research by integrating force measurement unit on the cutting tool to measure cutting forces. For example, Harbin Institute of Technology has developed a smart cutting tool by embedding a tiny triaxial force measurement system into a structure optimized cutting tool [10]; North University of China developed a micro-sensor for cutting force measurement based on Ni-Cr alloy[17]; University of Birmingham proposed a smart cutting tool based on piezoelectric film[18]. As the research of smart cutting tool is still in its infancy, there are a lot of work worth exploring. This paper proposes a smart cutting tool based on Micro-Electro-Mechanical System (MEMS) strain gauge. Its structure composite, working principle and performance test are introduced in following sections.

2. Materials and Methods

2.1. Structure design

According to problems of present cutting force dynamometers that have been mentioned above, and taking the requirement of cutting force measurement in intelligent machining, a smart cutting tool for cutting force measurement should fulfill following principles: ① good compatibility, the developed smart cutting tool can be easily installed on CNC lathe without affecting its machining performance; ② fine portability, the smart cutting tool should be structure simple and potable for application; ③ high accuracy, it needs high accuracy and good sensitivity for monitoring small cutting force variation.

Figure 1 depicts a schematic view of the designed smart cutting tool, it is basically the same with an ordinary cutting tool except that there are four strain gauge bonding slots distributed in the front of tool shaft. MEMS strain gauges are packaged in the strain gauge bonding slots and are organized into Wheatstone bridges for measuring cutting forces in X and Y directions. A signal cable channel is designed for placing power and signal wires. A shielded cable is adopted for signal transmitting in order to avoid electromagnetic interference.

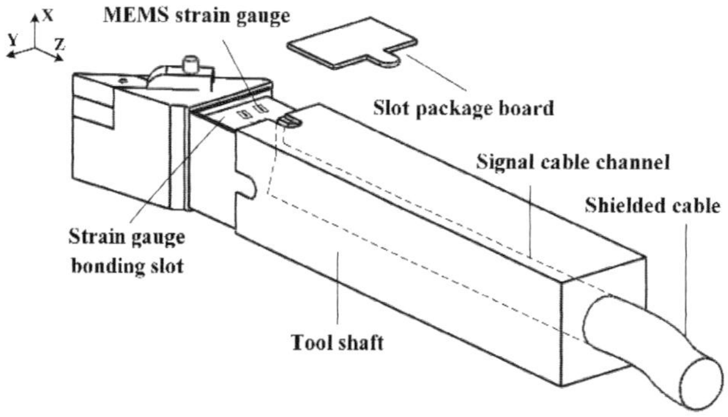

Figure 1. Smart cutting tool structure

2.2. MEMS strain gauge

The strain gauge used in this research is fabricated by MEMS technique, because MEMS strain gauge possesses the advantages of small size, large resistance, high sensitivity and batch manufacturing, which is good for improving the integration degree and measuring accuracy of smart cutting tool.

The MEMS strain gauge used in this paper is a commercial product of Anhui Tianguang Sensor Co. Ltd., Bengbu, China, as shown in figure 2. The strain gauge looks like a sandwich, its top and bottom layers are polyimide insulation, and the middle layer is MEMS strain gauge devices, including piezoresistor, gold wire, metal pad and metal pin. The MEMS strain gauge's resistance, sensitive coefficient and size are 1000Ω, 150±5% and $3\times0.20\times0.04mm^3$, respectively. Figure 2 also illustrates

IOP Conf. Series: Journal of Physics: Conf. Series **986** (2018) 012016 doi:10.1088/1742-6596/986/1/012016

fabrication process of the strain gauge: ①boron ions are implanted into single crystal silicon bar in order to form p-type piezoresistor material; ②p-type piezoresistor material is cut into small slices; ③ small pieces of piezoresistor material is ground for next step; ④the ground piece of silicon material is cut into smaller slices as a single resistor; ⑤each resistor is etched to appropriate size and a gold layer is patterned to form the contact area; ⑥ the resistor is bonded between two pieces of polyimide film and the resistor is linked to metal pad by gold wire, two pins are finally bonded to the metal pad.

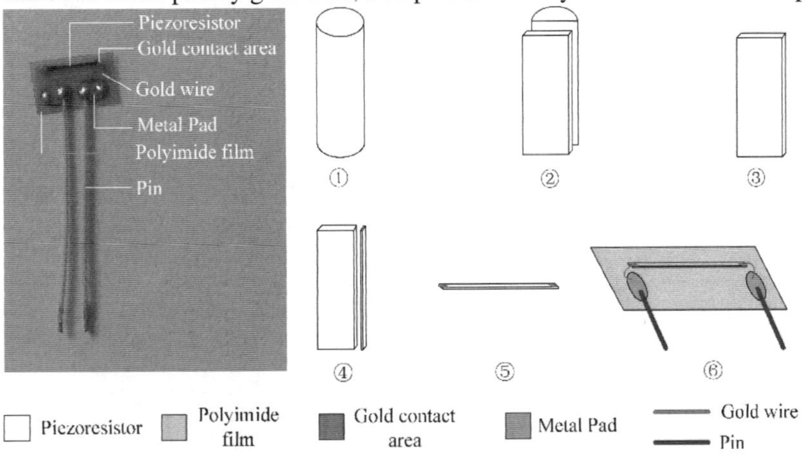

Figure 2. Schematic view of MEMS strain gauge and its fabrication process

2.3. MEMS strain gauge bonding and Wheatstone bridge organization

When X (or Y) direction cutting force component is applied to the tool tip, each two symmetrical slots of the 4 strain gauge bonding slots receive opposite surface stress. According to this principle, Wheatstone bridge circuit is adopted for cutting force measurement in order to improve sensitivity. Eight pieces of MEMS strain gauge are packaged symmetrically in bonding slots and two groups of Wheatstone bridge are organized for cutting force measurement in X and Y directions, as shown in figure 3. A layer of silica gel is coated on top of each MEMS strain gauge in order to isolate it from steam and other corrosive substances. Power and signal cable for Wheatstone bridge is collected into the signal cable channel, and then it is connected to data acquisition equipment by shield cable.

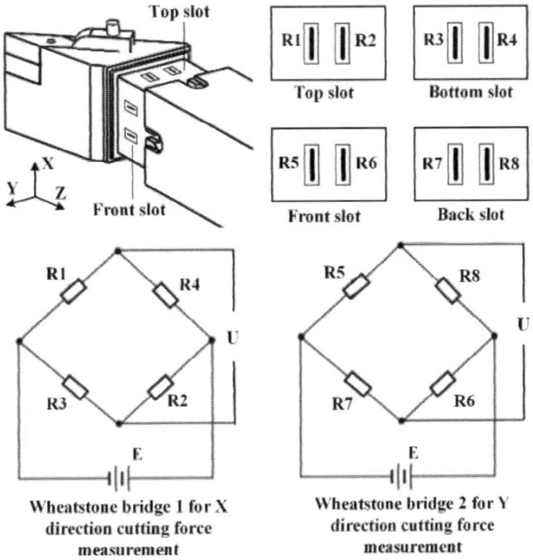

Figure 3. Strain gauge location and Wheatstone bridge organization method

It is important to note that: (1) all strain gauge should be placed symmetrically in the packaging slot, which is helpful for reducing measurement accuracy that caused by positioning error; (2) each strain gauge should be placed as close as possible to the slot's symmetrical axis in order to make it stay on the neutral layer of surface stress under the effect of cross force, which can reduce cross-interference error.

2.4. Static calibration test

Static calibration test was carried out to verify practical performance of the developed smart cutting tool and to obtain its static characteristic indexes. Figure 4 depicts the experiment setup for static calibration. During the test, the smart cutting tool was fixed on the base of an electro-mechanical universal testing machine (EMUTM, type UTM6104) and was powered by 5V DC. The EMUTM was pre-set to apply standard-static force on the tool tip in X and Y directions, respectively. Each calibration cycle includes a loading process and an unloading process. During loading process, the standard-static force applied to the tool tip rises from 0 N to 260 N with an integral of 20 N in each step; then it falls from 260 N to 0 N by 20 N in each step during unloading process. At least 3 cycles of calibration test were needed for each direction, and the test results were averaged for eliminating random error. The output of two Wheatstone bridges were recorded by two sets of precision digital multimeter (type FLUKE 8846A).

① Precision digital multimeter ② Power supplier
③ Electro-mechanical universal testing machine ④ Smart cutting tool

Figure 4. Picture of static calibration system

3. Results and Discussion

Figure 5 depicts the output voltage signal versus the input force signal of the smart cutting tool in X and Y directions' calibration test, respectively. It is apparent that the developed smart cutting tool possesses fabulous linear property and desirable cross-interference error. After further data processing, main static characteristic indexes are obtained and listed in table 1.

The developed smart cutting tool owns satisfying static properties, its sensitivity is 0.1 mV/N, which means it is sensitive to small variation of cutting force and is suitable for cutting force measurement in high speed or precision machining. The cross-interference error are 1.97% and 3.03% in X and Y directions, respectively. Thus, each measuring circuit is able to measure a single direction's cutting force component while hardly disturbed by cutting force component in another direction. Furthermore, its linearity error (0.87%, 1.21%) and hysteresis error (1.08%, 1.31%) are acceptable except repeatability error. The repeatability error may be affected by bonding material's stability, which can be reduced by optimizing MEMS strain gauge bonding process.

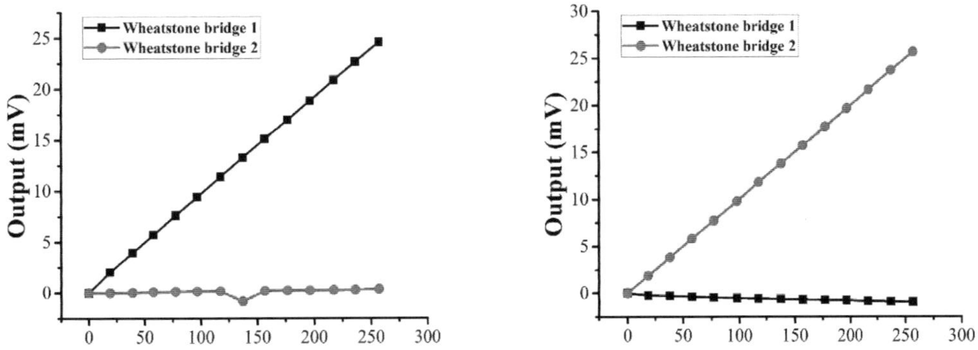

Figure 5. (a) X direction calibration result; (b) Y direction calibration result

Table 1. Static characteristic indexes of the smart cutting tool.

Wheatstone bridge	Sensitivity (mV/N)	Linearity (%)	Hysteresis (%)	Repeatability (%)	Accuracy (%)	Cross-interference (%)	
						Bridge 1	Bridge 2
Bridge 1	0.1	0.87	1.08	3.05	3.35	/	1.97
Bridge 2	0.1	1.21	1.31	2.74	3.27	3.03	/

4. Conclusion

This paper proposes a smart cutting tool based on MEMS strain gauge, which has fully considered the compatibility and reliability of its installation on CNC machine. MEMS strain gauges are bonded in the specially designed slots and are organized into Wheatstone bridge for cutting force measurement. Static calibration test proves that:

(1) The smart cutting tool possesses favorable sensitivity (0.1mV/N), which is capable of measuring small cutting force in high speed and precision machining process;

(2) Its accuracy error is around 3%, which is promising to be reduced by optimizing MEMS strain gauge bonding process.

(3) Cross-interference (1.97% and 3.03%) is acceptable for application, the developed smart cutting tool is able to measurement cutting force components in X and Y direction respectively while hardly interfering each other.

5. Acknowledgments

You Zhao designed the sensor and analyzed the data, Yulong Zhao Contributed guidance for sensor design and experiment, Yiwei Shao, Tengjiang Hu, Qizhang and Xiaohui Ge performed the experiment. In addition, this research is supported by the National Natural Science Foundation of China (Grant No. 51705408), the Fundamental Research Funds for the Central Universities (xjj2017017), and the Changjiang Scholars and Innovative Research Team in University of China (No. IRT_14R45), the National Science Fund for Distinguished Young Scholars (Grant No. 51325503), the National Natural Science Foundation of China (Grant No. 51421004).

References

[1] Teti R, Jemielniak K and O'Donnell G 2010 *CIRP. Ann. Manuf. Tech.* **59** 717-39
[2] Jain R, Rathore J and Gorana V 2016 *Proc. Int. Conf. on CARs & FoF* (Kolaghat, West Bengal India) 13-21
[3] Rehorn AG, Jiang J, and Orban PE 2005 *Int. J. Adv. Manu. Tech.* **26** 693-10
[4] Audy J 2006 *J. Zhejiang Uni. Sci.* **7** 1781-89

[5] Information on https://www.kistler.com
[6] Ren Z, Zhang J and Jia Z 2014 *Adv. Mater. Res.* **945** 2191-94
[7] Bingyao Cheng. Development of tool changeable piezoelectric dynamometer on turning, master's degree thesis of Dalian University of Technology. 2013
[8] You Z, Yulong Z and Chaohui Wang *2016 Sen. Act.* A. *Phys.* **237** 119-27
[9] You Z, Yulong Z and Songbo L 2015 *Sen.* **15** 7969-84
[10] Xiao CW, Ding H and Li WD 2012 *Proc. Preci. Engi. Nano.* (Aspen, Colorado, USA) **516** 373-77

Configuration of Air Microfluidic Chip for Separating and Grading Respirable Dust

Xiaofeng Zhu[1], Yiting Jia[1], Jianhai Sun[2,*], Peiyue Zhao[2], Jinhua Liu[1], Yanni Zhang[1], Zhanwu Ning[1]
[1]Beijing Municipal Institute of Labour Protection, Beijing, 100054, China
[2]State key laboratory of transducer technology, Institute of Electronics, Chinese Academy of Sciences, Beijing, 100190, China
[*]Correspondence: jhsun@mail.ie.ac.cn; Tel: +86-10-58887183

Abstract: Particulate matter (PM) is a category of airborne pollutants, and fine particles that have a diameter of 2.5 μm (PM2.5) or smaller are especially damaging to human health because of their ability to penetrate deep into our respiratory system, Therefore, Monitoring of PM is very important. In this work, an air micro- fluidic PM sensor based on MEMS was proposed, and numerical model of the sensor was simulated accurately. The sensor was able to separate particles according to their sizes, and then transports and deposits the selected particles using thermophoretic precipitation onto the surface of a microfabricated mass-sensitive film bulk acoustic resonator (FBAR), precisely weighing and providing the concentration of PM. The PM sensor has double stage separation function, and the primary separator can separate the particles with size of less 10 μm from the particles, and the secondary can separate particles with size of less 2.5 μm from the particles.
Keywords: Air microfluidic chip; Film bulk acoustic resonator; Particulate matter; Electrophoretic force

1. Introduction

With the development of industrialization, the living environment of human beings is suffering from the most serious air pollution [1-19], which was seriously affected people's health and normal operation of society. The main composition of fog and haze is sulfur dioxide, nitrogen oxides and particulate matter (Inhalable particles, the aerodynamic equivalent diameter are less than 10μm) [5-15].The inhalable particles can be deposited in the respiratory tract, lung and other parts due to small size. The smaller particle size of the PM is, the deeper the PM entered into the respiratory tract. It is well known that the PM with diameter of 10 μm are deposited in the upper respiratory tract, and the PM with 5 μm diameter can enter into the deep part of the respiratory tract, and the PM with diameter less than 2 μm can be easily penetrated deep into bronchioles and alveoli, which was easy to cause cancer and other diseases. Negative effects of PM on health recently prompted the Beijing Municipal Environmental Protection Bureau to formulate new rules for the safe exposure limits of PM. Therefore, high precision PM monitors are very needed to monitor the total concentration of PM and the concentration of various sizes of PM in real time. Recently, a variety of PM detectors have emerged in the market, some of which are based on light-scattering principles, and some are based on the principle of weighing. Although PM sensors based on light-scattering can collect particle count data and infer the particle concentration based on assumptions about particle density and size distribution, reducing their accuracy and applicability. Therefore, it is of great significance to develop a rapid on-site detection technique for inhalable particles [6], which will be very conducive to understanding of the formation mechanism of fog and haze. In this work, an air microfluidic PM sensor based was proposed, and the numerical model of the sensor was simulated accurately. The PM sensor has double stage separation function, and the primary separator can separate the particles with size of less 10 μm from the particles, and the secondary can separate particles with size of less 2.5 μm from the particles.

2. Configuration of the air microfluidic chip

Figure 1 shows a configuration of the air microfluidic PM sensor. Air containing PM enters inlet of the microfluidic sensor and immediately flows into an inertial size separator, and fine particles were directed to the thermophoretic deposition region where a temperature gradient induced across the microfluidic channel removes particles from the air stream by thermophoresis. The particles removed from the air stream are deposited on the exposed surface of a mass-sensing film bulk acoustic resonator. The rate of particle deposition, corresponding to the rate of change in the resonant frequency of the FBAR is proportional to the concentration of the particles in the airstream.

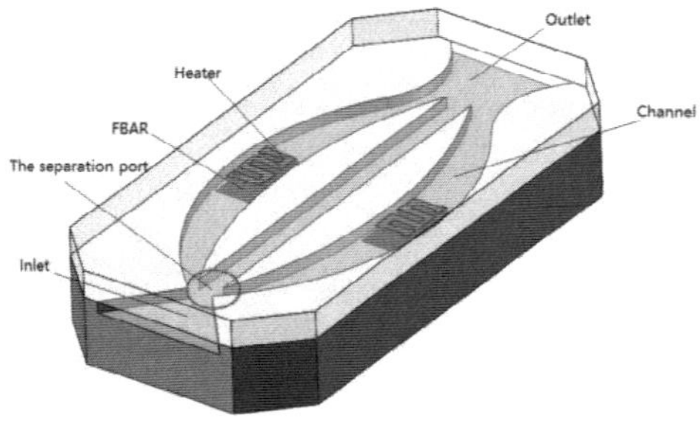

Figure 1.Configuration of air microfluidic chip.

It is well known that the inhalable particles can be affected by electrophoretic force at the gas separation port because of temperature gradient induced across the microfluidic channel. Where d is the particle diameter, η is the dynamic viscosity of the gas, δT is the temperature change, ρ is the density of particles, H is the molecular adjustment coefficient.

$$F = -\frac{9\pi d\eta^2 H(\delta T)}{2\rho gT} \quad \ldots\ldots \quad (1)$$

The smaller the particle size of inhalable particles is, the smaller the inertia is. In the role of electrophoretic force, the small particles turn to both sides of the channel, however, the large particles, directly outflow from the middle channel due to large inertia. Selection of particle size for inhalable particles can be expressed as follows:

$$d = -\frac{9\eta W^2 Dn}{\rho QC} \quad \ldots\ldots \quad (2)$$

Where W and D are the channel width and depth of the microfluidic chip, respectively, Q is the airflow velocity of the inlet and C the correction factor. The particle size of the PM can be graded by controlling these factors such as W and D as well as Q .

3. Experimental Section

3.1. Simulation of PM separation

In order to evaluate the separation performance of the chip, tobacco smoke was acted as the sensor prototype for analyzing the collection efficiency of PM under different flow rates (low flow rate and high flow rate) and different power (low power and high power). In the simulation analysis, the low flow rate is set as 5 ml/min, and high flow rate is set as 15 ml/ min, low loading voltage is set as 5 V, and high loading voltage is set as 15V. Fig. 2 and Fig.3 show the separation of PM (tobacco smoke) with different particle sizes from the air microfluidic chip. The flow rate of the sensor is 15 ml/min and the loading voltage is15 V. Fig. 4 shows the collection rate of PM with different particle sizes through the air microfluidic sensor. Fig.2 and Fig.3 show the Separation of inhalable particles with different diameter using air microfluidic chip. As we can see that the small particles turn to both sides of the channel, and the large particles, directly outflow the middle channel. The collection rate of PM with 5 microns was more than 90%.

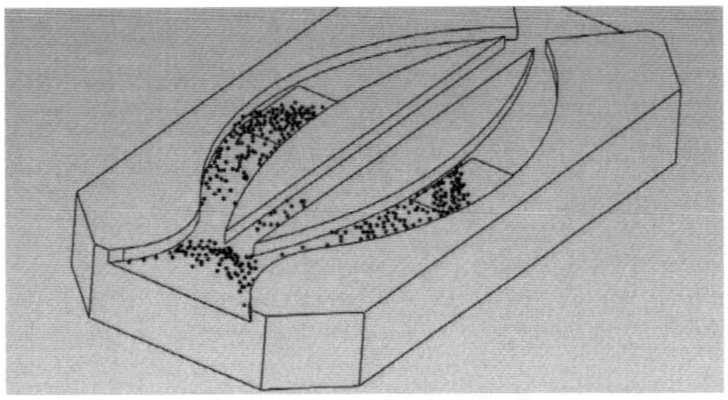

Figure 2. Separation of tobacco smoke with diameter less than 2.5 μm.

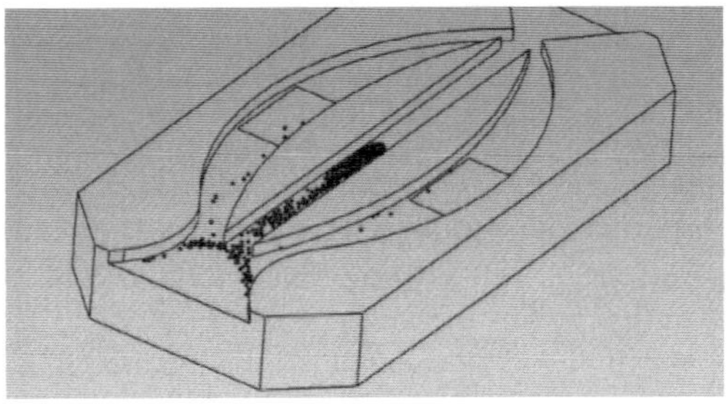

Figure 3. Separation of inhalable particles with large diameter of 10 μm.

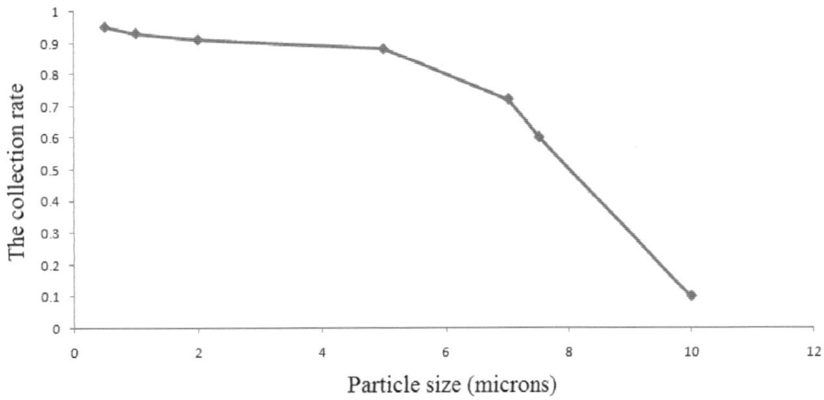

Figure4. Collection rate of PM for different particle size.

Fig. 5 show the sensitivity coefficient of PM for tobacco smoke while at two flow rates through the air microfluidic channel and two power settings to the thermophoretic heater. One of the analyses is that the heater of the sensor is loaded with 5V and 15V voltage at low flow rate of 5 ml/min, the heater of the sensor is loaded with 5V and 15V voltage at high flow rate of 10 ml/min, and the heater of the sensor is loaded with 5V and 15V voltage at high flow rate of 15 ml/min, respectively. As we can see from the result, the sensitivity of the PM sensor is dependent of the flow rate and loading voltage. Higher flow-rate and higher loading voltage imply higher sensitivity.

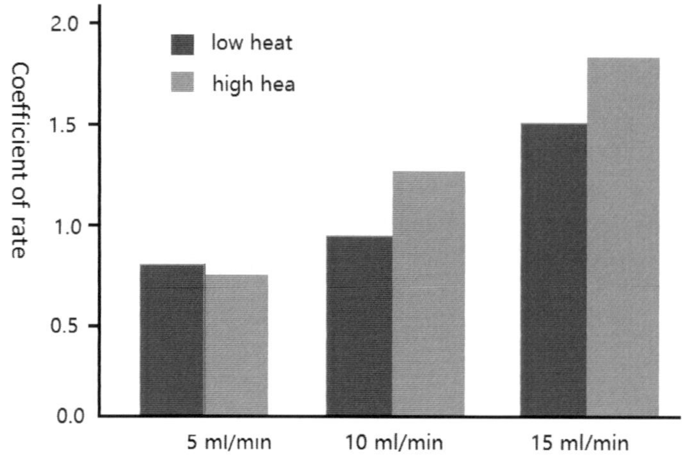

Figure 5. The sensitivity of the sensor exposed to tobacco smoke under low and high power.

3.2. Double-stage separation sensor

In this work, in order to accurately monitor the concentration of PM with various particle sizes, a miniaturized double stage separator was designed based on the same principle of single stage separator. The separator can separate the PM under 10μm from the primary separator and get its concentration by FBAR. Then the separated PM in the air was transported into the secondary separator, and the particles of 2.5 μm and below were separated from the particulates below 10 μm by the separator. Finally, the concentration of PM2.5 and the concentration of PM between 10 μm and 2.5 μm were obtained through FBAR, and Fig.6 shows the separation numerical diagram of PM using the double stage separator.

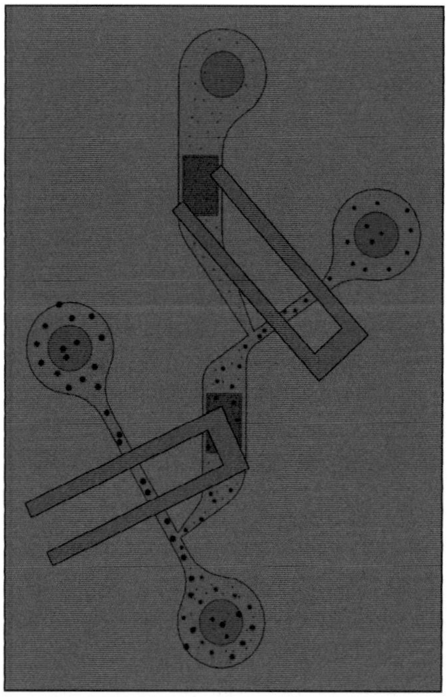

Figure 6 .Separation numerical diagram of PM using the double stage separator.

4. Conclusion

In this work, an air microfluidic PM sensor based on MEMS for monitoring PM was proposed. The performance of air microfluidic chip was influenced by the flow rate and loading voltage, Furthermore, the performance was also influenced by other factors, such as the location of the heater, the area occupied by the heater, etc. In addition, a miniaturized double stage separator was designed for monitoring the concentration of PM with various particle sizes based on the same principle of single stage separator.

Acknowledgment

The authors greatly acknowledge the financial support from national key research and development program of china under Grant numbers: 2016YFC07006 and 2016YFF02045. The authors greatly acknowledge the financial support from the National Science Foundation of China under Grant numbers: 11574219, 61774157, 61176112.

References

[1] Sydbom A, Blomberg A, Parnia S, Stenfors N, Sandstrom T, Dahlen SE 2001 *Eur Respir J* 17 733.
[2] Anderson JO, Thundiyil JG, Stolbach A 2012 *J Med Toxicol* 8166.
[3] Ozis F, Singh M, Devinny J, Sioutas C 2004 *Journal of the Air & Waste Management Association* 54 (8) 935.
[4] Andersen ZJ, Stafoggia M, Weinmayr G, Pedersen M, Galassi C 2017 *Environ Health Perspect* 348 (348) 7412.
[5] Araujo JA, Barajas B, Kleinman M, Wang X, Bennett BJ 2008 *Circulation Research* 102 (5) 589.
[6] Li N, Sioutas C , Cho A ,Schmitz D , Misra C 2003 *Environ Health Perspect* 111(4) 455.
[7] Campbell A, Oldham M,Becaria A, Bondy SC, Meacher D 2005 *Neurotoxicology* 26 (1) 133.
[8] Barakathaddad C, Elliott S , Pengelly D 2013 *Journal of Environmental Protection* 4 (8) 779.
[9] Nasari MM, Szyszkowicz M, Chen H, Crouse D, Turner MC 2016 *Air Quality Atmosphere & Health* 9 (8) 1
[10] Choi YS, Eom SY , Park H, Ali SF, Lantz-Mcpeak SM 2014 *Molecular & Cellular Toxicology* 10 (3) 245.
[11] Fox JR, Cox DP, Drury BE, Gould TR, Kavanagh TJ 2015 *Air Quality Atmosphere & Health* 8 (5) 507.
[12] Lankoff A, Brzoska K, Czarnocka J, Kowalska M, Lisowska H 2017 *Environmental Science & Pollution Research* 24 (23) 1.
[13] Klein SG, Cambier S,Hennen J, Legay S, Serchi T 2017 *Particle & Fibre Toxicology*14 (1) 7.
[14] Forchhammer L, Loft S, Roursgaard M, Cao Y, Riddervold IS, Sigsgaard T, Moller P 2012 *Toxicol Lett* 209 121.
[15] Pope CA, Burnett RT, Turner MC, Cohen A, Krewski D, Jerrett M, Gapstur SM, Thun MJ 2011 *Environ Health Perspect* 119 1616.
[16] Schwarze PE, Totlandsdal AI, Lag M, Refsnes M, Holme JA, Ovrevik J 2013 *Biomed Res Int* 685142.
[17] Nemmar A, Hoet PH , Vanquickenborne B, Dinsdale D, Thomeer M 2002 *Circulation*105 (4) 411.
[18] Li N, Xia T, Nel AE 2008 *Free Radic Biol Med* 44 1689.
[19] Nemmar A, Hoylaerts MF, Hoet PHM, Nemery B 2004 *Toxicol Lett* 149 243.

Simulation of micro/nano electroporation for cell transfection

Guocheng Zhang[1,2], Na Fan[1,2], Hai Jiang[1,2], Jian Guo[1,3,*], Bei Peng[1,2,*]

[1] School of Mechatronics Engineering, University of Electronic Science and Technology of China, Chengdu, Sichuan Province, P. R. China
[2] Center for Robotics, University of Electronic Science and Technology of China, Chengdu, Sichuan Province, P. R. China
[3] School of Mechanical Engineering, University of South China, Hengyang, Hunan Province, P. R. China
*Correspondence: jguo@uestc.edu.cn (J.G.); beipeng@uestc.edu.cn (B.P.).

Abstract. The 3D micro/nano electroporation for transfection has become a powerful biological cell research technique with the development of micro-nano manufacturing technology. The micro channels connected the cells with transfection reagents on the chip were important to the transmemnbrane potentical, which directly influences the electroporation efficiency. In this study, a two-dimensional model for electroporation of cells was designed to address the effects of channels' sizes and number on transmembrane potential. The simulation results indicated that the transmembrane potential increased with increasing size of channels' entrances. Moreover, compared with single channel entrance, the transmembrane potential was higher when the cells located at multiple channels entrances. These results suggest that it IS required to develop higher micro manufacturing technology to create channels as we expected size.

1. Introduction

Electroporation is widely adopted in cell transfection that uses external electric field to create transient pores in the cell membrane, which contributes to the delivery of biomolecules into cells. When a high voltage is applied, the original hydrophobic pores occur. Afterward, these pores in the cell membrane change into from the transient hydrophilic pores, thereby increasing its electrical conductivity and membrane permeability. The formation of these hydrophilic pores depends mainly on the strength of the electric field [1].

The conventional bulk elecporation has several significant disadvantages: the entire cells need to be suspended in the buffer and have to be exposed to very high voltages, which will causes high fatality rate. To overcome these problems, the Espinaosa group [2] invented nanofountain probe electroporation device for a individual adherent cell transfection with high viability at lower voltage. Afterward, they utilized the porous carbonate membrane to develop a microfluidic device for high-throughput cell transfection [3]. However, the pores in the carbonate membrane were randomly distributed, which leaded to a non-uniform pore-cell contact. Chang group [4-6] developed a 3D nanochannel array chip for high-throughput cell electroporation with high uniformity and dosage control. The 3D nano electraoporation silicon chip was a uniform and intense nanochannel array with a radius of 650 nm, which was made by projection photolithography and deep reactive ion etching technique. The experimental results indicated that the nanochannel electroporation chip showed a 20 fold improvement in uniformity and dosage control. However, limited by the microfabrication technology and the material anisotropic, the etched channels on the chip could not present uniform morphology. The cross sections of channels most had trapezoid shape. Moreover, the cells during transfection experiment may locate at multiple channels' entrances. These uncertain factors would result in different transmembrane potential difference and cause trouble in uniformity and dosage, even transfection validity.

To stress these issues, we performed infinite element analysis to study the effects of channels' sizes and channels' number on the transmembrane potential. The simulation results indicated that the bigger sizes resulted in higher transmembrane potential on condition that the cells were under safe potential difference. What's more, the transmembrane potential was higher when the cells located at multiple channels entrances compared with single channel entrance.

2. Theory and simulation

2.1 Electroporation theory

After applying an electric field in the cell culture medium, ions move to the surface of the cell membrane. The accumulated ions create an electric field in the cell membrane. The potential difference between the inside and outside of the cell membrane is called the transmembrane potential, which is proportional to the external electric field and the diameter of the cell. For a spherical cell as shown in Figure 1(a), the transmembrane potential can be described by the **Schwan** equation:

$$TMP = 1.5rE_0 \cos\theta \tag{1}$$

where the r is the cell radius, E_0 is the applied electric field intensity, and θ is the angle between the electric field direction and normal direction of the cell membrane surface point [7]. The transmembrane potential will increase when imposing to a long pulse voltage (ms) and the threshold transmembrane potential is in the range of 0.2~1 V [8]. Then the transient hydrophilic pores occur. Remove the pulse voltage, the cell membrane will be repair, there is no mechanical damage, a phenomenon called reversible electroporation. However, when a very high external electric field is applied in bulk electroporation, the most cells permanently deform and the cell membrane cannot be repaired, resulting in cell lysis. Unlike bulk elctroporation, the micro/nano electraoporation shown in Figure 1(b) creates a localized electric field on the cell membrane area connected to micro/nano channel, resulting in pores of cell membrane at low voltages. So the cell viability can be improved greatly compared with bulk electroporation.

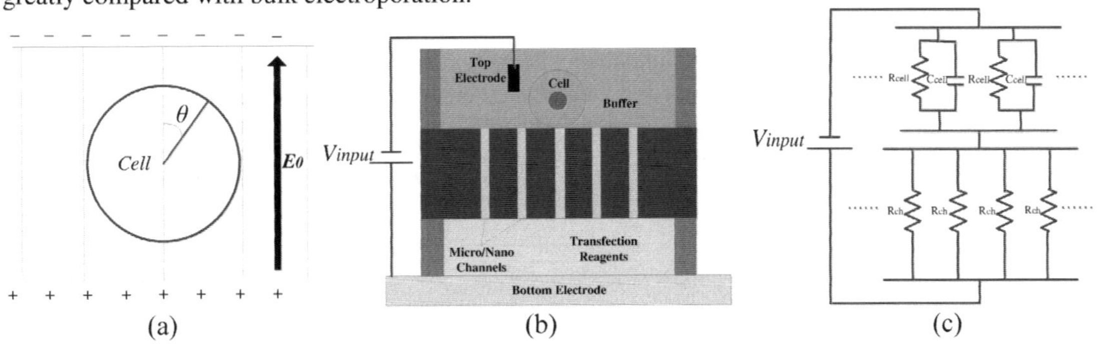

Figure 1. (a) Schematic of a spherical cell in the electric field; (b) Schematic of channels connected to a single cell at the top and transfection reagents at the bottom; (c) The corresponding circuit diagram of (b).

2.2 Modeling

To investigate the transmembrane potential at low voltage in micro/nano electroporation, we used Laplace equation $\nabla(\sigma\nabla V)=0$ to calculate in COMSOL with appropriate boundary conditions and geometrical model. In the partial differential equation, σ is the electrical conductivity and V is the input electrical potential [9]. In the experiments, the input voltage was applied on the glass slide connected with transfection reagents and the other electrode was submerged into the cell culture buffer. To satisfy the boundary condition, we applied input voltage at the bottom boundary while the top boundary was electrically grounded. And the other boundaries were set electrically insulated. The cell was modeled as a circle of 10 μm in diameter enclosed by a 5nm layer. The relevant schematic is shown in Figure 2, where l is membrane channel length, h is the gap between the cell and the channel, r is the culture chamber diameter and w is the micro/nano channel width. In this simulation, the electrical conductivity of cell membrane, cytoplasm, and buffer were respectively set to 5×10^{-7}, 0.2, 0.8 Sm^{-1} [10].

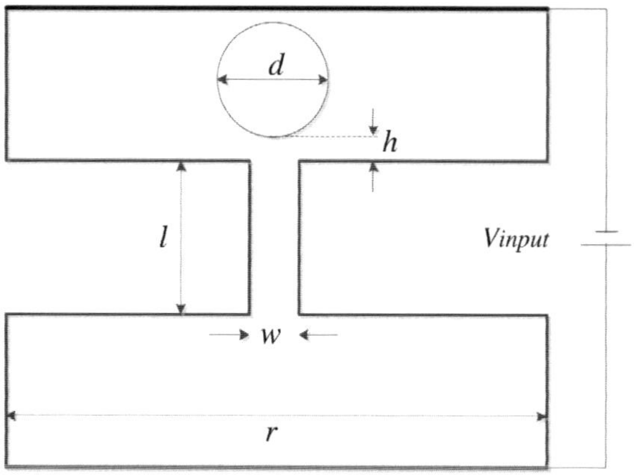

Figure 2. The schematic of simulation model and relevant dimension parameters in simulation.

3. Results and discussion

3.1 Effect of channels' sizes on the transmembrane potential

Due to different manufacturing techniques and anisotropic materials, the etched channels were not absolutely straight. As shown in Figure 3, we performed electrostatic numerical simulation with channels of different sizes to analyze transmembrane potential. It was found that the potential droped quickly in the micro/nano channels. Therefore, the cell viability was improved due to that only a small potential difference was located at a tiny area of the cell membrane. We predicted that the resistance in narrow channels was larger than that in wide channels. As predicted, the transmembrane potential drop increased with increasing width of channel entrances which side of channel the cells located at. Moreover, the transmembrane potential was lower than 1 V with a width of less than 0.5 μm at the input voltage of 10 V. The results suggested that it is required higher input voltage to achieve effective electroporation when the channel had a very small size.

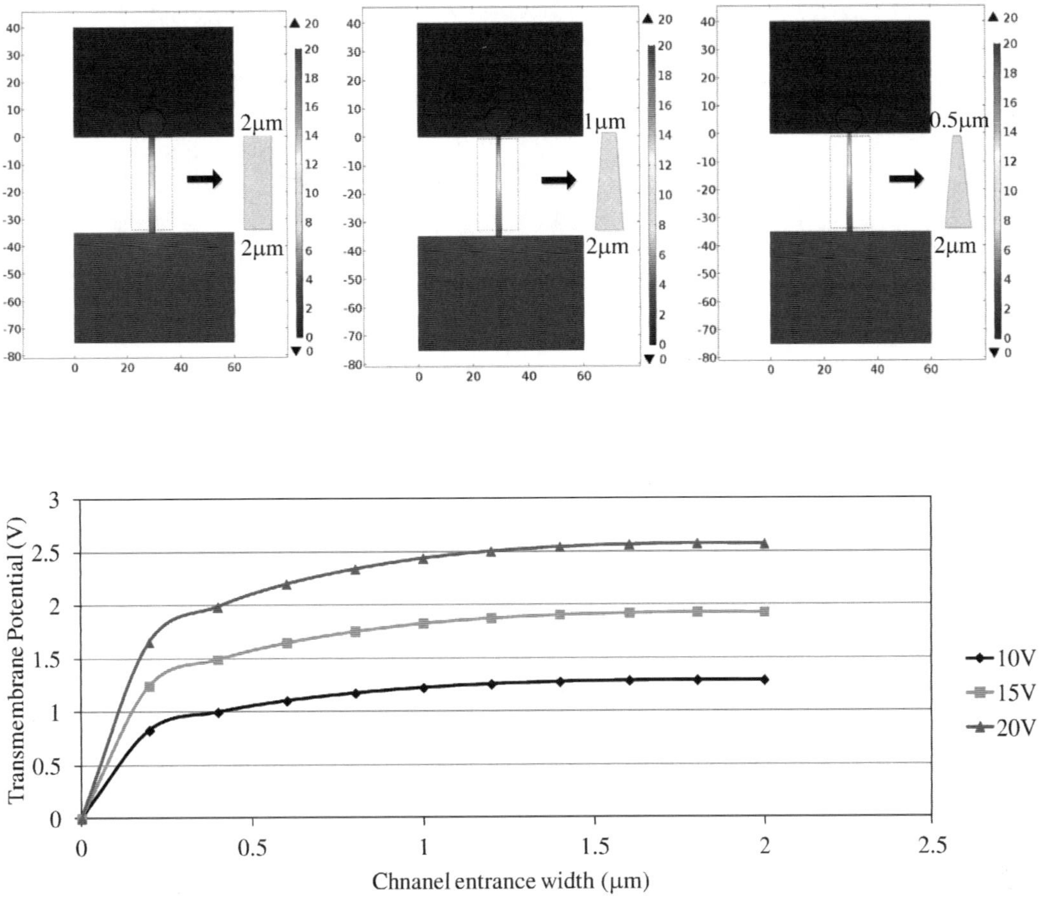

Figure 3. Numerical simulation of electroporation of different channels' sizes at input voltage of 10, 15, 20 V.

3.2 Effect of channels' number on the transmembrane potential

It was not certain how many channels cells locate at in actual micro/nano electroporation experiments. So in order to identify if the number of channels the cells placed on could affect the transmembrane potential, we set respectively one, two, three channels in our simulations. The width of rectangle channel was 1 μm and the spacing distance between two adjacent channels was 2 μm in our simulation. We could see from Figure 4, the highest transmembrane difference all occurred at the nearest area from the channel. The resistance were influenced directly by the number of channels. The potential dropped slower with mutiple channels. Therefore, the multiple channels could result in higher potential difference than single channel. Despite effective electroporation could be more likely happen at a higher transmembrane potential, it may cause trouble in dosage control. This was because the potential difference could influence the diffusion efficiency of transfection reagents.

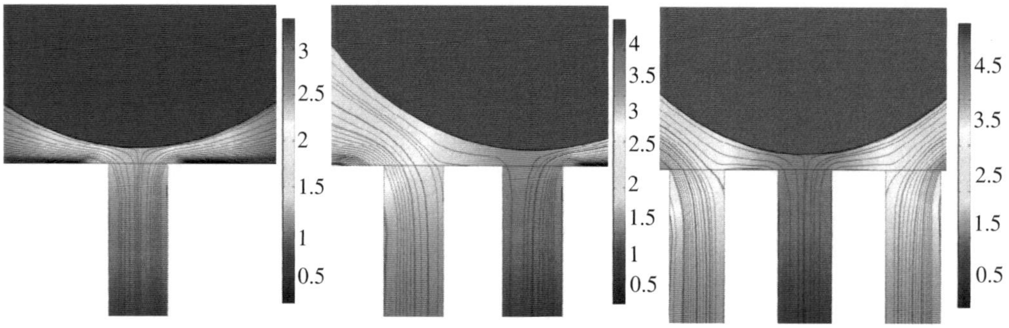

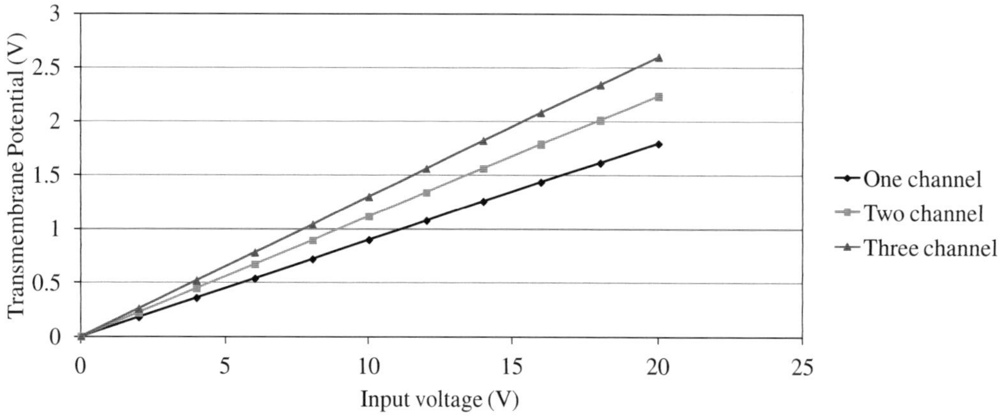

Figure 4. Numerical simulation of electroporation of different channels' number.

4. Conclutions

Toward the development of high-through micro/nano electroporation for transfections of the cells, we have emphasized the importance of channels' sizes and number on microporous chip to effective electroporation. It was found that the size and spacing distance between two adjacent channels affect the tansmembrane potential significantly. The results suggested that it is required higher input voltage to achieve effective electroporation when the channel had a very small size. Moreover, the more channels could result in higher potential difference. These findings suggest us we need more precise micro manufacturing techniques to create the channels on the chip as we designed.

5. Acknowledgements

The authors would like to acknowledge the partial supports provided by the China Postdoctoral Science Foundation Grant (No. 2016M590873), National Natural Science Foundation of China (No. 31771078, No. 51575090), Project of International Science and Technology Cooperation and Exchange of Sichuan (No. 2017HH0072), Fundamental Research Funds for the Central Universities (No. ZYGX2014Z004, No. ZYGX2016KYQD118 and No. ZYGX2015J084), and the National Youth Top-Notch Talent Support Program.

6. References

[1] Krassowska W, Filev P D and Krassowska W 2007 *Biophys. J.* ***92*** 404
[2] Kang W, Yavari F, Jolandan M M and Espinosa H D 2013 *Nano Lett.* ***13*** 2448-57
[3] Kang W, Giraldovela J P, Nathamgari S S and Espinosa H D 2014 *Lab on A Chip* ***14*** 4486-4495
[4] Chang L, Bertani P, Gallegoperez D, Gallegoperez D and Yang Z 2015 *Nanoscale* ***8*** 243
[5] Chang L, Gallegoerez D, Chiang C and Lee L J 2016 *Small* ***12*** 5971-5980

[6] Aderhold J, Davydov V Yu, Fedler F, Klausing H, Mistele D, Rotter T, Semchinova O, Stemmer J and Graul J 2001 *J. Cryst. Growth* **222** 701
[7] DeBruin K A and Krassowska W 1999 *Biophys. J.* *77* 1225-33
[8] Wang M Y, Orwar O, Olofsson J and Weber S G 2010 *Anal. Bioanal. Chem.* ***397*** 3235-48
[9] Agarwai A, Zudans I, Weber E A, Olofsson J, Orwar O and Weber S G 2007 *Anal. Chem.* ***79*** 3589-96
[10] Kotnik T, Bobanovic F and Miklavcic D 1997 *Bioelectrochem. Bioenerg* ***43*** 285-91

Effect of Slice Error of Glass on Zero Offset of Capacitive Accelerometer

R Hao[1], H J Yu[1], W Zhou [1], B Peng[1 *] and J Guo[1 2 *]

[1] Department of Mechatronics Engineering, Center for Robotics, University of Electronic Science and Technology of China, Chengdu, Sichuan, P.R. China
[2] School of Mechanical Engineering, University of South China, Heng yang Hunan421000, People's Republic of China
[*]Corresponding author:jguo@uestc.edu.cn, beipeng@uestc.edu.cn.

Abstract. Packaging process had been studied on capacitance accelerometer. The silicon-glass bonding process had been adopted on sensor chip and glass, and sensor chip and glass was adhered on ceramic substrate, the three-layer structure was curved due to the thermal mismatch, the slice error of glass lead to asymmetrical curve of sensor chip. Thus, the sensitive mass of accelerometer deviated along the sensitive direction, which was caused in zero offset drift. It was meaningful to confirm the influence of slice error of glass, the simulation results showed that the zero output drift was 12.3×10-3 m/s2 when the deviation was 40μm.

1. Introduction

The capacitance accelerometer adopted symmetrical structure and the die-adhesive-substrate packaging method, the curve of accelerometer caused by the different thermal expansion coefficient [1]. Only the asymmetric curve of accelerometer could cause the output zero drift, Peng P studied the thermal drift of accelerometers with the consideration of the adhesive overflow, which indicated that the adhesive overflow leaded to 10% increase of thermal drift compared with accelerometers without adhesive overflow [2]. Lai L S studied the void in DG-3S (An electronic packaging adhesive) which also leaded to accelerometer zero output drift [3]. If the zero output drift of the accelerometer neened to further reduced, the slicing error in the package process should be considered yet, study on slice error of glass had been mentioned by some papers [4]. This paper focused on slice error of glass and discussed the influence on the accelerometer zero output when slice error of glass rang from 0μm to 40μm.

Slicing process was a step in MEMS device packaging, each chip unit was separated along the scratches on the glass, if glass scratches had deviation that the position of sensitive mass also had deviation yet. Chip and glass was adhered on ceramic substrate by DG-3S, the adhesive curing process needed to keep temperature of 60℃ in incubator for two hours, then the temperature cooled down to the 20℃, because of the different thermal expansion coefficient, the three-layer structure curved in cooling process. As a result, the slice error of glass will leaded asymmetric curve of chip, which made the output zero offset.

2. Simulation

ABAQUS software was adopted in the Simulation, and half model of actual structure had been used in figure 1. The photo of micro-capacitance accelerometer structure was in figure 2, which was amplified 25 times, difference of d_1 and d_2 range from 0μm to 40μm had been set in simulation respectively, point A was the center of material contraction, the Young's modulus, Poisson ratio and thermal expansion coefficient of materials was showed in table 1. According to the packaging process, temperature parameter had been set from 60℃ to 20℃.

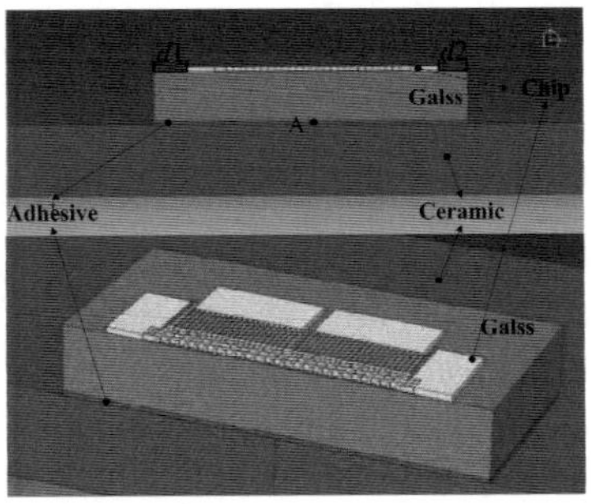

Figure 1 The ABAQUS model of micro capacitance accelerometer structure, point A was the contraction center.

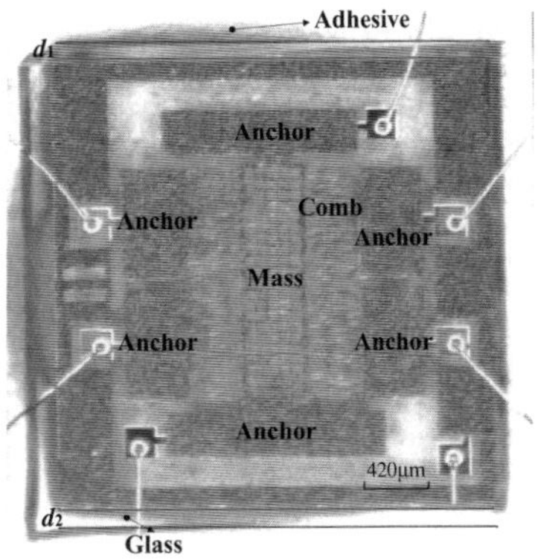

Figure 2. The micro capacitance accelerometer photo, the slice error of glass was marked by red line.

Table 1. Material properties for simulation

Material	Young' s modulus (GPa)	Poisson ratio	TCE
Silicon	160	0.22	2.6×10-6
DG-3S	4	0.37	6×10-5
Glass	62.7	0.2	3.25×10-6
Ceramic	360	0.22	6.5×10-6

3. Calculation

Each pair capacitance plate displacements of multiple nodes had been extracted from simulation for differential capacitance calculation. and the displacement of accelerometer showed in figure 3. Position changes of capacitance plate included displacement and angle [2], in figure 4. C_l and C_r expressed the left and right comb total capacitance respectively.

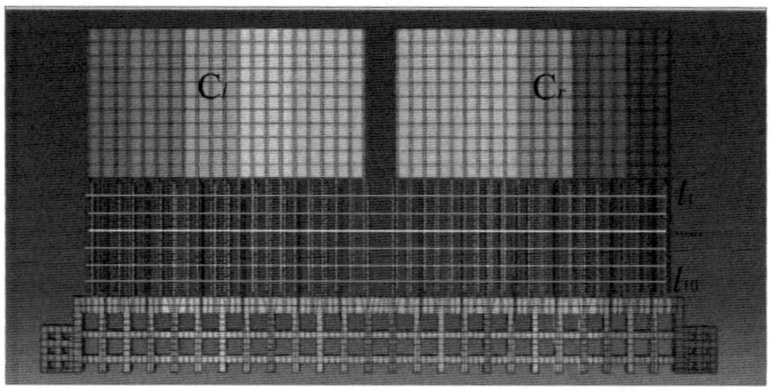

Figure 3 The capacitance plate deformation, C_l and C_r express the left and right comb total capacitance respectively.

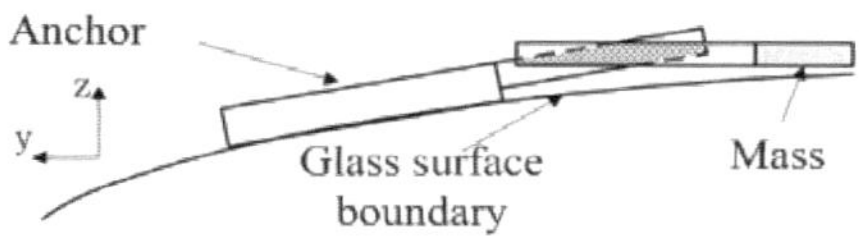

Figure 4 The schematic diagram of comb deformation, the shadow represents the effective common area of capacitance

Basic formula for capacitance calculation expressed in equation (1), which calculated the left and right total capacitance plate, the effective common area S was the shadow part in figure 4, which needed to extract the displacement of points form the simulation data, one method was provided to calculate the total capacitance [2].

$$C_i = \frac{\varepsilon\varepsilon_0 S}{d_i} \tag{1}$$

Where ε is relative dielectric constant of air, ε_0 is dielectric constant, ε_0=8.86e^{-12}F/m, S is the effective common area of capacitance plate, d_i expresses the capacitance comb narrow gap. The results of C_l, C_r under slice error of glass was shown in figure 5.

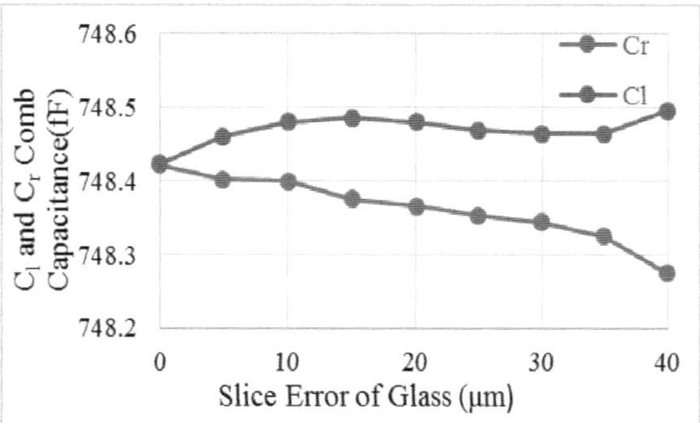

Figure 5 The left comb capacitance and right comb capacitance at different slice error of glass.

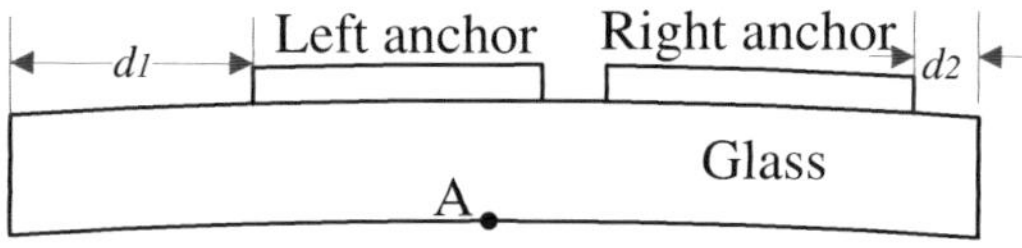

Figure 6 The deformation of sensor chip and glass, the point A is the contraction center.

There was two reasons of the change of capacitance, firstly, the decrease of apacitance comb plate gap was caused by material constriction when temperature reduced, secondly, the decrease of capacitance plate effective common area was caused by accelerometer chip curved. When $d_1 > d_2$, the left anchor more closer to the contraction center than right anchor, the main reason for the change of capacitance was material contraction, which caused capacitance increase, while the right comb far away from contraction center, the main reason for the change of capacitance was curve of accelerometer chip, which caysed capacitance decrease, in figure 6. Hence because of the two ways of the capacitance changes caused complex capacitance variation, which leaded to zero offset of capacitive accelerometer. The formula of zero output expressed in equation (2).

$$V_{\text{out}} = VG\frac{C_l\text{-}C_r}{C_l + C_r} \tag{2}$$

Where C_l and C_r is the left comb totle capacitance and right totle comb capacitance respectively, V is the amplitude of the open-loop voltage, G is the circuit gain coefficient. The acceleration output formula expressed in equation (3).

$$S = VG\frac{mg}{kd} \tag{3}$$

Where V is the amplitude of the open-loop voltage, G is the circuit gain coefficient, m is effective mass, g is the acceleration of gravity, k is total stiffness of the folded beams, d is capacitance plate gap, and the result of micro-capacitance accelerometer zero offset under slice error of glass range from 0μm to 40μm showed in figure 7.

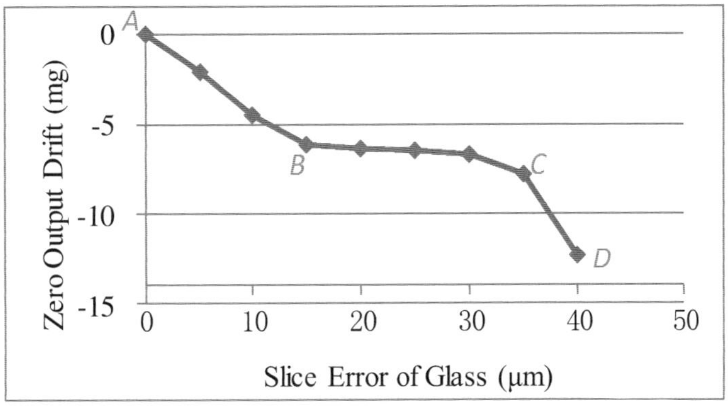

Figure 7 The effect of deviation on the zero output drift at different slice error of glass.

4. Conclusion

According to simulation and calculation, the onclusion was as follows: the accelerometer zero output drift was 12.3×10^{-3} m/s^2, if slice error of glass was 40 μm. In figure 7, in section AB, the main reason of the accelerometer zero output drift was material contraction, the change rate of C_l larger than the change rate of C_r, in the section BC, the main reason of the accelerometer zero output drift was capacitance plate effective common area changes, the change rate of C_l and C_r reached a balance, while in the section CD, the capacitance changes was caused by plate effective common area greater than the capacitance changes caused by material contraction, the accelerometer zero output drift was increased. The slice error of glass had no effect on the zero output drift of the accelerometer at nonsensitive direction, which shows in figure 2 $S_1 > S_2$.

The slice error of glass broke the balance of the change rate of C_l and C_r at the sensitive direction and caused the accelerometer zero output drift.

Acknowledgment

This paper was supported by the China Postdoctoral Science Foundation Grant (No. 2016M590873).

References

[1] Peng P, He X P, Du L M, Zhou W, Yu H J & Peng B. 2015. The influence of die-attach adhesives on the packaging of mems accelerometer. *Key Engineering Materials*, 645-646, 533-537.

[2] Peng P, Zhou W, Yu H J, Peng B, Qu H & He X. 2016. Investigation of the thermal drift of mems capacitive accelerometers induced by the overflow of die attachment adhesive.IEEE *Transactions on Components Packaging & Manufacturing Technology*, 6(5), 822-830.

[3] Zhou W, Li F, Yu H, Qu H, Hao R & Wang D. 2017. Influence of adhesive non-uniformity on zero offset of micro accelerometer. *International Journal of Modern Physics B*, 1750242.

[4] Zhou W, Li B, Peng B, Su W & He X. 2012. Prediction of gap asymmetry in differential micro accelerometers. *Sensors*, 12(6), 6857-6868.

[5] Zhou W, Li B, He X, Su W & Zhang F. 2009. Calibration study of force-balanced micro accelerometer based on least squared method and genetic algorithm. *International Conference on Mechatronics and Automation* (pp.1357-1361). IEEE.

[6] Lishchynska M, O'Mahony C, Slattery O, Wittler O & Walter H. 2007. Evaluation of packaging effect on mems performance: simulation and experimental study. *IEEE Transactions on Advanced Packaging*. 30(4), 629-635.

[7] Zhou W, Yu H, Peng B, Shen H, He X & Wei S. 2013. Modeling the microstructure curvature of boron-doped silicon in bulk micromachined accelerometer. *Materials*, 6(1), 244-254.

IOP Conf. Series: Journal of Physics: Conf. Series **986** (2018) 012020 doi:10.1088/1742-6596/986/1/012020

Numerical Simulation of Liquid Metal RF MEMS Switch Based on EWOD

Tingting Liu[1], Yang Gao[2], Tao Yang[1] and Huihui Guo[1]

[1]Robot Technology Used for Special Environment Key Laboratory of Sichuan Provinc e, Southwest University of Science and Technology, Mianyang, 621010, China
[2]Institute of Electronic Engineering, China Academy of Engineering Physics, Mianyang, 621900, China

Email: liutingting@swust.edu.cn

Abstract. Conventional RF MEMS switches rely on metal-to-dielectric or metal-to-metal contacts. Some problems in the "solid-solid" contact, such as contact degradation, signal bounce and poor reliability, can be solved by using "liquid-solid" contact. The RF MEMS switch based on liquid metal is characterized by small contact resistance, no moving parts, high reliability and long life. Using electrowetting-on-dielectric (EWOD) way to control the movement of liquid metal in the RF MEMS switch, to achieve the "on" and "off" of the switch. In this paper, the electrical characteristics and RF characteristics of RF MEMS switches are simulated by fluid mechanics software FLUENT and electromagnetic simulation software HFSS. The effects of driving voltage, switching time, dielectric layer, hydrophobic layer material and thickness, switching channel height on the RF characteristics are studied. The results show that to increase the external voltage to the threshold voltage of 58V, the liquid metal began to move, and the switching time from "off" state to "on" state is 16ms. In the 0~20GHz frequency range, the switch insertion loss is less than 0.28dB, isolation is better than 23.32dB.

1. Introduction

Previously, semiconductor devices like PIN diodes and FETs have been extensively used as switching devices in many Radio Frequency (RF) systems due to fast working speed, low voltages and have high power handling. However, they suffer from high insertion loss, low isolation and inferior linearity [1]. Since 1990's, Radio Frequency Micro-Electro-Mechanical Systems (RF MEMS) technology have been introduced in novel switches. The RF MEMS switches have the good operating characteristics with low insertion loss, high isolation and low cost of fabrication [2-5], and there is a very good application prospect in the test system, satellite communications, radar systems as well as the military field [6-7]. Conventional RF MEMS switches rely on metal-to-dielectric or metal-to-metal contacts. The contact reliability of solid-solid contacts becomes more important for MEMS switches as surface plays an increased role at microscale [8]. The contact degradation due to arcing, welding and material transfer, and the limited contact area due to the roughness of these surfaces, easily lead to contact bounce and limit the operational life of these devices [9]. These problems in solid-solid contact can be solved by using liquid-metal contact. The liquid metal is liquid at room temperature, characterized by good electrical conductivity, thermal conductivity and liquidity, and has been used as the contact material of switch in MEMS. The liquid metal in RF MEMS switch conforms almost perfectly over the CPW transmission line, which results in high on-state capacitances. By controlling the movement of liquid metal based on Electrowetting-on-dielectric (EWOD) effect, the RF MEMS switch is in the "on" state or "off" state. EWOD is well established as a fluid manipulation technique in such areas as lab-on-a-chip, visible light optics, and displays, RF electronics and electromagnetics [10-14]. Thus, the basic principle of liquid metal RF MEMS switches is to control the movement of liquid metal by controlling the applied voltage to change the wettability of liquid metal and solid surface.

In the paper, the characteristics of liquid metal RF MEMS switch are studied mainly from two aspects of the electrical and RF characteristics. The numerical simulations of RF MEMS switch are carried out by fluid mechanics software FLUENT and electromagnetic simulation software HFSS, the flow characteristics of mercury in RF MEMS switch, the driving voltage of switch and the influence of various structural parameters on RF performance are studied.

2. Structure and Modelling

2.1. Working Principle of EWOD-Based Liquid Metal RF MEMS Switch

2.1.1. Driving Principle of Droplet

The micro droplet control method based on EWOD has been studied in recent years. The basic idea is to control the droplet movement by the change of surface tension [15]. The EWOD system is usually composed of electrodes, hydrophobic layer, dielectric layer, micro droplet, as shown in Figure 1. When the droplet is located on both electrodes covered with the dielectric layer and the hydrophobic layer, the left electrode is unactuated and the contact angle of the left droplet remains unchanged. When the external voltage is applied to the right electrode, the contact angle of the right side will be reduced and the wettability of the micro droplet will be increased to spread to the right. Since the radius of curvature of the surface of the left and right sides of the micro droplet does not coincide, the droplet will move to the right when the internal pressure difference of the droplet caused by the change in the radius of curvature is greater than the received resistance.

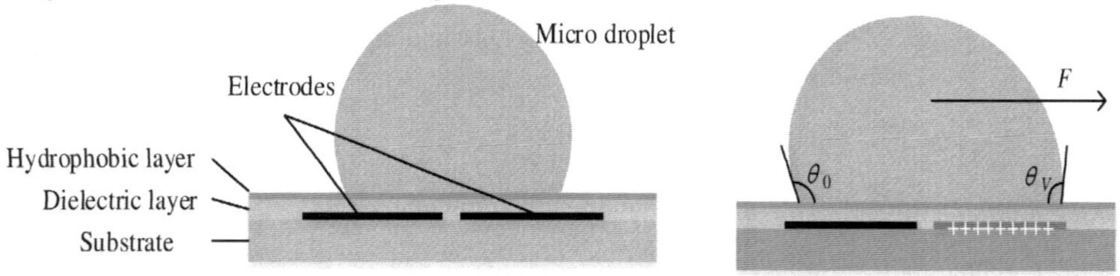

(a) no voltage is applied (b) applied voltage to the right electrode

Figure 1. Principle of droplet control based on EWOD

When the micro droplet is located on the rectangular driving electrode, according to the Young-Lippmann equation, the driving force F can be defined as [16-17] :

$$F = \frac{\varepsilon_0 \varepsilon_r}{2 \times d} \times V^2 \times L_{eff} \tag{1}$$

Where ε_0 is the vacuum dielectric constant, ε_r is the relative dielectric constant, V is the driving voltage, d is the thickness of the dielectric layer, and L_{eff} is the length of the droplet in the direction perpendicular to the driving force when the droplet is in contact with the driving electrode.

2.1.2. Structure and Principle of Liquid Metal RF MEMS Switch

The liquid metal RF MEMS switch utilizes the EWOD effect to drive the liquid metal droplet movement and control the status of the RF signal. RF MEMS switch is composed of the silicon substrate, glass cover, liquid metal droplet, CPW transmission line, driving electrode, the upper electrode, block, hydrophobic layer and dielectric layer. The overall structure is illustrated in Figure 2. In RF MEMS switch, the liquid metal is always in contact with the upper electrode to ensure that the liquid metal has a suitable bias voltage. The external voltage is applied to the upper electrode and the driving electrode, and becomes the main power source for driving the liquid metal droplet.

When the voltage is not applied and the liquid metal is on the CPW transmission line, the large capacitance to the ground is formed and the RF signal from the signal line conducts to the ground. Thus, the switch is in the "off" state. When the driving force is greater than the resistance, the liquid metal starts to move to the right. It moves to leave the CPW transmission line, the liquid metal is not in direct contact with the signal line, and the RF signal transmit to the on state with smaller loss. Two blocks are used to limit the movement of liquid metal in the switch.

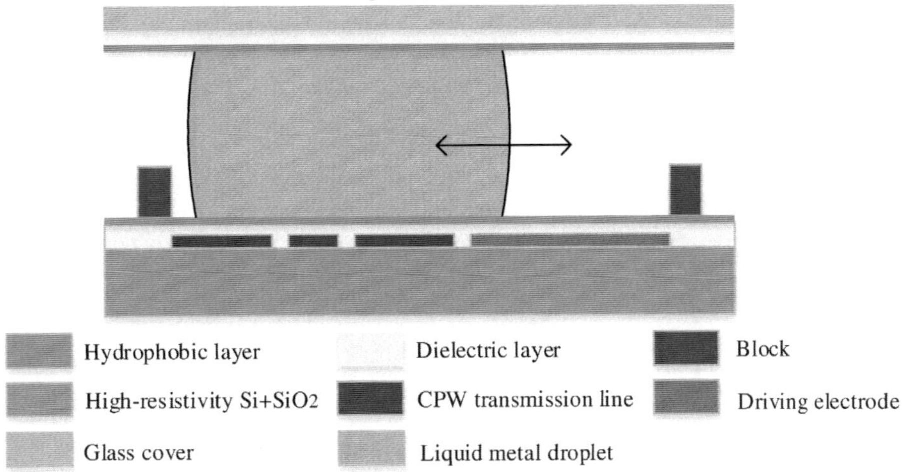

Figure 2. EWOD-based RF MEMS switch overall structure diagram

2.2. *Modelling of RF MEMS Switch*

In this paper, the research of characteristics of liquid metal RF MEMS switch is carried out mainly from two aspects of the electrical characteristics and RF characteristics. The numerical simulation model is established by using the numerical simulation software Fluent, and the driving voltage and switching time are simulated to study the flow characteristics of liquid metal droplet in RF MEMS switch. The "on" state and "off" state models of the switch are established by the electromagnetic simulation software HFSS. The impact of the dielectric layer, the thickness of the hydrophobic layer and the height of the switching channel on RF performance such as the return loss, insertion loss and isolation are studied.

2.2.1. *Numerical model of RF MEMS switch by FLUENT*

Liquid metal is influenced by multi-physical field coupling such as the electrostatic attraction, surface tension and viscous force in the whole movement process. After the upper electrode is earthed and the driving electrode is actuated, the droplet began to move right because of electrostatic attraction. If the upper electrode and driving electrode are applied the same voltage, the droplet and the CPW transmission line are subjected to electrostatic attraction, and the droplet and the driving electrode are driven by repulsing force, causing the droplet to move to the left. In this paper, the flow characteristics of liquid metal droplet in the switch are simulated by FLUENT. In the simulation, the volume of fluid (VOF) method and the continuum surface force (CSF) model are used. The user defined functions (UDF) are also used to realize the setting of boundary conditions such as driving force and contact angle. The numerical simulations are focused on the minimum driving voltage and switching time of RF MEMS switch.

The RF MEMS switch numerical model is shown in Figure 3. In order to facilitate the observation of the movement position of the liquid metal droplet in the RF MEMS switch, the numerical model established only contains the fluid region and the solid region below the switch. The CPW line and the driving electrode can be clearly seen from the numerical model. The size of the entire switch model is 1360μm×1220μm×600μm, and the size of the CPW line is 50μm/80μm/50μm. The discretization of the numerical model is realized by using the hexahedral mesh. According to equation (1), the UDF of driving force is written and loaded in the momentum source terms of FLUENT software. After applying the external voltage, the contact angle of the droplet on the dielectric layer changes, so it is necessary to set different contact angle hysteresis at different motion positions, and write the UDF of contact angle hysteresis. The FLUENT software calculates the value of the driving force and the contact angle at each iteration and returns it to the FLUENT solver.

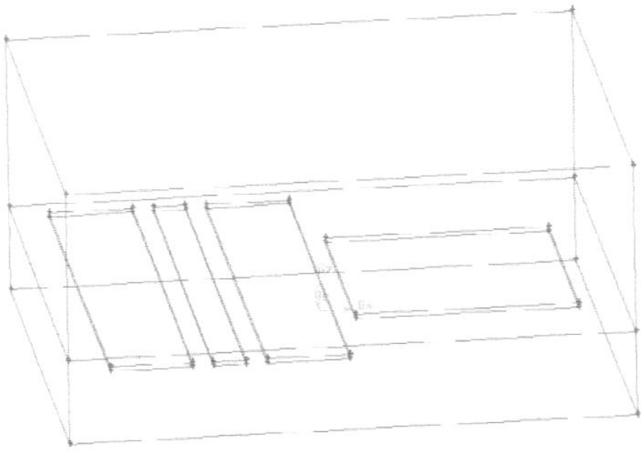

Figure 3. Numerical simulation model of RF MEMS switch

The contact angle of the liquid metal on the surface of the hydrophobic layer changes due to the applied external voltage. The external voltage V and the contact angle θ satisfy the Young-Lippmann equation [18]:

$$\cos\theta = \cos\theta_0 + \frac{\varepsilon_0 \varepsilon_r}{2 \times d \times \sigma_{lv}} \times V^2 \tag{2}$$

Where θ is the contact angle when the applied voltage is V, θ_0 is the static contact angle when there is no applied voltage, and σ_{lv} is the surface tension between the droplet and the gas. It can be seen from the Young-Lippmann equation that the contact angle θ after applying the voltage decreases as the applied voltage V increases. However, as the voltage increases, the contact angle finally tends to a fixed value. This phenomenon is called the contact angle saturation. According to the literature [19], when the voltage increased to 90V, the contact angle of the liquid metal droplet on the silicon dioxide surface is 80° when the phenomenon of contact angle saturation occurs. In the RF MEMS switch, the liquid metal droplet is always in contact with the CPW line and the driving electrode at the same time, and the contact angle (right side) of the contact portion with the driving electrode decreases due to the applied external voltage. In numerical simulations, FLUENT software invokes the UDF function at each iteration to read the x-direction size of each grid of liquid metal droplet. If the droplet moves above the driving electrode, set the contact angle according to equation (2).

2.2.2. Numerical model of RF MEMS switch by HFSS

The influence of the switch parameter including the material and thickness of the dielectric layer, the thickness of the hydrophobic layer and the height of the switching channel on RF performance are analyzed by HFSS software. RF MEMS switch adopted silicon as the substrate material and the relative dielectric constant is 11.9. The material of dielectric layer is silicon dioxide and the relative dielectric constant is 3.8. The material of the CPW transmission line and driving electrode is gold and the conductivity is 4.1×10^7S/m. The geometry parameters of RF MEMS switch are shown in Table 1. The switch simulation models are established in the HFSS software, as shown in Figure 4.

Table 1. The geometry parameters of the switch

	Symbol	Size	Material
width of substrate	length	1020μm	Silicon
height of substrate	ground	200μm	Silicon
interval of signal line and ground line	g	50μm	Gold
width of signal line	signal	80μm	Gold
thickness of the thermal oxidation silicon dioxide layer	t	200nm	Silicon dioxide
thickness of CPW line	t1	1100nm	Gold
thickness of dielectric layer	t2	300nm	Silicon dioxide
thickness of dielectric layer	t3	10nm	Teflon

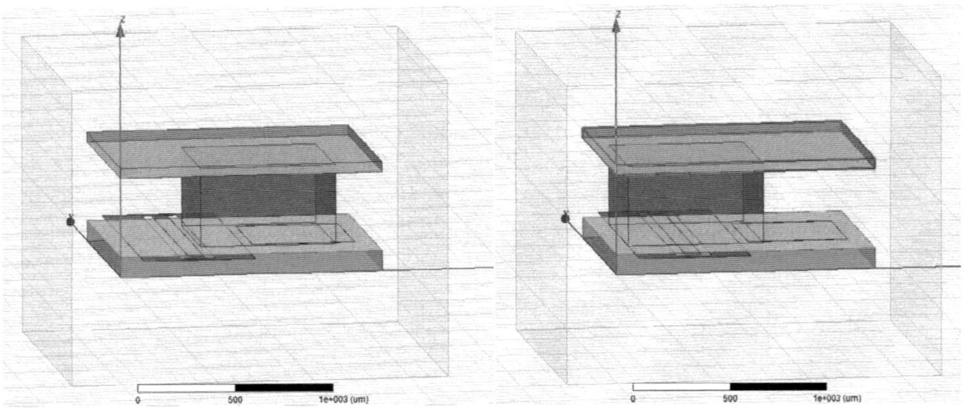

(a) "on" state (b) "off" state

Figure 4. The simulation models of RF MEMS switch in HFSS software

3. Results and Discussion

3.1. Simulation Results of Electrical Characteristics by FLUENT

Set the external voltage as 80V, the flow characteristics of liquid metal in the switch are shown in Figure 5. When the time is 0ms, the droplet is located on the entire CPW line, and the switch is in the "off" state. After 4.7ms, the droplet is driven by the external voltage to completely leave the signal line, and the switch was in the "on" state. If the initial position of the switch is in the "on" state, the droplet moves toward the CPW line direction with the action of external voltage. After 2.8ms, the entire droplet relocated on the CPW line, and the switch is in the "off" state.

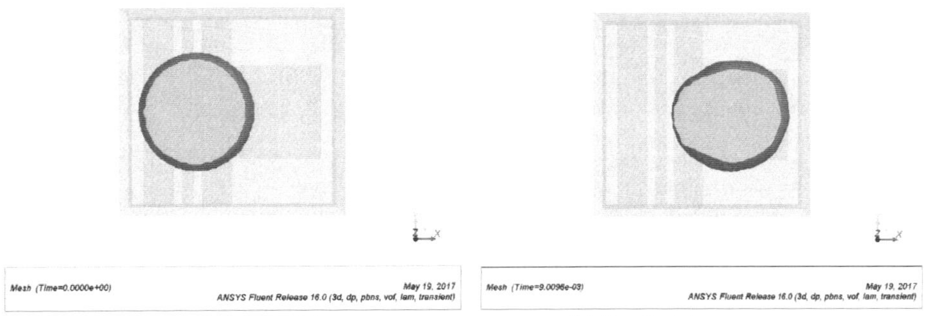

(a) "off" state (b) "on" state

Figure 5. The flow characteristics of liquid metal in RF MEMS switch

Table 2 shows the contact angle under different driving voltages and the switching time required for the switch from "off" to "on". The simulation results show that when the driving voltage is less than 58V, the driving force is too small and the droplet does not move. The switch is still in the off-state. The larger the driving voltage, the faster the droplet moves, and the less time is required. When the driving voltage increased to 90V, the contact angle saturation of the liquid metal droplet on the surface of the silicon dioxide appears, and the contact angle is 80°.

Table 2. The contact angle and the switching time from "off" to "on" under different driving voltage

Driving voltage（V）	Contact angle of right side（°）	Switch time（ms）
58	107.8	16.5
60	106.1	16.1
70	97.3	12.6
80	87.3	4.7

3.2. Simulation Results of RF Characteristics by HFSS

The S parameters of the switch in the "on" and "off" state are shown in Figure 6 and Figure 7 respectively. The simulation results show that the insertion loss S_{21} is -0.037~-0.282dB and the return loss S_{11} is -47.33~-19.30dB in the frequency range of 0~20GHz when the switch is in the "on" state. The isolation is better than 23.32dB when the switch is in the "off" state.

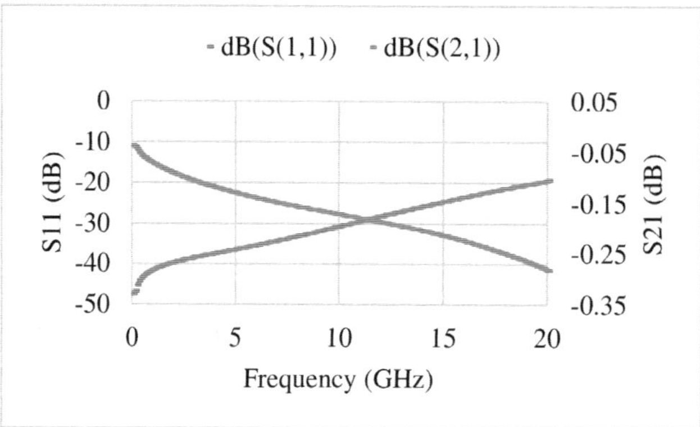

Figure 6. S parameter of the switch in the "on" state

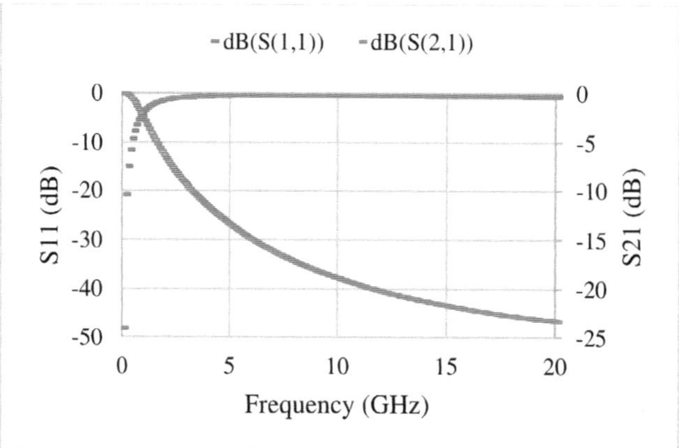

Figure 7. S parameter of the switch in the "off" state

3.2.1. *Effect of the Thickness of Dielectric Layer on RF Performance of the Switch*

When the thickness of the dielectric layer is 300nm, 400nm and 500nm respectively, the variations of the insertion loss and isolation with the thickness of the dielectric layer are shown in Figure 8.

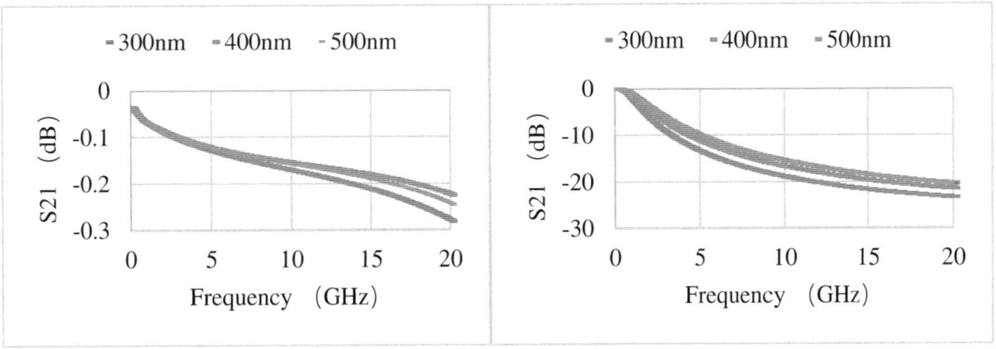

(a) Insert loss S_{21} (b) Isolation S_{21}

Figure 8. Variations of S_{21} parameter with thickness of dielectric layer

It can be seen from Figure 8 that the isolation of the switch is -23.32dB when the frequency is 20GHz and the thickness of dielectric layer is 300nm, and the isolation of the switch is -20.40 dB when the thickness is 500nm. The isolation of RF MEMS switch is related to the capacitance of "off" state. The thinner the dielectric layer, the better the isolation and the greater capacitance of the switch in the "off" state. Therefore, reducing the thickness of the RF MEMS switch dielectric layer can effectively improve the isolation of the switch. However, reducing the thickness of the dielectric layer may lead to the acceleration of the breakdown in the dielectric layer. There is a minimum thickness of the dielectric layer. The breakdown field strength of the silicon dioxide is about 2MV/cm. Considering the contact angle saturation phenomenon of liquid metal droplet occurs with the external voltage of 90V, so the working voltage should be lower than 90V and the dielectric layer thickness should be greater than 180nm to ensure that the dielectric layer will not be punctured by the field. The results also show that the thickness of the dielectric layer has not much impact on the insertion loss. The insertion loss of the switch is mainly the loss of the transmission line. Since the size of the CPW does not change, the thickness of the dielectric layer of the switch has little effect on the insertion loss of the switch.

3.2.2. *Effect of the hydrophobic layer thickness on RF performance of the switch*

In the RF MEMS switch, hydrophobic layer formed on the dielectric layer is used to increase the initial contact angle between the liquid and solid surfaces [20], which facilitates the movement of liquid metal droplet in the switch. When the thickness of the hydrophobic layer is 7.5nm, 10nm and 12.5nm respectively, the resulting insertion loss and isolation are shown in Figure 9.

It can be seen from Figure 9 that the insertion loss is -0.27dB, -0.28dB and -0.25dB at the frequency of 20 GHz and the thickness of the hydrophobic layer are 7.5 nm, 10 nm and 12.5 nm, respectively, and the isolation is -23.21dB, 23.31dB and -22.95dB. The hydrophobic layer is used to adjust the contact angle between the droplet and the solid surface, and the hydrophobic layer has little effect on the insertion loss and the isolation.

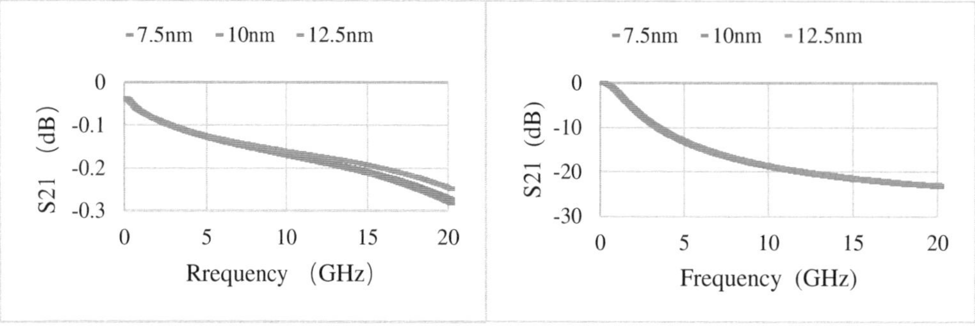

(a) Insert loss S_{21} (b) Isolation S_{21}

Figure 9. Variation of S_{21} parameter with thickness of hydrophobic layer

3.2.3. Effect of dielectric layer material on RF performance

In order to study the effect of different dielectric layer materials on the RF characteristics, SiO_2 and Si_3N_4 are used as the dielectric layer materials, and their dielectric constants are 3.8 and 7.0 respectively. The insertion loss and isolation achieved from the numerical simulation analysis of HFSS software are shown in Figure 10.

It can be seen from Figure 10 that the isolation of the switch with a dielectric layer made of Si_3N_4 has been significantly improved in the frequency range of 0~20GHz, and the isolation of the switch is -25.68dB at 20GHz, while the isolation of the switch made of SiO_2 material is -23.10dB at 20GHz. The use of high dielectric constant material effectively improved the capacitance of the switch in the "off" state. The higher the capacitance in the "off" state, the greater the isolation of the switch. The dielectric layer material has little effect on the insertion loss, since the change of dielectric layer material has no impact on the loss of the CPW transmission line, so that the insertion loss of the switch is substantially unchanged. From the RF performance of the switch, the use of high dielectric constant material can more effectively improve the RF performance of the RF MEMS switch.

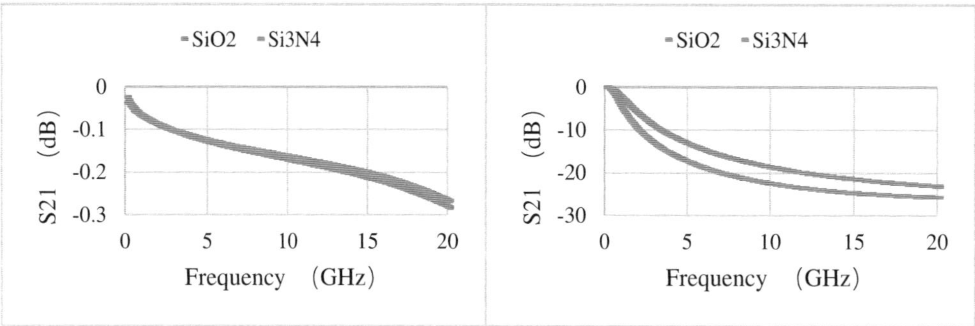

(a) Insert loss S_{21} **(b)** Isolation S_{21}

Figure 10. Variation of S_{21} parameter with different dielectric layer material

3.2.4. Effect of channel height on RF performance

Mercury droplet is in contact with the upper and lower electrodes, and the volume of droplet is related to the microchannel height. The microchannel height increases and the volume of mercury droplet increases accordingly. Considering the microchannel height of 400μm, 425μm and 450μm respectively, the insertion loss and isolation of the switch are shown in Figure 11.

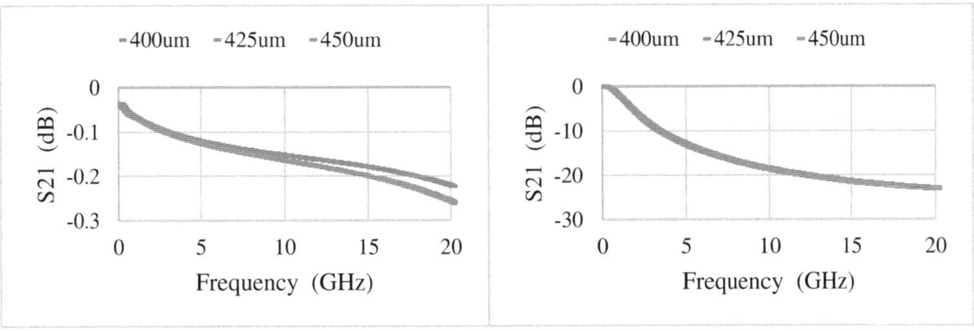

(a) Insert loss S_{21} **(b)** Isolation S_{21}

Figure 11. Variations of S_{21} parameter with different microchannel height

As can be seen from the figure, the microchannel height has little effect on the insertion loss and isolation. At the frequency of 20 GHz, the insertion loss corresponding to the microchannel height of 400μm, 425μm and 450μm is -0.22dB, -0.26dB and -0.26dB, respectively, and the isolation is -23.11dB, -22.86dB and -23.21dB.

4. Conclusions

Liquid metal gradually being applied in the MEMS devices with its excellent physical properties. Liquid metal RF MEMS switch control the movement of liquid metal to achieve the "on" and "off" of the switch through the EWOD method. In addition to liquid metal, the switch has no moving parts. The switch has the advantages of reliable contact, low contact resistance and so on. In this paper, the electrical characteristics and RF characteristics of liquid metal RF MEMS switches are studied by numerical simulation. When the external voltage is too small, the electrowetting force is less than the contact angle hysteresis force, and the liquid metal does not move. When the external voltage is increased to the threshold voltage of 58V, the liquid metal starts to move, the switching time of the switch from the "off" state to the "on" state is 16ms. The greater the external voltage, the faster the switching speed. In the frequency range of 0~20GHz, the insertion loss of the switch in the "on" state is -0.037~-0.282dB, and the return loss is -47.33~-19.30dB. The isolation in the "off" state is better than 23.32dB. Parametric simulation studies show that the dielectric layer thickness and material have a great influence on the RF performance.

Acknowledgments: This work was supported by National Natural Science Foundation of China (No.61640403 and No.61501384), Major Project of Education Department in Sichuan Province (No.17ZA0402), and Doctoral Scientific Research Foundation of Southwest University of Science and Technology (No.15ZX7156).

References

[1] Mirzajani, H.; Ilkhechi, A.K.; Zolfaghari, P.; Azadbakht, M.; Aghdam, E.N.; Ghavifekr, H.B. Power efficient, low loss and ultra-high isolation RF MEMS switch dedicated for antenna switch applications. 2017. *Microelectron. J. 69*, pp 64–72.

[2] Yeonsu, J.; Sungchan, K.; Hyeon Cheol, K.; Kukjin, C. An RF MEMS switch with a differential gap between electrodes for high isolation and low voltage operation. 2011. *J. Micromech. Microeng. 21,* pp 1-9.

[3] Agarwal, S.; Kashyap, R.; Guha, K.; Baishya, S. Modeling and analysis of capacitance in consideration of the deformation in RF MEMS shunt switch. 2017. *Superlattices Microstruct. 101*, pp 567-574.

[4] Angira, M.; Rangra, K. A novel design for low insertion loss, multi-band RF-MEMS switch with low pull-in voltage. 2016. *JESTECH. 19*, pp 171-177.

[5] Yanqing, Z.; Lei, H.; Qin, M.; Qingan, H. Novel DC-40 GHz MEMS series-shunt switch for high isolation and high power applications. 2014. *Sens. Actuators, A. 214*, pp 101-110.
[6] Chakraborty, A.; Gupta, B.; Sarkar, B.K. Design, fabrication and characterization of miniature RF MEMS switched capacitor based phase shifter. 2014. *Microelectron. J. 45*, pp 1093–1102.
[7] Kundu, A.; Sethi, S.; Mondal, N.C.; Gupta, B.; Lahiri, S.K.; Saha, H. Analysis and optimization of two movable plates RF MEMS switch for simultaneous improvement in actuation voltage and switching time, 2014. *Microelectron. J. 41*, pp 257–265.
[8] Sen, P.; Kim, C.J. A liquid-solid direct contact low-loss RF micro switch. 2009. *J. Microelectromech. Syst. 18*, pp 990–997.
[9] Sen, P.; Kim, C.J. Microscale liquid-metal switches-A review. 2009. *IEEE T. Ind. Electron. 56*, pp 1314–1330.
[10] Jain, V.; Devarasetty, V.; Patrikar, R. Effect of electrode geometry on droplet velocity in open EWOD based device for digital microfluidics applications. 2017. *J. Electrostat. 87*, pp 11–18.
[11] Diebold, A.V.; Watson, A.M.; Holcomb, S.; Tabor, C.; Mast, D.; Dickey, M.D.; Heikenfeld, J. Electrowetting-actuated liquid metal for RF application. 2017. J. Micromech. Microeng. *27*, pp 025010.
[12] Krogmann, F.; Monch, W.; Zappe, H. Electrowetting for tunable microoptics. 2008. *J. Microelectromech. Syst. 17*, pp 1501–1512.
[13] You, H.; Steckl, A.J. Three-color electrowetting display device for electronic paper. 2010. *Appl.Phys.Lett. 97*, pp 023514.
[14] Nelson, W.C.; Kim, C.J. Droplet actuation by Electrowetting-on-Dielectric (EWOD): a review. 2012. *J. Adhes. Sci. Technol.*, *26*, pp 1747-1771.
[15] Jain, V.; Devarasetty, V.; Patrikar, R. Effect of electrode geometry on droplet velocity in open EWOD based device for digital microfluidics applications. 2017. *J. Electrostat. 87*, pp 11-18.
[16] Ni, Q.; Capecci, D.E.; Crane, N.B. Electrowetting force and velocity dependence on fluid surface energy. 2015. *Microfluid Nanofluid. 19*, pp 181-189.
[17] Chang, J.; Pak, J.J. Twin-plate electrowetting for efficient digital microfluidics. 2011. *Sens. Actuators, B. 160*, pp 1581-1585.
[18] Chen, L.; Bonaccurso, E. Electrowetting—From statics to dynamics. 2014. *Adv. Colloid Interface Sci. 210*, pp 2-12.
[19] Ke, L. Preparation and wetting of porous silicon. Master's degree, 2012. Shenyang aerospace university, Shenyang.
[20] Jiangang, W.; Ruifeng, Y.; Xuefeng, Z.; Litian, L. Droplet actuating chip using the electrowetting-on-dielectric method. 2006. J Tsinghua Univ (Sci&Tech). *46*, pp 1341-1344.

IOP Conf. Series: Journal of Physics: Conf. Series **986** (2018) 012021 doi:10.1088/1742-6596/986/1/012021

Surface micromachined MEMS deformable mirror based on hexagonal parallel-plate electrostatic actuator

Wenying Ma[1], Changwei Ma[1] and Weimin Wang[2,3]

[1] College of Communication Engineering, Chengdu University of Information Technology, Chengdu 610225, China
[2] State Key Laboratory of Optical Technologies on Nano-Fabrication and Micro-Engineering, Institute of Optics and Electronics, Chinese Academy of Science, Chengdu 610209, China

[3] weiminwang@qq.com

Abstract. Deformable mirrors (DM) based on microelectromechanical system (MEMS) technology are being applied in adaptive optics (AO) system for astronomical telescopes and human eyes more and more. In this paper a MEMS DM with hexagonal actuator is proposed and designed. The relationship between structural design and performance parameters, mainly actuator coupling, is analyzed carefully and calculated. The optimum value of actuator coupling is obtained. A 7-element DM prototype is fabricated using a commercial available standard three-layer polysilicon surface multi-user-MEMS-processes (PolyMUMPs). Some key performances, including surface figure and voltage-displacement curve, are measured through a 3D white light profiler. The measured performances are very consistent with the theoretical values. The proposed DM will benefit the miniaturization of AO systems and lower their cost.

1. Introduction

Deformable mirrors (DM) are the most important device of adaptive optics (AO) technology [1,2]. They can compensate wavefront distortions which are induced by atmospheric turbulence, ocular aberration, etc. Miniature DMs based on microelectromechanical system (MEMS) technology, or MEMS DMs, have many advantages compared to conventional DMs, such as low cost, low energy consumption and high actuator count [3,4].

The most conventional structure of DMs is a continuous facesheet supported by a periodic arrangement of actuators, whose deformation will deform the continuous facesheet and correct the phase of incident light. The arrangement of actuators mainly has two geometries, hexagon and square. Hexagonal geometry has less actuator count and smaller fitting error when the two geometries have the same period [5]. Unlike conventional DMs, MEMS DMs adopt square geometry more universally.

Surface micromachining and bulk micromachining fabrication processes are two main MEMS processes. Surface micromachining has the higher integration, better IC compatibility and reliability. In this paper, we proposed a surface micromachined MEMS DM based on hexagonal electrostatic actuator. The remaining sections of the paper are organized as follow: Section 2 is the structural design and analysis of the DM. The fabrication and test are described in Section 3 and a conclusion is detailed in Section 4.

2. Design and analysis

The hexagonal and square geometry arrangements for DM are shown in Figure 1 and d is the pitch. Three parameters, stroke, bandwidth, and actuator coupling, mainly determine the performance of MEMS DMs. Among them, the stroke is determined by the fabrication processes for an electrostatic actuator. Thus the stroke is not need to be considered in the structural design. The displacement of facesheet under different voltages can be solved through the balance between electrostatic force and elastic restoring force. These two force can be expressed as [6]

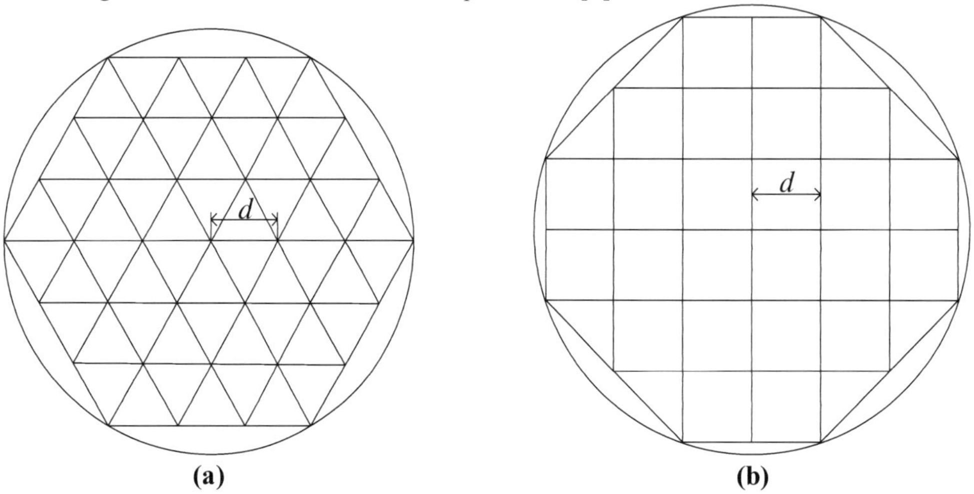

Figure 1. (a) Hexagon; (b) Square.

$$F_{\text{elec}} = \frac{\varepsilon_0 S V^2}{2(g-x)^2}, \tag{1}$$

$$F_{\text{elas}} = k_{\text{a}} x, \tag{2}$$

where ε_0, S, V, g, k_{a}, and x are permittivity of vacuum, area of actuator electrode, driving voltage, gap between actuator electrodes, spring constant of actuator, and displacement of facesheet, respectively.

Another parameter, bandwidth, is also not need to be considered because that a bandwidth of several hundred or one thousand Hz is totally enough for most AO systems. On the contrary, the optimum value of actuator coupling is 5%~12% [7]. Therefore the actuator coupling should be analyzed and designed carefully. The actuator couplings for square and hexagonal geometries can be expressed as

$$C_{\text{Rect}} = \frac{1}{1+4k_{\text{a}}/k_{\text{m}}}, \tag{3}$$

$$C_{\text{Hex}} = \frac{1}{1+6k_{\text{a}}/k_{\text{m}}}, \tag{4}$$

where k_{m} is spring constants of facesheet. Obviously, hexagonal geometry has a lower actuator coupling compared with square geometry when they have the same spring constants of actuator and facesheet. The spring constant of facesheet can be expressed as [8]

$$k_{\text{m}} = \frac{24EI}{d^3}, \tag{5}$$

where I and E are moment of inertia of the section of facesheet and Young's modulus, respectively. When there is a residual tensile stress σ in the facesheet, the spring constant should be modified to [9]

$$k_{\mathrm{m}} = \frac{8EI}{d^3 \frac{u - \tanh u}{u^3}} \approx \frac{24EI}{d^3} + \frac{12N}{5d}, \tag{6}$$

where $u = \frac{d}{2}\sqrt{\frac{N}{EI}}$, $N = \sigma(1-v)dt$, and v are the Poisson's ratio of the facesheet. Thus it can be seen that the residual stress produces an extra term in the equation. Moreover, generally facesheet is a multilayer membrane (silicon + metal) to increase the surface finishment and reflectivity. So an effective EI_{eff} product should be used to replace the one in above equations. N also should be replaced by the sum of residual stress of every layer. Based on the design and process parameters [10,11] listed in Table 1, the spring constant of facesheet can be computed as k_{m}=19.74N/m.

Table 1. Designed parameters of the facesheet.

Symbol	Quantity	Value
d	actuator period	408μm
E	Young's modulus of material silicon	158GPa
v	Poisson's ratio of material silicon	0.22
t	thickness of silicon	1.5μm
σ	residual stress of silicon	-10MPa
E_{metal}	Young's modulus of material metal	78GPa
v_{metal}	Poisson's ratio of material metal	0.42
t_{metal}	thickness of metal	0.52μm
σ_{metal}	residual stress of metal	50MPa

Based on k_{m} and the optimum value of C_{Hex}, we can compute k_{a} using equation (4) and the result shows that it should be located in 24.13N/m~62.52N/m. For tuning k_{a} accurately, the supporting beam of the actuator is designed as Figure 2(a). Here b is a tunable parameter and b=0 represents the conventional fixed-fixed beam.

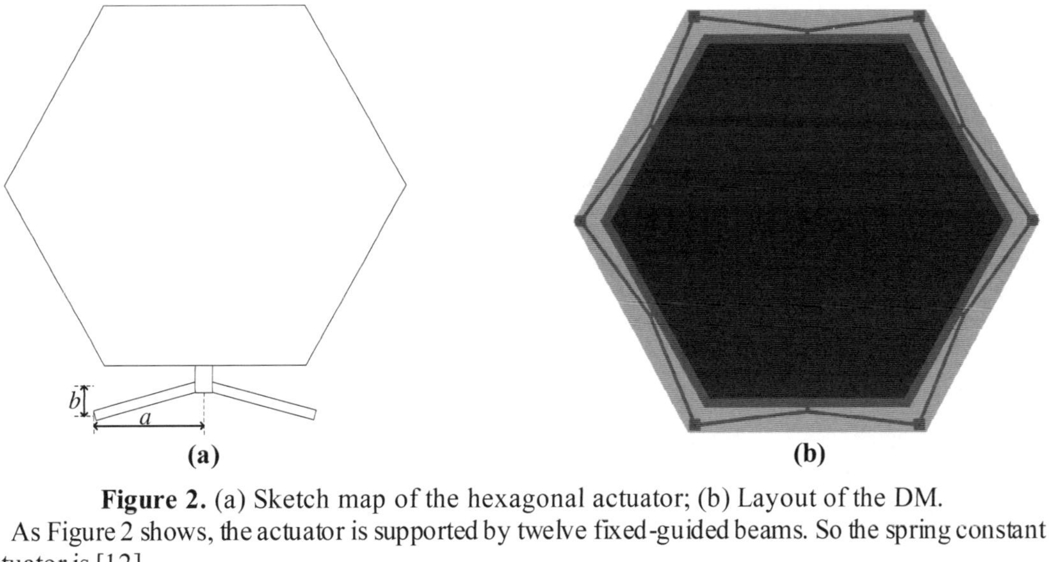

(a) (b)

Figure 2. (a) Sketch map of the hexagonal actuator; (b) Layout of the DM.

As Figure 2 shows, the actuator is supported by twelve fixed-guided beams. So the spring constant of actuator is [12]

IOP Conf. Series: Journal of Physics: Conf. Series **986** (2018) 012021 doi:10.1088/1742-6596/986/1/012021

$$k_{\mathrm{a}} = \frac{12kP}{2\tan\frac{kl}{2} - kl}, \tag{7}$$

where $P = \sigma_{\mathrm{act}}\left(1 - v_{\mathrm{act}}\right) w t_{\mathrm{act}}$, $k = \sqrt{\frac{P}{E_{\mathrm{act}} I_{\mathrm{act}}}}$, and $l = \sqrt{x^2 + y^2}$. E_{act}, v_{act}, I_{act}, t_{act}, and σ_{act} are Young's modulus, Poisson's ratio, moment of inertia, thickness and residual stress of actuator. All these parameters are listed in Table 2.

Table 2. Designed parameters of the hexagonal actuator.

Symbol	Quantity	Value
E_{act}	Young's modulus of the hexagonal actuator, silicon	158GPa
v_{act}	Poisson's ratio of the hexagonal actuator, silicon	0.22
t_{act}	thickness of actuator	2μm
σ_{act}	residual stress of actuator	-10MPa
w_{act}	width of beam	3μm
a	horizontal length of the skew beam	111.5μm
b	vertical length of the skew beam	12.25μm
g	initial gap between actuator electrodes	2μm
L_{upper}	side length of the upper electrode of actuator	210μm
L_{lower}	side length of the lower electrode of actuator	200μm

Based on the parameters listed in above table, we can obtain that k_{a}=26.22N/m, and satisfy the optimum range.

Considering actuator coupling, the elastic restoring force should be modified to

$$F_{\mathrm{elas}} = k_{\mathrm{a}} x + k_{\mathrm{m}}\left(x - C_{\mathrm{Hex}} x\right). \tag{8}$$

3. Fabrication and Characterization

3.1. Fabrication

A DM prototype including 7 element is fabricated by a commercial MEMS surface process PolyMUMPs [10] in Run 90. The SEM picture and micrograph of the DM are shown in Figure 3.

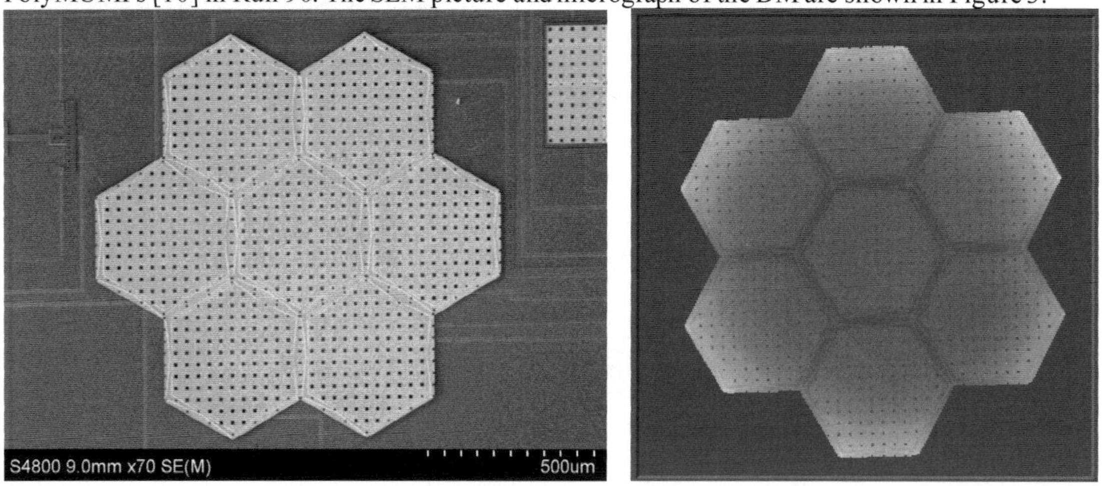

(a) **(b)**

Figure 3. SEM picture (a) and micrograph (b) of the prototype.

Due to the uncertainty of the process, some fabrication parameters are different from the reported ones listed in above tables. Table 3 list these different parameters provided by the foundry [10], except for the last row, which is measured by us.

Table 3. Different structural parameters compared to above two tables.

Symbol	Quantity	Value
t	thickness of silicon	1.52μm
σ	residual stress of silicon	-6MPa
t_{metal}	thickness of metal	0.49μm
σ_{metal}	residual stress of metal	28MPa
σ_{act}	residual stress of actuator	-6MPa
g	initial gap between the two actuator electrodes	4.53μm

3.2. Characterization

3.2.1. Surface profile. The characterization of the prototype is measured by a white light 3D profiler (Zygo NewView 7300). Figure 4(a) shows the surface figure of the DM prototype. As the figure shows, except the central element, all elements are bending upwards due to the residual stress. Thus the initial gap between the two actuator electrodes is increased to 4.53 μm, as shown in Table 3. In the succeeding work we should avoid this phenomena.

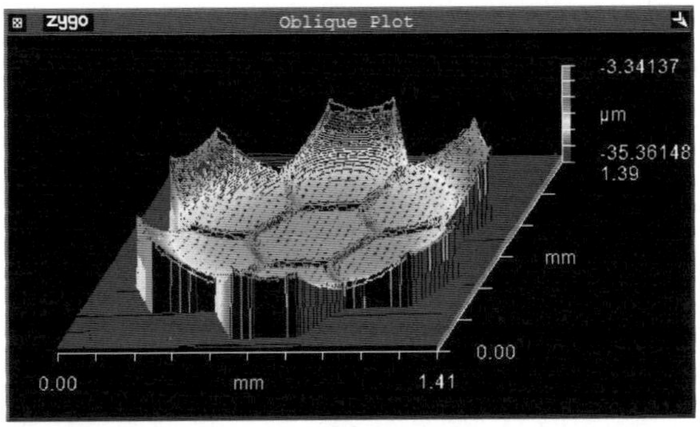

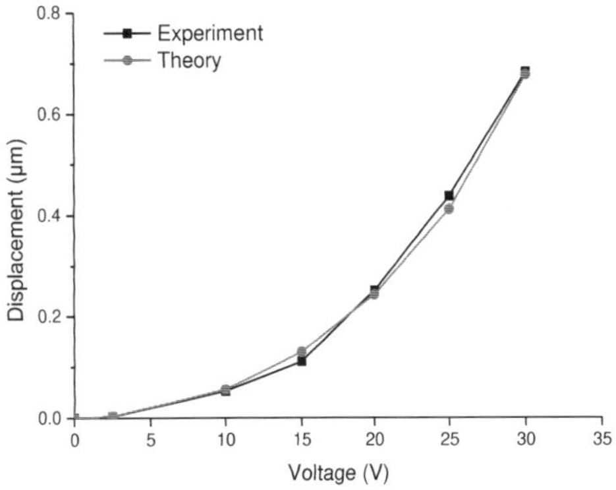

(a) **(b)**

Figure 4. (a) Measured surface profile of the DM; (b) Theoretical and experimental results (displacement of the drived actuator versus driving voltage) of the DM.

3.2.2. Device performance. Through measuring the surface of the prototype under different voltages, the displacement of the prototype is measured and plotted in Figure 4(b). At the same time the displacement is calculated through above equations and plotted in Figure 4(b), too. The theoretical performances are consistent with the measured values. A stroke, i.e., maximum displacement, of 0.68μm is obtained under a driving voltage of 30V.

4. Conclusions

A novel DM based on hexagonal parallel-plate electrostatic actuator is proposed. The actuator coupling of the device is analyzed and computed. A prototype of the DM is fabricated and measured. The computed and measured results of the displacement are consistent very much. In the future the surface figure of the presented DM should be improved to promote its application.

Acknowledgments
This research is supported by National Natural Science Foundation of China (NSFC) (11403029), Science & Technology Department of Sichuan Province (2017GZ0329), and Youth Innovation Promotion Association CAS (2014346).

References

[1] Devaney N, Dalimier E, Farrell T, Coburn D, Mackey R, Mackey D, Laurent F, Daly E and Dainty C 2008 *Appl. Optics* **47** 6550-62
[2] Vdovin G, Loktev M and Simonov A 2005 *Proc. SPIE* **5894** 58940B
[3] Manzanera S, Helmbrecht M A, Kempf C J and Roorda A 2011 *Biomed. Opt. Express* **2** 1204-17
[4] Cornelissen S A, Bierden P A, Bifano T G and Lam C V 2009 *J. Micro/Nanolith. MEMS MOEMS* **8** 031308
[5] Jiang W, Ling N, Rao X and Shi F 1991 *Proc. SPIE* **1542** 130-7
[6] Senturia S D 2001 *Microsystem Design* (New York: Kluwer Academic Publishers) p 131
[7] Pearson J E and Hansen S 1977 *J. Opt. Soc. Am.* **67** 325-33
[8] Hardy J W 1998 *Adaptive Optics for Astronomical Telescopes* (New York: Oxford University Press) p 187
[9] Frisch-Fay R 1962 *Flexible Bars* (London: Butterworth & Co. Limited) p 85
[10] Information on http://www.memscap.com
[11] Liu C 2006 *Foundations of MEMS* (New York: Pearson Education Inc.)
[12] Young W C and Budynas R G 2002 *Roark's Formulas for Stress and Strain* (New York: McGraw-Hill) p 229

Microsacle PolySilicon Hemispherical Shell Resonating Gyroscopes with Integrated Three-dimensional Curved Electrodes

Xuye Zhuang [1,3], Binggen Chen [1], Xinlong Wang [1], Lei Yu [1], Fan Wang [1] and Shuwen Guo [2]

[1] East China Institute of Photo-Electronic IC, Bengbu 233033, China
[2] School of Electronic Information, University Soochow, Suzhou 215000, China

[3] E-mail:zxye8888@163.com

Abstract. A novel approach for fabrication of polysilicon hemispherical resonator gyroscopes with integrated 3-D curved electrodes is developed and experimentally demonstrated. The 3-D polysilicon electrodes are integrated as a part of the hemispherical shell resonator's fabrication process, and no extra assembly process are needed, ensuring the symmetry of the shell resonator. The fabrication process and materials used are compatible with the traditional semiconductor process, indicating the gyroscope has a high potential for mass production and commercial development. Without any trimming or tuning of the n=2 wineglass frequencies, a 28 kHz shell resonator demonstrates a 0.009% frequency mismatch between two degenerate wineglass modes, and a 13.6 kHz resonator shows a frequency split of 0.03%. The ring-down time of a fabricated resonator is 0.51 s, corresponding to a Q of 22000, at 0.01 Pa vacuum and room temperature. The prototype of the gyroscope is experimentally analyzed, and the scale factor of the gyro is 1.15 mV/°/s , the bias instability is 80 °/h.

1. Introduction

Micro-Electro-Mechanical Systems (MEMS) gyroscopes are electromechanical devices that measure rate or angle of rotation. They have been adopted in a large variety of applications, such as smart phone, car navigation, electronic toy and industrial robot [1]. A wide variety of surface and bulk fabrication approaches for MEMS gyroscopes have been investigated over the past decades, and various micromachined resonators are developed [2-4]. Among these vibrating elements, the shell-type gyroscopes are the most popular and favorable ones owing to their high axial symmetry and high possibility to operate in whole angle mode.

Shell type gyroscopes of multifarious shapes and different materials are developed. Polysilicon hemispherical resonators were fabricated using isotropically-etched molds in silicon, and for the one with a shell thickness of 700 nm and a diameter of about 1.2 mm, a Q factor around 8500 at a resonant frequency of 6.7 kHz was observed, and for the one with a long anchor , a Q factor of 40400 was obtained [5]. The gyroscope based on this kind of resonator showed a scale factor of 8.57 mV/°/s [6]. A prototype of polysilicon hemispherical resonator with integrated spherical electrodes was fabricated using isotropic silicon etching molds and ICP etching process [7]. The resonator possessed several attractive characters, including large transduction area, small and uniform capacitive gap, no assembly process and high axis symmetry. The radial deviation of the hemispherical shell was less than 1.22 μm for the 500 μm radius shell (0.24%), promising a small frequency split. Silicon oxide hemispherical resonators were produced and Q factors of resonators with 500 μm in diameter and 1μm in thickness were in excess of 10000 when operated in 50mT vacuum. The separation in the degenerate frequencies of wineglass modes was 5 Hz at a resonant frequency of 22 kHz [8]. Cylindrical resonators made of microcrystalline diamond were batch-fabricated using diamond deposited by hot filament chemical vapor. The Q factor of the fabricated resonator was 528000 at 19 kHz wineglass modes with the ring-down time 8.9 s [9]. Frequency split less than 1Hz was demonstrated on borosilicate glass wineglass structures, and Q factors above 1 million were demonstrated on fused silica wineglass structures [10]. Fused silica wineglass resonators with high Q factors were fabricated though a simultaneous process of micro blow-torching and microwelding. The resonator had a shell radius/height of about 2.8 mm, and a stem diameter of 1 mm. At < 10 μTorr vacuum, the frequency of the wineglass mode was 22.6 kHz, the value of the Q factor was 2.55 million [4]. For the gyroscope with a fused silica Birdbath Resonator demonstrated, a bias stability of 0.0391 deg/hr was obtained using a force-rebalanced control architecture without temperature control or additional compensation [11].

Microscale hemispherical resonator gyroscopes have exhibited several intriguing characters and are considered to be a viable solution for inertial navigation applications. During its way to commercial market, one of the most serious challenges is to make its fabrication process be compatible with traditional semiconductor process. In this way, the gyroscope will be suitable to mass production and the cost can be reduced. Gyroscopes with assembled electrodes face the problem of requiring extra fabrication steps which increases process complexity and cost, making a bottle-neck in batch-fabrication of the devices at wafer level. Some gyroscopes equipped integrated alternatives by sputtering and etching metal electrodes around the shell or doping some areas of the silicon handle wafer. With these techniques, electrodes as thick as several microns can be fabricated, but they do not take the advantage of the large transduction area that the use of 3-D spherical and hemispherical shells provides. Integrated 3-D glass blown electrodes and high aspect ratio silicon electrodes for spherical shell gyroscopes have been successfully demonstrated [9, 10]. However, the curved walls of the shell-type resonators make it challenging to obtain small, uniform capacitive gaps across the height of the structure.

An alternative wafer-level approach for the batch micromachining of polysilicon hemispherical resonators equipped with integrated 3-D curved silicon electrodes is introduced. Gaps between the electrodes and the shell resonator were shaped by a sacrificial layer, ensuring a small, uniform width of the gaps. Assisted by the 3-D curved electrodes, which are self-aligned located around the resonator and have a similar profile with the shell, the gyroscope owns a large initial capacitance. Low voltage (the amplitude of ~mV) is needed to drive these resonators, making them suitable candidates for very low-power applications.

2. Structure Design

Structure of the designed gyroscope is illustrated schematically in Figure 1. It is a three-layers sandwich structure, including a upper cap layer, a middle silicon layer, and a bottom glass layer. Among them, the middle silicon layer is the gyro's main structure layer, it consists of a polysilicon hemispherical shell resonator and dozens of discrete three-dimensional electrodes. The number of the electrodes is variable according to the electrical connection design. The electrodes can be divided into two groups according to their functions. One group is named as functional electrodes which used for electrical driving, sensing and tuning the resonator. The other group is dominated supporting electrodes. All the electrodes in this group joined together to form a solid pedestal to support and provide electrical access to the resonator. Trenches between these electrodes are used as air passages when pumping vacuum. Metal pads are sputtered and patterned on the top surface of the electrodes. A shallow cave is fabricated in the upper cap glass layer, which is used to cover the shell resonator and make sure the resonator can vibrate freely and safely. Dozens of through holes are drilled in this layer and are aligned with the pads on silicon wafer. When the gyroscope chip is encapsulated in a ceramic package, these holes are the aisles for gold lines to reach the pads. The bottom glass layer is just a flat glass wafer whose function is to protect the resonator from being contaminated.

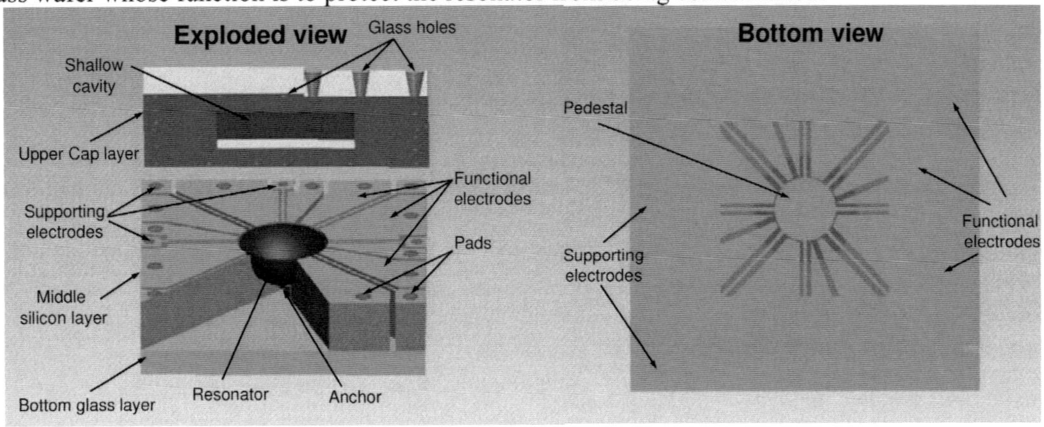

Figure 1. 3D diagram of the polysilicon hemispherical gyroscope with integrated 3-dimensional electrodes.

The gyroscope offers several attractive features. Firstly, the resonator is made of polysilicon. Polysilicon is a very common material used in semiconductor industry. It is easy to be handled and processed, so the cost of the gyroscope is reduced. Secondly, the initial capacitance of the gyroscope is large. By using integrated 3-D curved electrodes, the transduction area of the gyro is greatly improved. Besides, the capacitive gaps between spherical electrodes and shell are formed by a thin sacrificial dioxide layer, the width of the gap can be very uniform and small. The driving voltage of the resonator is as low as less than 30 mV.

3. Fabrication Process

The fabrication process is schematically illustrated in Figure 2. The resonators are fabricated on a 400 μm low-resistivity p-type single-crystal <111> silicon wafer. First, a hemispherical silicon cave is formed using isotropic etching in hydrofluoric acid \ nitric acid \ acetic acid (HNA) system (Figure 2(a)). The mask of the HNA etching is a composite membrane which consists of a 0.2 μm thermal SiO_2 layer and a 2 μm low-stress LPCVD silicon nitride layer. Then the mask layer is stripped and a bottom cave is ICP etched at the bottom of the hemispherical silicon cave. Subsequently, a blanket sacrificial thermal oxide layer is grown on the wafer and an anchor mold is fabricated by stripping a part of sacrificial oxide from the bottom cave's bottom, as shown in Figure 2(b). The anchor mold is the bridge which connects the hemispherical shell resonator and the pedestal mechanically and electrically. A 1-2 μm low-stress polysilicon shell layer is conformally deposited on the oxide layer. Then the polysilicon is doped with boron and annealed. After that, the shell layer on the top surface of the wafer is removed using chemical mechanical polishing (Figure 2(c)).

IOP Conf. Series: Journal of Physics: Conf. Series **986** (2018) 012022 doi:10.1088/1742-6596/986/1/012022

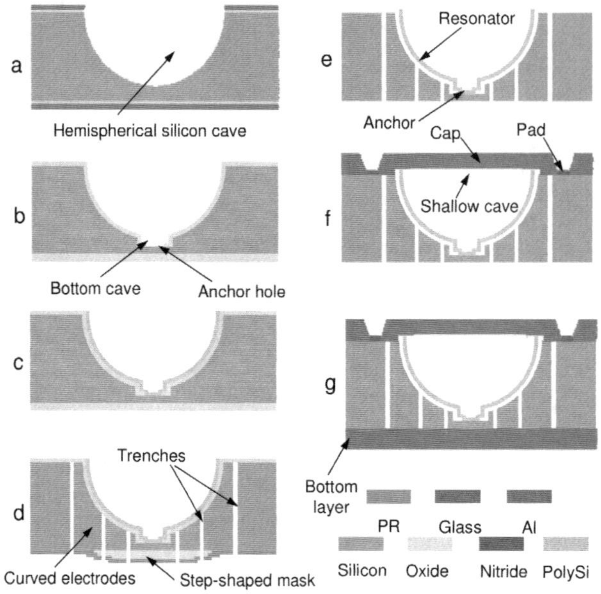

Figure 2. Fabrication process steps for the polysilicon hemispherical gyroscope with integrated 3-dimensional curved electrodes.

Subsequently, three-dimensional curved electrodes are ICP etched from the bottom side of the wafer using the step-shaped silicon oxide mask [12], as shown in Figure 2(d). The shell is then released in vapor hydrofluoric (VHF). Capacitive gaps between the resonator and the electrodes are formed simultaneously at this step (Figure 2(e)). Figure 3 shows a close-up scanning electron microscope (SEM) of the 1.7 μm gap between the resonator shell and electrodes. The capacitance between the resonator and each electrode is about 0.4 pF. The upper cap glass layer with through holes and shallow caves is bonded to the wafer using silicon-glass anodic bonding. Metal pads are patterned on the electrodes using sputtering with a mask. The mask is carefully aligned to make sure the pads are just located in the bottom of these holes and directly connected with the electrodes (Figure 2(f)). Finally, the bottom glass layer is bonded to the stack (Figure 2(g)).

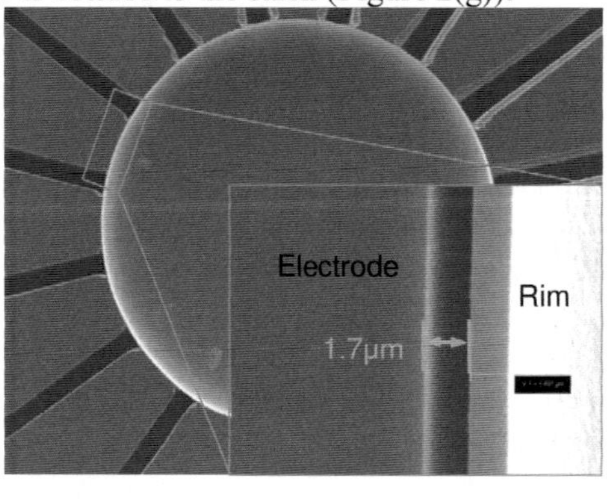

Figure 3. SEM image of the gap between the shell and electrode (gap size = 1.7 μm).

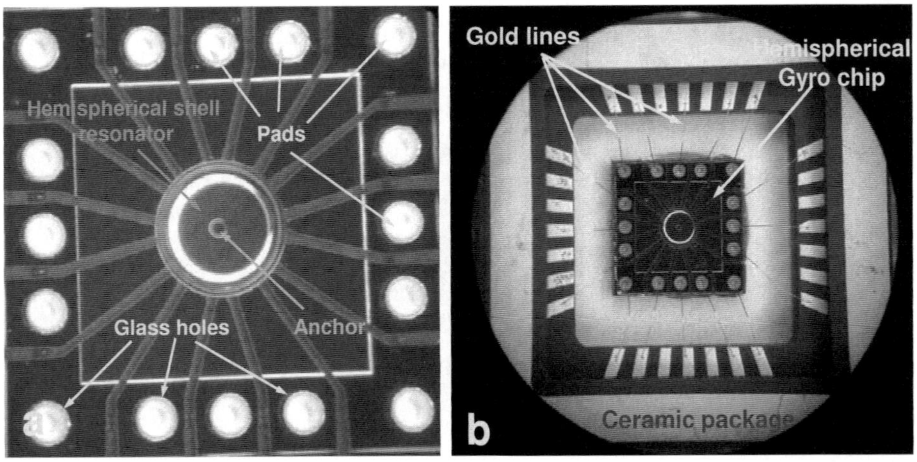

Figure 4. Photograph of a gyroscope chip (a) and a completed gyroscope chip wire bonded into a ceramic package (b).

Figure 4 shows a finished resonator chip (Figure 4(a)) fabricated using the proposed approach and a chip packaged in a ceramic package (Figure 4(b)). It shows clearly that the metal pads are just under the glass holes and attached to the electrodes firmly. Compared with the pads used in [7], the electrical connection is 100% successful in here. The pads fabricated using the method introduced in [7] often fail because of the break of the metal layer, which usually occurs at the corners between glass holes and silicon electrodes. Figure 5 is the SEM picture of the polysilicon hemispherical shell resonator whose 3-D curved electrodes are stripped away. The resonator is anchored to the pedestal created by the supporting electrodes, and the functional electrodes stretch into the cave deeply (about 160 μm), ensuring a large transduction area.

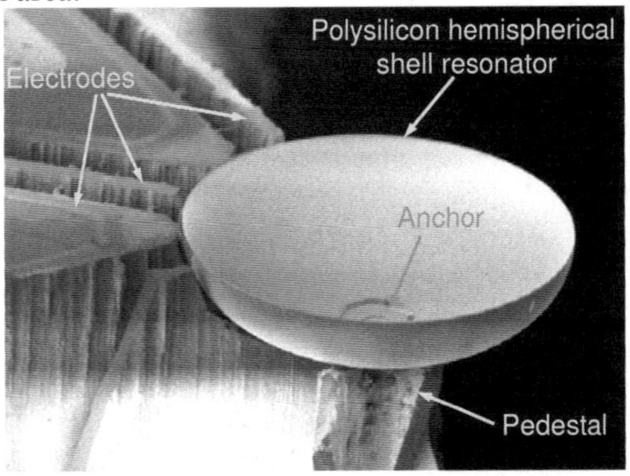

Figure 5. SEM micrograph of a 0.78 mm diameter hemispherical shell resonator whose electrodes have been stripped away.

4. Experimental results

A series of experiments have been performed to characterize the fabricated polysilicon hemispherical shell resonators, in which the resonator n=2 wineglass vibration is measured using laser vibrometer Ploytec MSA-500 and Zurich Instrument's HF2LI lock-in amplifier respectively. The laser vibrometer testing setup is shown in Figure 6. The device is fixed in a sealed chamber and actuated using the power that supplied through the electrical wires. A glass window is located on the top of the chamber, so the laser beam can translate into and out of the chamber. The gyro chip is mounted on the platform using glue with an angle in order that the laser beam can be focused on the edge of the shell where the vibration is much more violent, and a strong signal can be obtained.

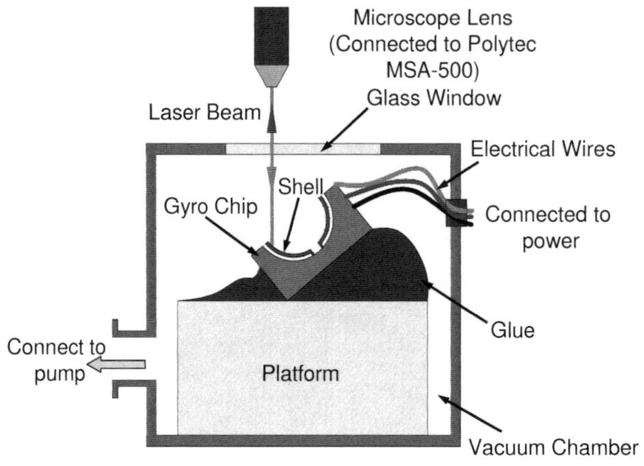

Figure 6. Schematics of the experimental testbed for characterization of polysilicon hemispherical shell resonating gyroscopes using a laser vibrometer.

After the chamber is pumped to vacuum, the resonator is actuated using electrostatic force and vibration information is obtained. Figure 7(a) is the frequency response of a polysilicon shell with a diameter of 0.78 mm and a thickness of 1.8 μm at 0.9 Pa and room temperature. The frequency of the n=2 wineglass mode are at 28.012 kHz and 28.0146 kHz, the frequency split is 2.6 Hz ($\Delta f_{n=2}$ / $f_{n=2}$ = 0.009%). The Q factors of these two frequencies are 14365 and 11435 respectively.

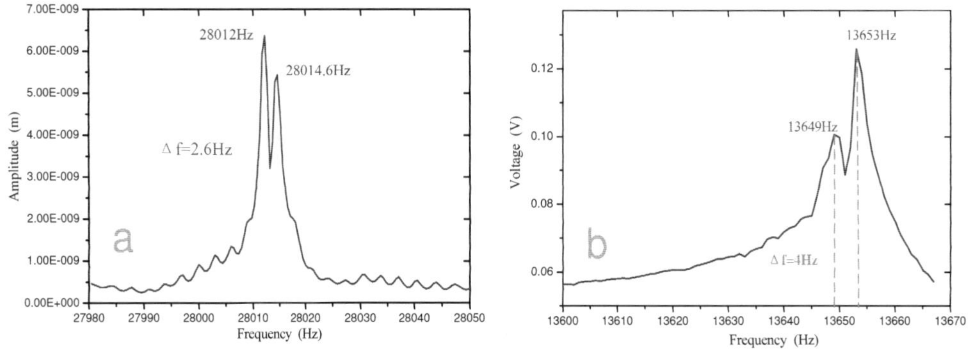

Figure 7. The frequency spectrum of a polysilicon shell with a diameter of 0.78 mm and a thickness of 1.8 μm (a), and a shell with a diameter of 1.3 mm and a thickness of 1.5 μm (b).

Frequency sweeps of a polysilicon hemispherical shell resonator with a diameter of 1.3 mm and a thickness of 1.5 μm are obtained using a Zurich Instrument's HF2LI lock-in amplifier. Figure 7(b) shows the frequency spectrum of the shell. The n=2 modes frequency are 13.649 kHz and 13.653 kHz respectively. The test is implemented at 0.02 Pa vacuum and room temperature, the driving voltage is 20 mV, the bias voltage is 400 mV. The as-fabricated frequency difference of these two modes is 4Hz ($\Delta f_{n=2}$ / $f_{n=2}$ = 0.03%) , which is larger than the one with a diameter of 0.78 mm. The main reason is that the symmetry of the shell is becoming difficult to be maintained in the fabrication process as the diameter increases.

The Q factor is measured in time domain by ring-down method. The ring-down time constant τ is related to the quality factor by $Q = \pi f \tau$. Time series data was exported to MATLAB for curve fitting to extract τ. A hemisphere ring-down time of 0.51 s (corresponding to a Q of 22000) at 0.01 Pa vacuum and room temperature is shown in Figure 8.

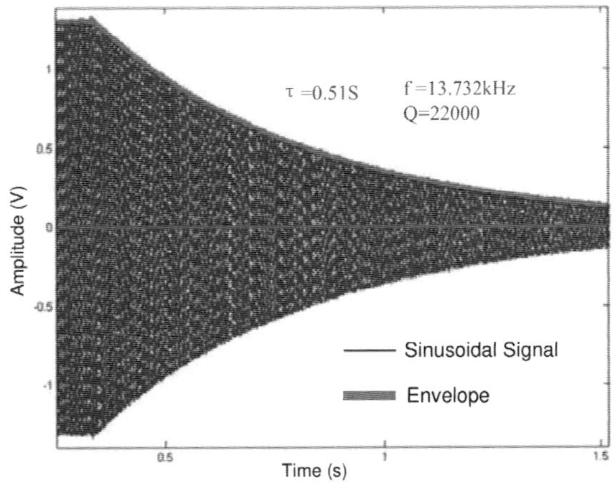

Figure 8. Ring-down plot of a polysilicon hemispherical shell resonator with a diameter of 1.3 mm and a thickness of 1.7 μm.

A prototype of the polysilicon hemispherical gyroscope has been successfully excited and sensed. The gyroscope without any trimming and tuning operates in an open loop mode. The measured scale factor is about 1.15 mV/°/s and bias instability (BS) is 80 °/h, as shown in Figure 9. The primal experimental results indicate a promising future of polysilicon hemispherical gyroscopes and motivate further research in improving the mechanical quality factor, optimizing the fabrication process, and improving the interface circuit.

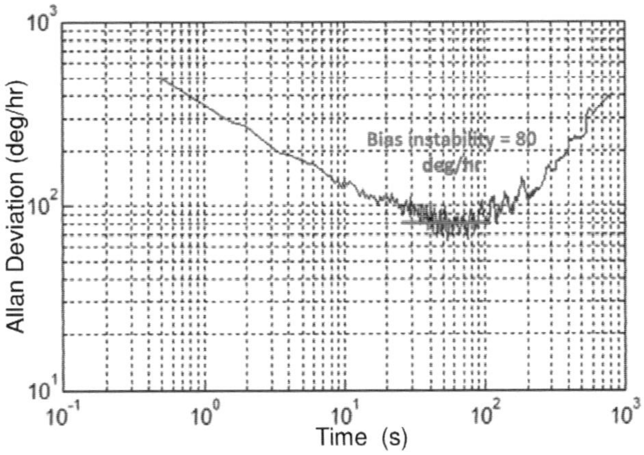

Figure 9. Measured Allan variance stability of a prototype of the designed polysilicon hemispherical gyroscope.

5. Conclusions

Micromachined polysilicon hemispherical shell resonating gyroscopes with integrated three-dimensional curved capacitive electrodes are fabricated and experimentally analyzed. 3-D curved electrodes are integrated as a part of the hemispherical shell resonator fabrication process, and are manufactured in company with the shell resonator in the same process, no extra assembly process is needed. The fabrication process and materials used are compatible with the traditional semiconductor process, indicating the gyroscope has a high potential for mass production and commercial development. Without any trimming or tuning of the n=2 wineglass frequencies, a 28 kHz shell resonator demonstrates a 0.009% frequency mismatch between two degenerate wineglass modes, and a 13.6 kHz resonator shows a frequency split of 0.03%. The Q factor of a fabricated resonator is measured using ring-down method. The ring-down time is 0.51 s, corresponding to a Q of 22000, at

0.01 Pa vacuum and room temperature. The prototype of the gyroscope is experimentally analyzed, and the scale factor of the gyro is 1.15 mV/°/s , the bias instability is 80 °/h. The process may inspire new classes of high-performance MEMS devices for inertial sensor applications.

Acknowledgments
The authors thank Wan Quan, Gu Honghua, Cheng Mengmeng, and Professor Xu Dacheng at Soochow University for the Zurich Instrument based testing support.

References
[1] Lammel G 2015 The future of MEMS sensors in our connected world *IEEE Int. Conf. on MEMS* (Estoril, Portugal) vol 2015 (New Jersey: IEEE) pp 61-64
[2] Senkal D, Ng E, Hong V, Yang Y, Ahn C H, Kenny T W and Shkel A M 2015 Parametric drive of a toroidal MEMS rate integrating gyroscope demonstrating < 20 PPM scale factor stability *IEEE Int. Conf. on MEMS* (Estoril, Portugal) vol 2015 (New Jersey: IEEE) pp 29-32
[3] Simon B R, Shkel A M 2012 Anti-phase mode isolation in tuning-fork MEMS using a lever coupling design *IEEE SENSORS* (Taipei, Taiwan) vol 3 (New Jersey: IEEE) pp 1-4
[4] Cho J Y and Najafi K 2015 A high-Q all-fused silica solid-stem wineglass hemispherical resonator formed using micro blow torching and welding *IEEE Int. Conf. on MEMS* (Estoril, Portugal) vol 2015 (New Jersey: IEEE) pp 821-824
[5] Peng S, Curtis L M, Xin G, Vahid T and Farrokh Ayazi 2014 *J. Microelectromech. S.* **23** 762
[6] Peng S, Vahid T, Curtis L M and Farrokh Ayazi 2014 *IEEE Sensors Journal* **15** 4974
[7] Xuye Z, Xinlong W, Lei Y, Pinghua L 2016 Micromachined silicon hemispherical resonators with self-aligned spherical capacitive electrodes *IEEE SENSORS* (Busan, South Korea) vol 2016 (New Jersey: IEEE) pp 1-4
[8] Pradeep P, Massood T A 2013 Fabrication and testing of hemispherical MEMS wineglass resonators *IEEE Int. Conf. on MEMS* (Taipei, Taiwan) vol 2013 pp 677-680
[9] Daisuke S, Chen Y, Amir H, Hadi N, Liwei L and David A H 2016 *Appl. Phys. Lett.* **108** 1157
[10] Senkal D, Ahamed M J, Askari S and Shkel A M 2014 *J. Procedia Engineering* vol **87** 1489
[11] Christopher B, Jae Y C 2017 Effect of drive-axis displacement on MEMS birdbath resonator gyroscope performance *INERTIAL* (Kauai, USA) vol 2017 (New Jersey: IEEE) pp 1-2
[12] Xuye Z, Lei Y, Xinlong W, Pinghua L, Dongfeng L 2016 *Opt. Precision Eng.* **24** 2746

BCB Bonding Technology of Back-Side Illuminated COMS Device

Y Wu[1], G Q Jiang[1], S X Jia[1] and Y M Shi[2]

[1]Nanjing Electronic Device Institute, China Electronics Technology Group, Nanjing, Jiangsu Province, China

[2]Key Laboratory of Infrared Imaging Materials and Detectors, Shanghai Institute of Technical Physics of the Chinese Academy of Sciences, Shanghai, China

wuyan_baade@163.com

Abstract. Back-side illuminated CMOS(BSI) sensor is a key device in spaceborne hyperspectral imaging technology. Compared with traditional devices, the path of incident light is simplified and the spectral response is planarized by BSI sensors, which meets the requirements of quantitative hyperspectral imaging applications. Wafer bonding is the basic technology and key process of the fabrication of BSI sensors. 6 inch bonding of CMOS wafer and glass wafer was fabricated based on the low bonding temperature and high stability of BCB. The influence of different thickness of BCB on bonding strength was studied. Wafer bonding with high strength, high stability and no bubbles was fabricated by changing bonding conditions.

1. Introduction

As a mature technology, spaceborne hyperspectral imaging technology is an efficient mean of ground target detection. Quantified hyperspectral imaging technology requires flat spectrum in response wave band of CMOS image sensor (CIS). However, wave peak and trough exist in spectral response of CIS because of reflection and diffraction of metal layer fabricated by normal CMOS process. Thus, studies focused on Back-side Illuminated CMOS (BSI) have become necessary (Figure 1.). BSI image sensor simplifies the incident light path greatly and avoids the reflection and diffraction of metal layer, which makes the planarization of spectrum response possible. With the enlargement of the scale of CMOS active pixel sensor (APS) chip and the decrease of the size of pixel units, BSI shows its advantage of huge structure and great performance, which includes great decrease of incident light path, small pixel units (<5μm) and improvements of spectrum response and duty ratio[1-3]. The fabrication of BSI device includes wafer bonding, which is the key process, substrate thinning, and sputtering of anti-reflection film. Based on 6 inch bonding technology, bonded wafers with high strength, high stability and no bubbles were fabricated by developing bonding process and improving surface quality of wafer.

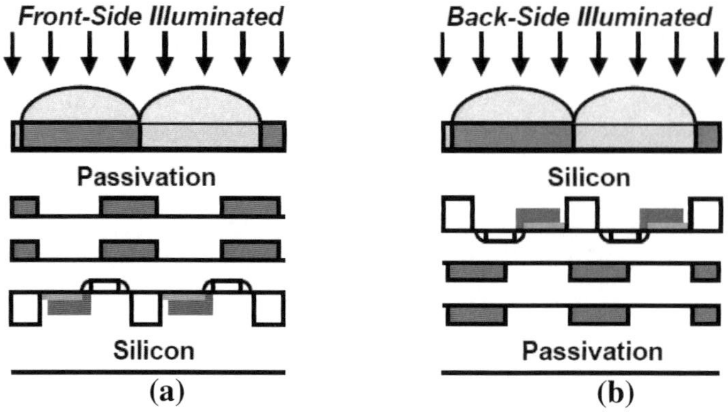

Figure 1. Front-side illuminated device (a) and back-side illuminated device (b)

2. Formatting the title, authors and affiliations

Photo BCB, which is a kind of organic polymer bonding material with advantages of high stability of process and great light transmission of BSI devices, is used to be the bonding material of BSI devices. The CYCLOTENETM 4000 Series of the Dow Chemical Company have the advantages of low bonding temperature and high stability. Thus, 4022-35 BCB was used as bonding material in our paper report. Figure 2 is the process of BCB bonding, including adhesion promoter coating, adhesion promoter baking, BCB coating, BCB baking, soft cure of BCB and bonding. The detailed bonding process is as follows:

1.Wafer cleaning by using megasonic cleaning with acetone and alcohol in order to remove organic particles.

2.Adhesion promoter coating by spinning AP8000, with a spin speed of 2000rpm, a spin acceleration of 1000r/s and a spin time of 30 seconds. The time of baking adhesion promoter is 2 minutes.

3.4022-35 BCB coating with a spin speed of 2000rpm, a spin acceleration of 1000r/s, and a spin time of 30 seconds.

4.Using T1100 to remove the BCB on the edge of wafer for 30 seconds. The rotation speed of wafer is 800rpm and the acceleration is 1000r/s.

5.Baking BCB for 2 minutes with a temperature of 110 degree.

6.Soft cure of BCB with a chamber pressure of 10^{-5} mBar and a heating rate of 8 degree per minute. Figure 3 is the heating curve of soft cure.

7.BCB bonding with a chamber pressure of 10^{-5}mBar, a heating rate of 8 degree per minute and a tool pressure of 5000 mBar (approximately 11000N). Figure 4 is the heating curve of bonding.

Plenty of water molecules exist on the wafer surface, which decreases the adhesion of BCB. Thus, adhesion promoter was used with AP8000 (used for most surfaces, including silicon oxide, silicon nitride, silicon oxynitride, aluminum, copper, titanium and chromium) before BCB coating and bonding. After adhesion process of wafer that increases the adhesion of BCB effectively, high bonding strength and stability are received. Because of plenty of wires on the wafer, the flatness of wafer is low and spin uniformity and step coverage are concerned when spinning BCB. In order to get a flat surface, a proper spin speed of 2000rpm was chosen. The thickness of BCB with different spin speed is as Figure 5 shows. The thickness of BCB was 4μm after soft cure with a spin speed of 2000rpm.

IOP Conf. Series: Journal of Physics: Conf. Series **986** (2018) 012023 doi:10.1088/1742-6596/986/1/012023

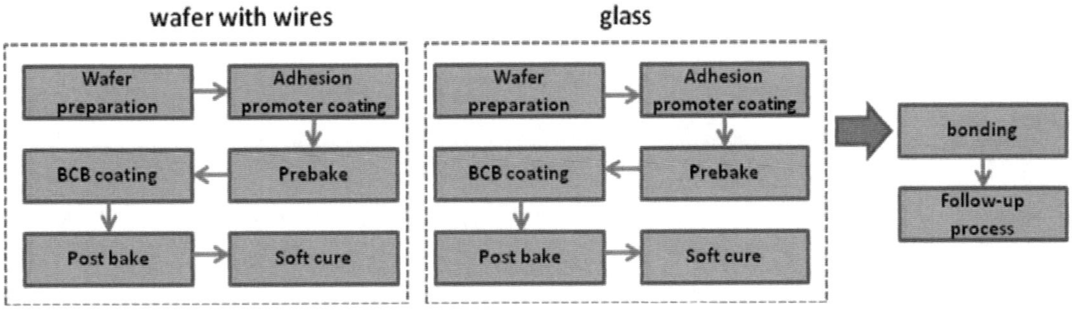

Figure 2. Process of BCB bonding

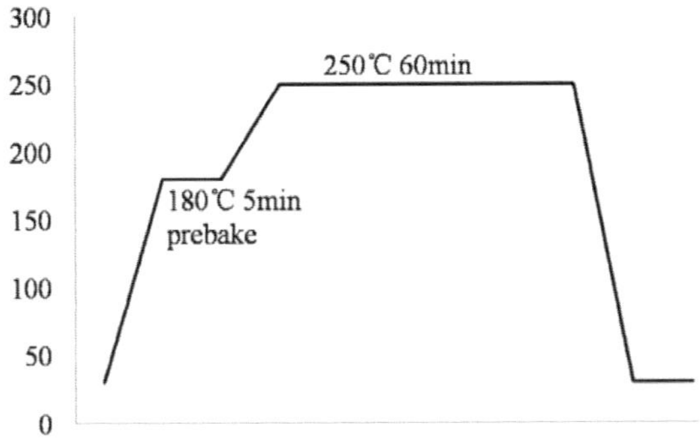

Figure 3. Temperature curve of soft cure.

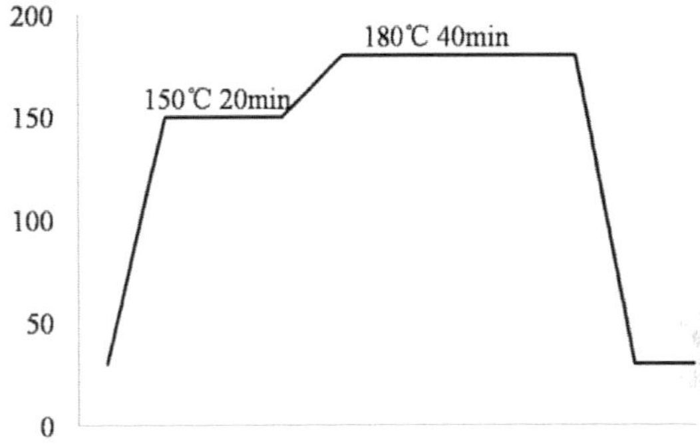

Figure 4. Temperature curve of bonding.

3. Optimization of bonding process

The process of low temperature wafer bonding is used to improve the flatness of bonding surface and make the disappearance of bubbles possible. Due to bonding temperature that higher than the temperature of soft cure, the viscosity of BCB become lower and the flowability of BCB stronger, which causes BCB bonded through Van der Waals force between wafers[4]. Very high pressure is applied to plate to enhance bonding strength and bonding stability.

Because of the existence of bubbles caused by these reasons that obvious difference exists between the refractive indexes of bubbles and BCB and the path of incident light includes the bonding area, huge changes of optical property appear when the device turns to work. The bubbles is mainly cause by three reasons: 1) When heating the plate the gas that BCB releases during heating has no way to escape because of the adhesion of wafers, then bubbles appear after bonding; 2) Due to the low flatness of wafer surface that caused by plenty of wires, the BCB between wafers aren't bonded completely within the limited deformation range; 3) The deformation of BCB becomes lower during bonding process because of high temperature of soft cure. We confirm that the first two points are the main reasons. For these reasons, solutions are provided as follows: 1) Optimization of bonding process was studied by adding EBR process that operated on wafers coated by BCB. Before wafer bonding, wafers are separated and heated for five to ten minutes at 180 degree by inserting spacers until gas escapes completely. 2) Through PECVD of silicon nitride and CMP process, the flatness of wafer is improved. 3) In order to get lower crosslinking ratio of BCB, lower temperature of soft cure is adopted, which causes the softening degree and deformation limit of BCB increased.

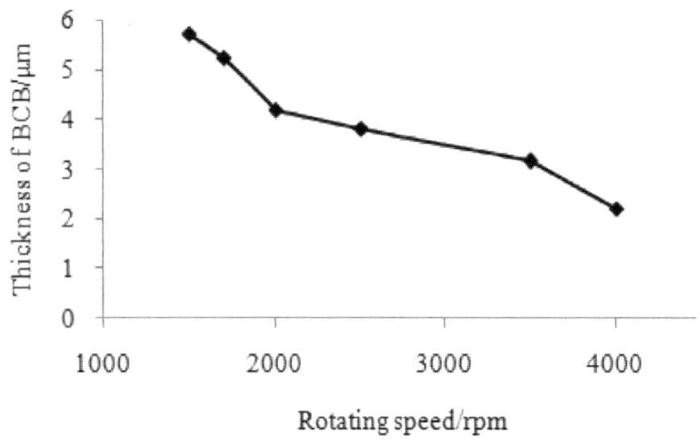

Figure 5. Thickness of BCB with different rotating speed.

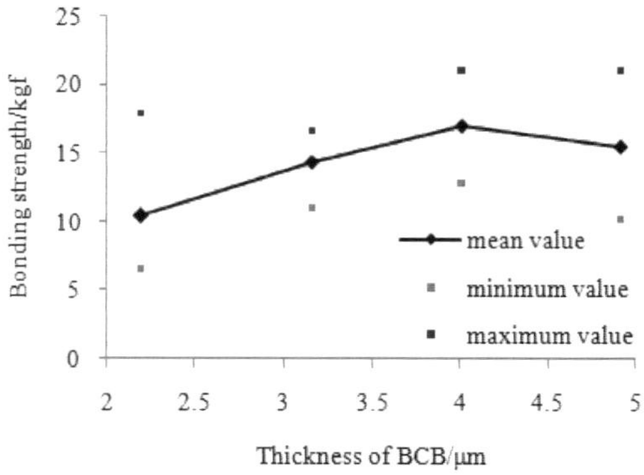

Figure 6. Bonding strength with different thickness of BCB.

4. Results and conclusions

Figure 6 is the bonding strength with different thickness of BCB(after soft cure). Bonding strength is improved obviously when increasing the thickness of BCB. When the thickness of BCB is 4μm, the mean value of bonding strength is greater than 15 kgf, and the maximum value of bonding strength is greater than 20 kgf. Even if the thickness of BCB is less than 2μm, the minimum value of bonding strength is greater than 5 kgf, which meets the requirements of applications. Figure 6 are the bonded BSI device and bonding area before and after the optimization of bonding process. The bubbles disappear through the optimization of bonding process, which improves the performance of BSI devices. In addition, reducing the quantity of bubbles increases the bonding area and then improves the bonding strength further.

5. References

[1] Pain B, Sun C, Vo P, Hancock B, Cunningham T J, Wrigley C, Toda R, White V, Banerjee A, and Misra D 2007 Wafer-Level thinned monolithic CMOS imagers in a bulk-CMOS technology *International Image Sensor Workshop Proceedings* pp158-61

[2] Pain B 2005 Fabrication and initial results for a back-illuminated monolithic APS in a mixed SOI/bulk CMOS technology *IEEE Workshop on CCD & AIS* pp102-4

[3] Fontaine R 2011 A Review of the 1.4 μm Pixel Generation *Proc of International Image Sensor Workshop* pp5-8

[4] Zheyao W 2014 Three-Dimensional Integration Technology *Tsinghua University Press* chapter 3

A Silicon Anode Material with Layered Structure for the Lithium-ion Battery

Dongliang Liao[1], Xuanlin Kuang[2],Jiangfeng Xiang[3]and Xiaohong Wang[4]
[1、2、4]Tsinghua University, Tsinghua National Laboratory for Information Science and Technology, China Institute of Microelectronics, Beijing, China
[3] GCL System Integration Technology Co. Ltd, Suzhou, China

Abstract. Although the theoretical capacity is 4200 mAh/g, it suffers from a large volume expansion (over 300%). We design a novel stratified rGO/CNT/Si structure by the vacuum filtration method. rGO is used to form a multilayer structure which is supported by CNT and the silicon nanopartarticles (SiNPs) are dispersed in the CNT gap. This structure can effectively protect the SiNPs from forming SEI in the first lithiation and de-lithiation process. The anode material owns the specific capacity of 1794.23mAh / g in the first cycle, and remains 1211.15mAh/g after 100 cycles.

Keywords Si anodes, Lithium-ion battery, stratified structure

1. Introduction

Rechargeable Li–ion batteries have been successfully commercialized, which has a widespread application, such as electric vehicles and energy storage systems for grid support [1]. So, much efforts have been devoted to study advanced energy storage technologies to order to enhance the performance of current Li–ion batteries [2,3]. To meet the ever increasing high-performance requirements, scientist need find a kind new anode materials which has high energy density and long cycle life.

Silicon reserve is very rich in nature, which is 26.4% ranked only second to oxygen. Recently, silicon has emerged as one of the most attractive high-capacity anode materials for it has the highest theoretical capacity (4200 mAh/g in a composite of $Li_{4.4}Si$) compare with conventional graphite anodes (theoretical capacity of 372 mAh/g) [4]. However, silicon has a large volume change (over 300%) during lithium-ion intercalation and deintercalation [5]. The volume change can lead to a series of problem for example, causing continuous consumption of the active material and Li-ion for reformation of solid electrolyte interphase (SEI) layers [6].

The lithium-plating mechanism of Silicon does not is very different from graphite with a layered structure for graphite-based materials, however silicon depends on the alloying and de-alloying of lithium ions in the lithiation and de-lithiation process, its electrode reaction can be written as the following formula in the charge and discharge process [7].

$$\mathrm{Si} + x\mathrm{Li}^{+} + xe^{-} \leftrightarrow \mathrm{Li}_x\mathrm{Si}$$

There is a huge change in volume when Silicon and lithium together form alloy [7]. To overcome volume problem, many researchers have done a lot of research. People has researched some nanostructure strategies, such as thin films [8], nanowires/nanotubes [9,10] and nanoparticles [11,12]. These special structure facilitate relaxation of the stress associated to the intercalation/deintercalation process and thus demonstrating high capacities in initial cycles. In addition, many researchers design a shell to protect the silicon a, the silicon particles wrapped in the shell inside. The shell could provide a void space for silicon's volume expansion and a stable shell to prevent electrolyte react with silicon directly was demonstrated in Si-SiO2 hollow tubes [13] and the Si-C yolk-shell [14,15]. The conductivity of pure silicon is very poor, carbon-based anode material has a good conductivity. Therefore, the Silicon - carbon composites which has a high-performance get more and more attention for lithium-ion anode material. The composite of silicon and carbon is gaining more and more attentions in the search of new high-performance Li-ion battery anodes. Much progress in terms of cycle stability has been made by designed silicon/carbon nanocomposite anodes from various carbon sources, such as graphene, PVDF, pitch, acetylene and polydopamine [16].

In summary, in order to solve the problem of volume expansion and conductivity of silicon, we have designed the anode material of rGO / CNT / Si with a layered structure as shown Figure.1. The layered structure can effectively protect the silicon so that the SEI film does not form on the surface of the silicon particles and uses a highly conductive graphite material to improve the electrical conductivity of the material.

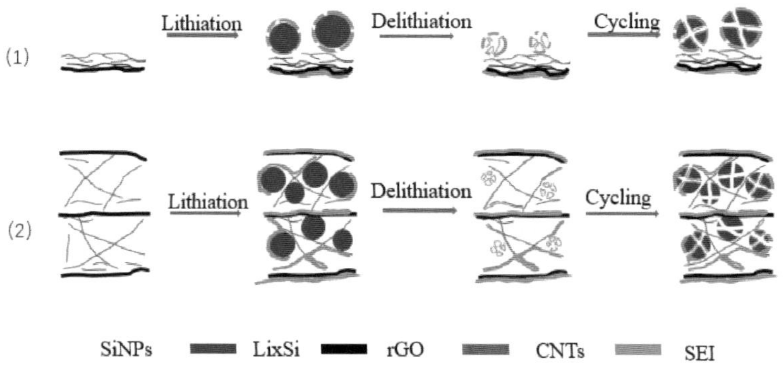

Figure 1. The schematic of lithiation and delithiation processes of different silicon-based materials (1) the anode material is fabricated by natural evaporation. The SEI films is continuously destroyed and meanwhile formed because the silicon has a huge volume expansion during the reaction process; (2) the anode material is fabricated by vacuum filtration. The layered structure protects the silicon, and prevents the SEI layer forming directly on the silicon surface.

IOP Conf. Series: Journal of Physics: Conf. Series **986** (2018) 012024 doi:10.1088/1742-6596/986/1/012024

2. Materials and Methods

Figure 2 describes the process of fabricating for the rGO/CNT/Si anode materials, the process is divided into four steps: the first step, fabricating the precursor solution of GO/CNT/SiNP s (the mass ratio of each substance is 38:12:50)——SiNPs are dispersed in deionized water containing a certain amount of GO and CNT, SiNPs are used as active material which provides main battery capacity, GO will form a layered structure to protect silicon and the CNT acts as a support role between the GO layers; the second step, the rGO/CNT/Si material is obtained by different ways, (a) the mixed solution of GO/CNT/Si is transferred to the culture vessel with a dropper and naturally evaporates the excess liquid (b) we deal with the compound of GO/CNT/Si by vacuum filtration; The third step, the compound requires a high-temperature sintering process in 900 °C about 2h (the process called as carbonation), in order to transform the GO into rGO; the fourth step, the surface of silicon particles has been oxidized to silica when it exposes to the air, so the SiO_2 need etch by using hydrofluoric acid. Finally, we fabricate the rGO/RGO/CNT/SiNPs composite.

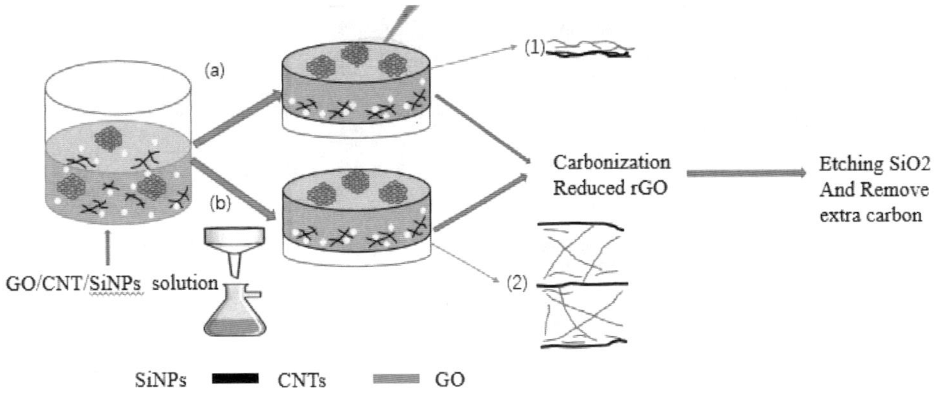

Figure 2 Fabricating the process of rGO/CNT /SiNPs composite anode by different ways(a) natural evaporation; (b) vacuum filtration.

3. Results and Discussion

The SEM of the rGO/CNT/Si composite is shown in Figure. 3 and Figure.4. Figure. 3(a) and Figure. 4(a) show morphological features of this material by natural evaporation. The layered structure can't been found in Figure. 4(a) which is a sectional view, so the rGO failed to encapsulate the silicon particles. From the figure.3(b) and figure.4(b), We can clearly find a layered structure for the composite which is manufactured by vacuum filtration, thus it can protect the silicon particles. This is a good proof to our design.

Figure 3. SEM images of the anode material (a) natural evaporation; (b) vacuum filtration.

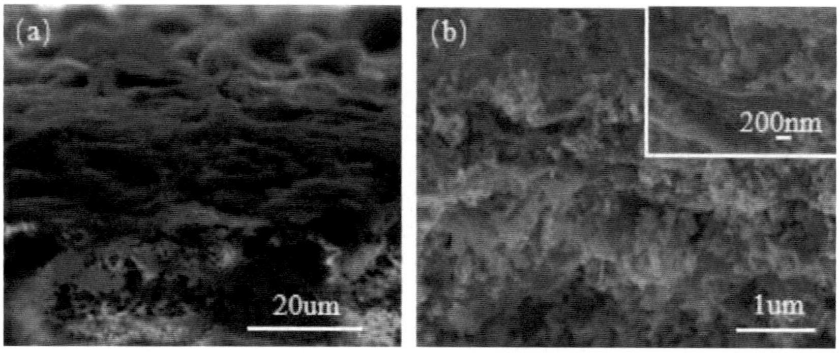

Figure 4. SEM images of the anode material for the sectional view (a) natural evaporation; (b) vacuum filtration.

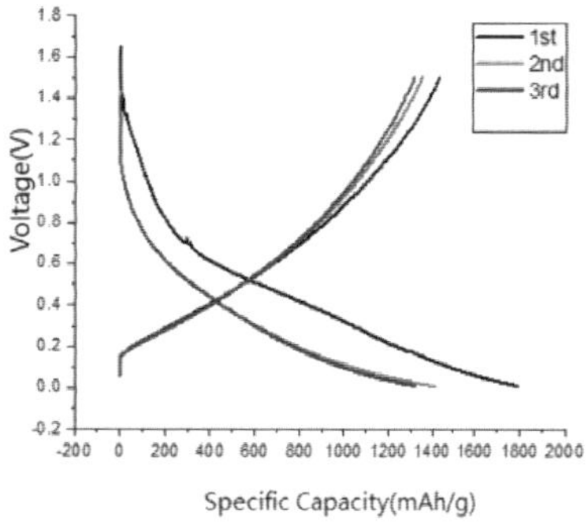

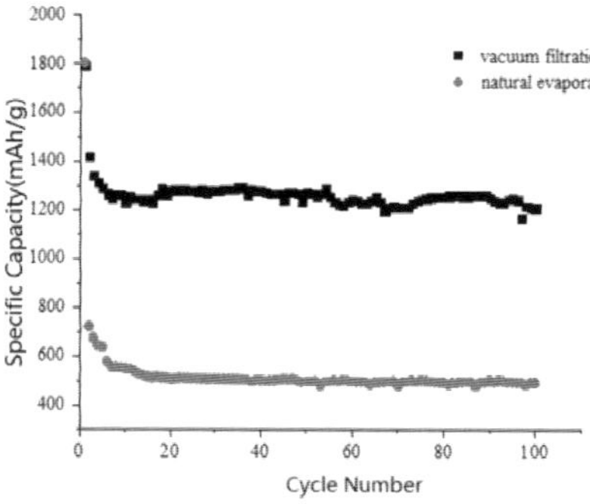

(a) (b)

Figure 5. (a) The profiles of the rGO/CNT/Si composite anode for different cycles on continuous current charge and discharge; (b) The specific capacity of rGO/CNT/Si composite anode under galvanostatic cycling tests.

We fabricate the two kinds materials into half-cell. Figure 5(a) shows the charge and discharge diagram of the layered structure in constant current for different cycles. The first cycle has a longer discharge plateau compared with others. This is because the effective silicon content is most at the first cycle, so the initial capacity is the highest and the process of discharge is the longest. The discharge plateau for the second and third is almost similar, which indicates that the battery structure has been stable and the layered structure has protected the SiNPs. The cycle performance of half-cell is shows in

IOP Conf. Series: Journal of Physics: Conf. Series **986** (2018) 012024 doi:10.1088/1742-6596/986/1/012024

Fig 5(b) for the two kinds materials. The first specific capacity is as high as 1804.12mAh / g for the first method, but it rapidly drops to 497.01mAh / g after the 100 cycles, and the capacity decays by 72.45%. This demonstrates that the rGO can't protect the silicon, the continuous formation of SEI layers consumes a mass of lithium in the silicon surface, which makes the battery capacity rapid decrease. The anode material is prepared by the second method, its first specific capacity is 1794.23mAh / g, and capacity falling to 1211.15mAh / g after 100 cycles, so the capacity still remains 67.50%. Thus, these can verify that the rGO / CNT / Si composite with layered structure not only effectively protects the silicon, but also greatly enhances the specific capacity.

The reason why the rGO/CNT/Si fabricating by vacuum filtration can form a layered structure lies in the high precipitation speed. During the natural evaporation, the graphene with great mass precipitates at the bottom and the silicon precipitates at the upper. However, for the vacuum filtration, the solvent is rapidly filtered. The compositor materials are synchronously precipitated, which can retain the layered structure.

4. Conclusions

To solve the volume expansion problem of silicon, we designed the rGO/ CNT/Si composites with stratified structure, which shows high capacity compared with graphite material. The experiment demonstrates the anode which has a later structure own the initial specific capacity of 1794.23mAh / g when the battery performs the first cycle, and remain 1211.15mAh / g after 100 cycles. Meanwhile the preparation process of material is simple and benefit for the large scale production.

Acknowledgments

The project is support by the National Natural Science Foundation of China (No. 61531166006), Suzhou-Tsinghua Leading Innovation Specific Project(No.2016SZ0302), and 863 program (No. 2015AA043502) of China. In addition, the majority experimental process is completed in Professor Xingpeng Qiu Department of Chemistry in Tsinghua University, so we are very grateful to them for their help.

References

[1] M. Armand, J.M Tarascon, 2008 J. Nature. vol 451 652–657

[2] J. M. Tarascon, M. Armand 2001 J. Nature. "Issues and Challenges Facing Rechargeable Lithium Batteries" vol 414 359-367

[3] V. Etacheri, R. Marom, R. Elazari, G. Salitra, D. Aurbach 2011 J. Energy Environ. Sci. vol 4 3243–3262

[4] A. S. Arico, et al 2005 J. Nature Materials. Nanostructured Materials for Advanced Energy Conversion and Storage Devices. vol 4 366-377

[5] B. A. Boukamp, G. C. Lesh, R. A. Huggins 1981 J. Electrochem. Soc. All-solid Lithium Electrodes with Mixed-conductor Matrix. vol 128 725-729

[6] S. Nadimpalli, et al 2012 J. Power Sources. "Quantifying Capacity Loss due to Solid-electrolyte-interphase Layer Formation on Silicon Negative Electrodes in Lithium-ion Batteries," vol 215 145-151

[7] Sharma RA, Seefurth RN 1976 J. Electrochem. Soc. vol 123 1763-1768

[8] T. Takamura, et al. 2004 J. Power Sources. "A Vacuum Deposited Si Film Having a Li Extraction Capacity over 2000 mAh/g with a Long Cycle Life". vol 129 96-100
[9] C. K. Chan, et al., R. A. Huggins, and Y. Cui 2008 J. Nature Nanotechnology. "High-performance Lithium Battery Anodes Using Silicon Nanowires". vol 3 31-35
[10] H. Kim, J. Cho 2008 J. Nano Letters. "Superior Lithium Electroactive Mesoporous Si@Carbon Core-Shell Nanowires for Lithium Battery Anode Material," 3688-3691.
[11] X. S. Zhou, et al 2012 J. Chem. Commun."Facile Synthesis of Silicon Nanoparticles Inserted into Graphene Sheets as Improved Anode Materials for Lithium Ion Batteries". vol **48** 2198-2200
[12] A. Magasinski, et al 2010 J. Nature Materials."High-performance Lithium-ion Anodes Using a Hierarchical Bottom-Up Approach". vol 9 353-358
[13] H. Wu, et al 2012 J. Nature Nanotech. "Stable cycling of double-walledsilicon nanotube battery anodes through solid-electrolyteinterphase control". 310-315
[14] N. Liu, et al 2012 J. Nano Lett. "A yolk-shell design for stabilized and scalable Li-ion battery alloy anodes". vol 12 3315-3321
[15] X. Z. Li. et al 2015.MEMS 2015. "Silicon anode supported by carbon scaffold for high performance lithium ion micro-battery" 130-133
[16] M. L. Terranova, et al 2013 J. J. Power Sources. "Si/C Hybrid Nanostructures for Li-ion Anodes: an Overview" vol 246 167-177

Design of a hybrid power system based on solar cell and vibration energy harvester

Bin Zhang[1], Mingxue Li[2], Shaoxuan Zhong[2], Zhichao He[1], Yufeng Zhang[2,*]
[1]Information Science Academy of China Electronic Technology Group Corporation, Beijing, China.
[2]MEMS Center, School of Astronautics, Harbin Institute of Technology, Harbin, Heilongjiang, China

*yufeng_zhang@hit.edu.cn

Abstract. Power source has become a serious restriction of wireless sensor network. High efficiency, self-energized and long-life renewable source is the optimum solution for unmanned sensor network applications. However, single renewable power source can be easily affected by ambient environment, which influences stability of the system. In this work, a hybrid power system consists of a solar panel, a vibration energy harvester and a lithium battery is demonstrated. The system is able to harvest multiple types of ambient energy, which extends its applicability and feasibility. Experiments have been conducted to verify performance of the system.

1. Introduction

Wireless sensor network (WSN) plays a more and more important role in many industries[1]. With the development of manufacture technology, the volume of wireless sensor node tends towards micromation[2]. Although micro wireless sensor node consumes tiny power, it has very limited lifetime powered by batteries[3][4]. Wireless sensor network usually consists of a large amount of nodes distributed in a broad area[5] which build up communication internet. Since sensor nodes generally use traditionally secondary battery as power supply, it costs prohibitively to replace sensor node batteries at intervals.

Using renewable resource, such as solar energy[6][7], wind[8], motion[9], and temperature difference[10] will significantly extend working life of sensor node and lower the maintenance cost. Since vibration energy and solar energy is ubiquitous in most area, they are good candidate for outdoor wireless sensor nodes[11][12]. However, renewable energy harvester is easily affected by actual working environment and weather, research of hybrid power source containing both generator and storage devices has become a hot point. Furthermore, power system combined with multiple energy types such as solar cell and thermoelectric generator (TEG)[13][14], fuel cell and lithium battery[15], solar photo voltaic and wind turbine[16], etc. Energy scavenging aiming at different form from environment will enhance environmental suitability of power system, which will also gain more efficiency of energy transformation.

In this paper, a hybrid power system consists of a solar panel, a vibration energy harvester and a lithium battery is demonstrated. System structure is firstly introduced. Then, regulator circuit of each component is proposed to realize functional integration of the system. Based on this, energy management strategy is designed to promote system's overall performance. Finally, experiments are conducted for verification.

IOP Conf. Series: Journal of Physics: Conf. Series **986** (2018) 012025 doi:10.1088/1742-6596/986/1/012025

2. Design of hybrid power management system

The power system includes two power generators, solar panel and vibration energy harvester, and an energy storage device, micro lithium battery. As output voltage of the power sources are improper for direct integration, corresponding converters are inherently necessary. A MSP430G2553 low power microcontroller is implemented in the system, in which proposed energy management strategy runs on it. System structure is presented in Figure 1.

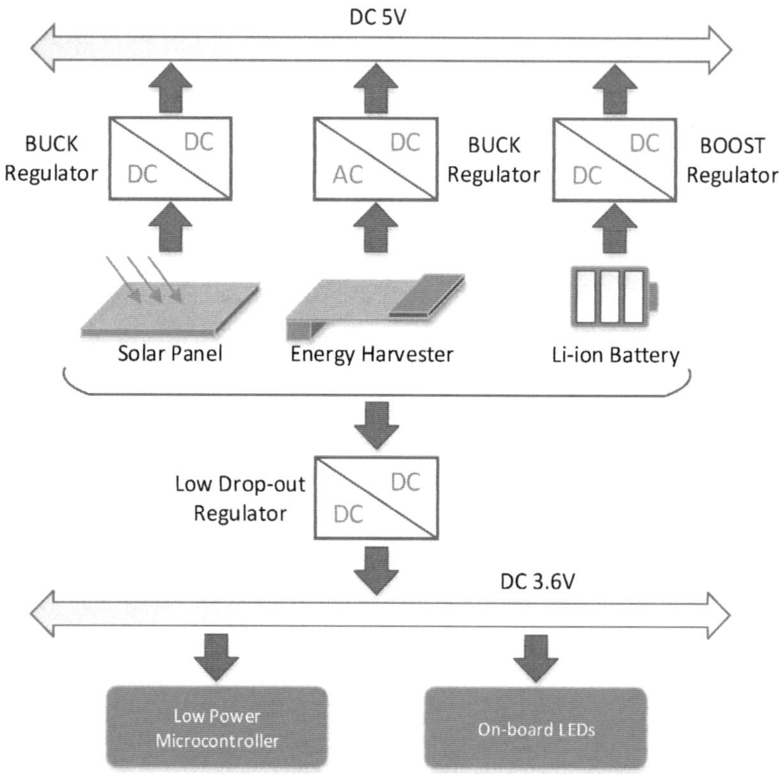

Figure 1. System energy flow route diagram

The power system generates two power rails. One is 5V direct current which is used as primary output power line, the second is 3.6V direct current which is for microcontroller and other on-board peripherals. Both power rails are driven by the power components.

2.1. Solar panel regulator circuit

Due to low manufacturing cost[17] and relatively high efficiency, poly-Si solar panel is used as photovoltaic component, of which the parameters are presented in Table 1. Low drop-out regulator (LDO) is implemented to obtain stable 5V output. LDO is suitable for situations with high input voltage, under which the energy cost is relatively low and peripheral circuit is very simple. Texus Instruments (TI) LP2985-5.0 is selected as the LDO circuit, which has a very low leakage current of 850μA.

Table 1. Solar panel parameters

Parameter Type	Value
Open circuit voltage	6.28V
Short circuit current	0.15A
Maximum power	750mW
Efficiency	17%
Size (length×width×thickness)	84×55×3mm

The enable port is connected to microprocessor, which controls system structure dynamicly. The output pin is connected to 5V power rail V_{out}, which supplies energy when LDO is enabled and enters high-resistance state when disabled.

2.2. Lithium battery regulator circuit design

When ambient solar energy is not sufficient for the system's requirement, additional backup power supply is necessary to stabilize the system.A small lithium battery is integrated in the system, with model of CR2032, capacity of 230mAh and nominal voltage of 3V. A high efficiency boost converter is implemented to normalize output voltage of the battery, with model of Texas Instrument TPS61041-Q1 and quiescent current of 28uA (1μA when shut down). The inductor selection is restrict within equations as

$$f_{s,max} = \frac{V_{in,min} \cdot (V_{out} - V_{in})}{I_{peak} \cdot L \cdot V_{out}} \tag{1}$$

$$f_{s,norm} = \frac{2I_{load} \cdot (V_{out} - V_{in} + V_d)}{I_{peak}^2 \cdot L} \tag{2}$$

where $f_{s,max}$ and $f_{s,norm}$ respectively denote the maximum switching frequency and nominal switching frequency, L represents inductance value, V_d stands for Schottky diode drop-out voltage and I_{peak} means the maximum switching current expressed as

$$I_{peak} = I_{lim} + \frac{V_{IN}}{L} \cdot T_{delay} \tag{3}$$

where T_{delay} is internal propagation delay time and I_{lim} is DC current limit. With formula (1)-(3) and parameters given by manufacturer, inductance value range of L can be calculated. Additionaly, L can be uniquely determined with the tradeoff between switching frequency and efficiency. Also, enable ports is connected to microcontroller for power management between output state and high impedance state.

2.3. Piezo vibration energy harvester regulator circuit design

Piezo vibration energy harvester is effective for low power wireless sensor devices. Its generated energy can be directly deposit in storage devices[18]. Piezo vibration energy harvester has a structure of cantilever with piezoelectric materials on both surfaces. When vibrationally excitated, response of the cantilever beam is periodic, and output voltage of the energy harvester is alternating. The energy harvester implemented in system is Mide V25W with mass tip attached which lowers the resonant frequency to 45Hz. As vibration amplitude increases, the output peak-to-peak voltage (V_{PP}) can reach over 50V with 1g excitation. Linear Technology LTC3588 with low drop-out Schottky rectifier circuit resistant and high-efficiency step-down switching regulator integrates is implemented. The required inductance value can be derived from the equation

$$L = \frac{V_{out} \cdot (V_{in} - V_{out})}{V_{in} \cdot \Delta I_L \cdot f_{s,norm}} \tag{4}$$

where ΔI_L is ripple current of inductor which values 30% of the maximum load current.

The Power Good signal is delivered to microcontroller when output reaches pre-programmed value of 5V (both D0 and D1 is tied to VIN2). Output enable function is realized by using PMOS with very low on resistance as switch.

2.4. Design of system internal power rail

A microcontroller with model of Texas Instruments MSP430G2553 is implemented to monitor states of each power source and executes respective control strategy. Energy management algorithm is programmed in the microcontroller, with very low current consumption and abundant low power mode (LPM). It is necessary to build up an internal power rail, since output voltage of power sources regulator is set to 5V, which exceeds the safe operation area of microcontroller and requires additional components for system to cold start with regulators disabled and microcontroller deactivated.

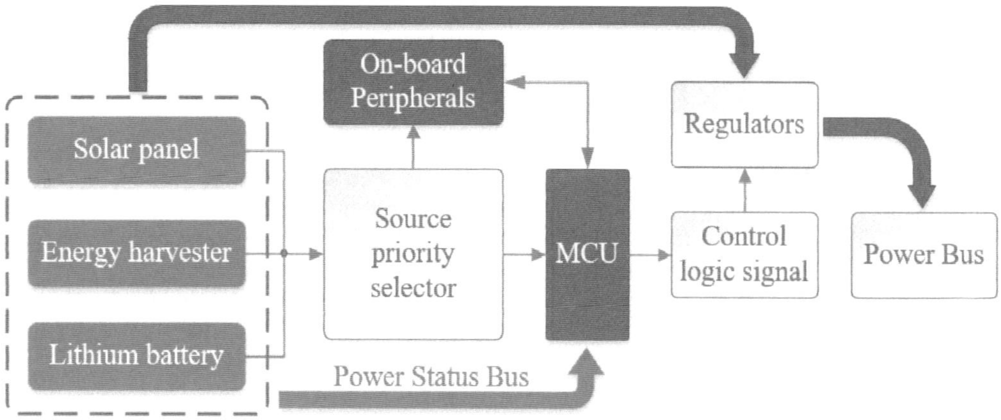

Figure 2 Schematic of internal power rail and system controller

The schematic of internal power rail is shown in Figure 2. The power selector is realized by TPS22933 with a 3.6V LDO integrated, which stabilizes the output voltage of solar panel and battery. As output voltage of vibration energy harvester is high and alternative, it is not used for cold start of microcontroller. Multi-channel integrated 10-bit analog-to-digital converter of the microcontroller is used to monitor output voltages of lithium battery and solar panel for the further control.

2.5. Energy management strategy

During operation, output voltages of all modules are monitored and controlled as either power sources or high-impedance ports. Monitoring is realized via the built-in 10-bit ADC of the microcontroller, and control is realized via the enable ports of polar panel, vibration energy harvester and lithium battery with the energy management strategy shown in Figure 4, microcontroller decides the operation mode of each power source.

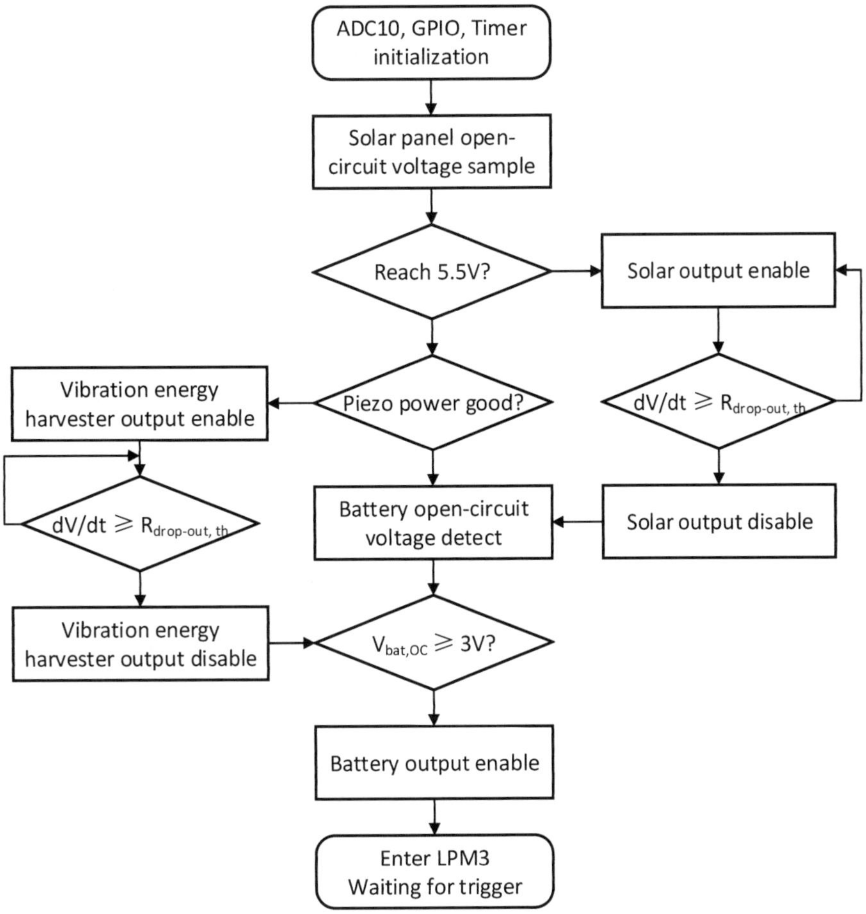

Figure 3. crocontroller working flow diagram

State of the vibration energy harvester can be obtained from the power good ports of LTC3588. Converting time of triple converting channels is about 12μs and interval period of sampling is set to 10ms, which are sufficient for double converting and data process. Interval of the converting is used to calculated change rate of output voltage which indicates availability of a power source.

Microcontroller detects open-circuit voltage of solar panel first. If the voltage reaches power good threshold, solar panel step-down regulator will be enabled and supply external load with 5V voltage. Meanwhile, output voltage variation is calculated within sampling cycle.If the voltage decreases rapidly and exceeds the threshold, it indicates that the power source is not sufficient for power load, and then solar panel converter will be shut down by setting the enable port to low (nearly ground voltage, 0V). In that case, even if piezo output shows power good state, lithium battery will be enabled as the piezo generater is far weaker than solar panel. On the other hand, when piezo is in resonance state, the energy collected is sufficient. Then the controller will enter low power mode (LPM) after finishing sampling and processing period which consumes several μA current and wait for next period or sudden power failing.

3. Results and discussion

3.1. Internal power rail and system cold start

Before output power rail is activated, microcontroller boots ahead and samples states of the power sources to ensure that the power system is capable to reach stabilization output. This procedure is called cold start during which power can be supplied by any part of the system sources, solar panel, vibration energy harvester or lithium battery. Operating waveform of the 3.6V internal power rail powered by solar panel at 45500Lux illumination intensity is shown in Figure 4(a).Detailed data is listed in Table 2. As shown, higher optical power input leads to a shorter system start time and a higher open-circuit voltage. Operating waveforms of the 3.6V internal power rail powered by vibration energy harvester under acceleration of 1.5g are shown in Figure 4(b). Detailed data is shown in Table 3. Because the output power of vibration energy harvester is much lower than solar panel, start time becomes longer. The power good signal appears to be square-wave pulse since the output power is not fully sufficient for stabilization. When acceleration is over 2g, the power good signal is stable high level, which means the generated power is larger than internal consumption.

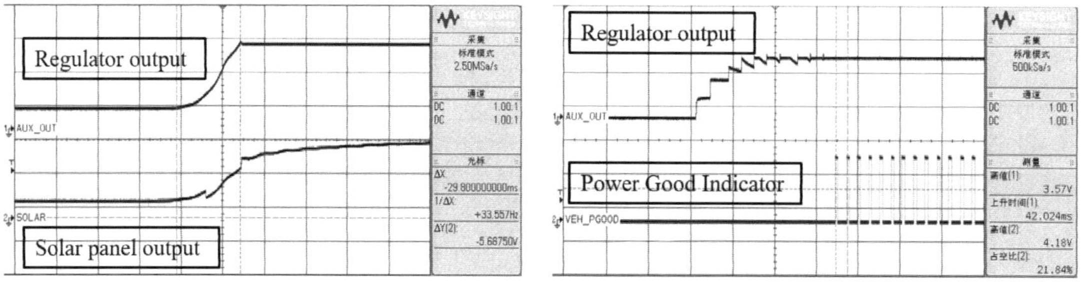

(a) (b)

Figure 4. Oscilloscope pattern of system internal power rail starts: (a) solar panel; (b) energy harvester

Table 2. Response time at different illumination intensity

illumination intensity/Lux	Start time /ms	Open-circuit voltage of solar panel/V
15000	45	5.71
20000	37	5.93
30000	32	6.03
38000	30.5	6.11
45000	29.8	6.22

Table 3. Response time at different acceleration stimulation

Acceleration	Start time /ms	Resonant frequency	PGOOD duty
1.0g	360		0%
1.5g	153	47Hz	21.8%
2.0g	114		100%

3.2. Output power rail and wireless sensor node demonstrate

The output result of primary power rail supplied by solar panel is presented in Figure 5. Efficiency reaches up to over 90% with the output power of 600mW and output current of 150mA. When the illumination intense is around the same as outdoor daytime environment (40000Lux), the maximum output power can reach 620mW which is powerful enough for most wireless sensors with microcontrollers during data transmitting. When vibration accelaretion is 2g, the vibration energy harvester gives a 4.7mW output power and 3.9mW actual output power with system convention efficiency of 83%, which can power up the internal microcontroller or CC2650 active mode without data transmitting. If ambient condition is changed dramaticly, backup battery maintains the output power until either vibration energy or solar energy back online. As shown in Figure 6, hybrid power system implemented in this work can power up Texas Instrument CC2650STK Bluetooth sensor node.

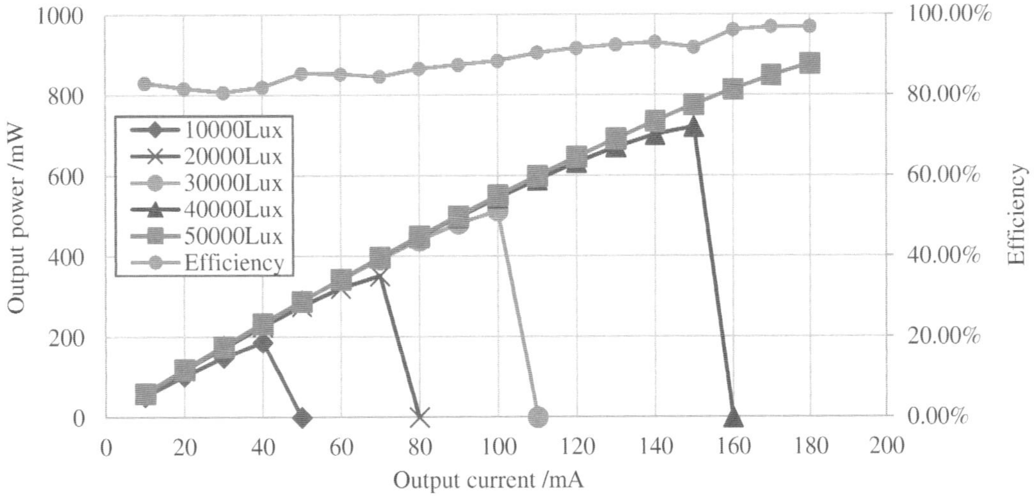

Figure 5. Output power and efficiency of 5V power rail

Figure 6. Demonstration of supplying wireless node with hybrid power system

4. Conclusions

A hybrid power system consists of a solar panel, a vibration energy harvester and a lithium battery is demonstrated in this paper. The charge stored in lithium battery is significantly economized due to the strategy of renewable power sources supplying the load preferentially. Ambient solar energy and vibration energy are converted into electric energy, cached in super-capacity and exported with high quality via the strategy of energy management system. The lithium battery works only as back-up power source to cover circumstance where solar cell and vibration energy harvester cannot gather energy from environment. Experiments show that designed system has a maximum power output of 620 mW, which is sufficient for most wireless sensor node applications. The power of vibration energy harvester can reach 4.7mW, which can be used to supply low-power sensors and system working at standby mode. The life of battery is prolonged with few active time during sensor node working cycle.

References

[1] Shaikh, Faisal K, Zeadally and Sherali 2016 *Renewable and Sustainable Energy Reviews* **55** 1041-1054
[2] Devaraj V, Dinil S, Diju M S, Vineeth S, Jose A, Jyothy M J and Neethu M J 2012 *Proceedings - ISPTS-1, 1st International Symposium on Physics and Technology of Sensors* 269-273
[3] Yanqiu L, Hongyun Y, Bo S and Yonghong S 2008 *IEEE Sensors Journal* **8** 678-681
[4] Akbari S 2014 *2014 Federated Conference on Computer Science and Information Systems* 987-992
[5] Amir E, Moataz A, Jamal A and Gelareh K 2013 *Proceedings of ASEE Northeast Section Conference 2013* 1588-1596
[6] López-Lapeña O, Penella M T and Gasulla M 2015 *IEEE Transactions on Industrial Electronics* **59** 1588-1596
[7] Javanmard N, Vafadar G and Nasiri A 2009 *2009 13th European Conference on Power Electronics and Applications*
[8] Tan Y K and Panda S K 2011 *IEEE Transactions on Power Electronics* **26** 38-50
[9] Lefeuvre E, Audigier D, Richard C and Guyomar D 2007 *IEEE Transactions on Power Electronics* **22** 2018-2025
[10] Dalola S, Ferrari M, Ferrari V, Guizzetti M, Marioli D and Taroni A 2009 *IEEE Transactions on Instrumentation and Measurement* **58** 99-107
[11] Elizabeth K R, Fred B, Romy F and Paul W 2011 *Smart Materials and Structures* **20**
[12] Raghunathan V, Kansal A, Hsu J, Friedman J and Srivastava M 2005 *4th International Symposium on Information Processing in Sensor Networks* **2005** 457-462
[13] Shang Y, Li Y, Yu H and Sun H 2007 *Proceedings of SPIE - The International Society for Optical Engineering* **6724**
[14] Yu H, Li Y, Shang Y and Su B 2008 *3rd IEEE International Conference on Nano/Micro Engineered and Molecular Systems* 196-201
[15] Jiang Z, Dougal and Roger A 2006 *IEEE Transactions on Industrial Electronics* **53** 1094-1104
[16] Dawoud S M, Lin X, Sun J, Mohsin Q K, Flaih F M F, Long P 2015 *Proceedings of 6th International Conference on Intelligent Control and Information Processing* 431-437, Reliability study of hybrid PV-wind power systems to isolated micro-grid
[17] He S and Sproul A B 2010 *Thin Solid Films* **519** 351–356
[18] Michael I F, S Faruque A, Onur B, Sondipon A, Arthur W L and Grzegorz L 2013 *Journal of Intelligent Material Systems and Structures* **23** 1505–1521

IOP Conf. Series: Journal of Physics: Conf. Series **986** (2018) 012026 doi:10.1088/1742-6596/986/1/012026

Research on Dust Concentration Measurement Technique Based on the Theory of Ultrasonic Attenuation

Yan Zhang [1,*], **Wenzhong Lou**[2], **Maohao Liao** [3]

[1,2,3] School of Mechatronical Engineering, Beijing Institute of Technology, Beijing, China

[*]Corresponding author's E-mail: zyatbit@126.com

Abstract. In this paper, a method of characteristics dust concentration is proposed, which based on ultrasonic changes of MEMS piezoelectric ultrasonic transducer. The principle is that the intensity of the ultrasonic will produce attenuation with the propagation medium and propagation distance, the attenuation coefficient is affect by dust concentration. By detecting the changes of ultra acoustic in the dust, the concentration of the dust is calculate by the attenuation-concentration model, and the EACH theory model is based on this principle. The experimental results show that the MEMS piezoelectric ultrasonic transducer can be use for dust concentration of 100-900 g/m^3 detection, the deviation between theory and experiments is smaller than 10.4%.

1. Introduction

In order to comprehensively and accurately detect the dust concentration information in the enclosed area, it is better to grasp the dust concentration situation, prevent the damage of excessive concentration to the operator's body, and the influence on the operation equipment and to forecasting and solve the high concentration of dust. To ensure personal safety and improve the quality of the environment play an important role [1].

At present, a variety of method for measuring dust concentration are studied at home and abroad, such as filter weighing method [2], ray method [3], piezoelectric crystal induction method, optical method [4], capacitance method [5], Charge sensing method [6] and ultrasonic method [7]. The weigh method is the most basic measurement method, but it cannot be use for continuous monitoring of dust concentration. Although the measurement of ray method is accurate, but it needs to compare with the dust sampling measurement, it is difficult to achieve on-line monitoring. For the use of low dust concentration on the ground occasions, it requires that the operator has an experience, on its use [8]. Microwave measurement of dust is in the experimental stage, optical method vulnerable to the impact of measurements [9], and the measurement is easy to make the optical system pollution, the need for frequent calibration, large maintenance; capacitance measurement of dust is safe and reliable, fast responding, etc., but its sensitivity is limited [10].

In recent years, the Los National Laboratory and the Chinese University of Shanghai Su Ming Xu team [11,12] by detected the complex wave acoustic which can be synchronized to obtain the acoustic velocity changes and acoustic attenuation phenomenon. Two kinds of acoustic characteristics were tested and analyzed; you can get more mixture of the data characteristics, to achieve the concentration and particle size distribution at the same time detection. Based on the principle of acoustic attenuation, this paper puts forward to the idea of using MEMS piezoelectric ultrasonic transducer to characterize

IOP Conf. Series: Journal of Physics: Conf. Series **986** (2018) 012026 doi:10.1088/1742-6596/986/1/012026

the measurement of dust concentration, then designed the dust concentration detection device, its structure is simple, the expense is low and the practicability is strong.

2. Ultrasonic method of principle

2.1. Theoretical model of ultrasonic concentration detection

The use of acoustic attenuation for concentration detection is mainly based on the acoustic waves in the media attenuation. The intensity of the acoustic wave decreases with the propagation distance increasing, it is called the attenuation of the acoustic. Modern acoustics think that absorption attenuation and scattering attenuation are the acoustic energy conversion and dissipation in the propagation medium, the attenuation factor contains the characteristic information of the propagation medium. Therefore, the dust concentration can be calculated by measuring the attenuation coefficient inversion. The principle of this method is shown in figure 1:

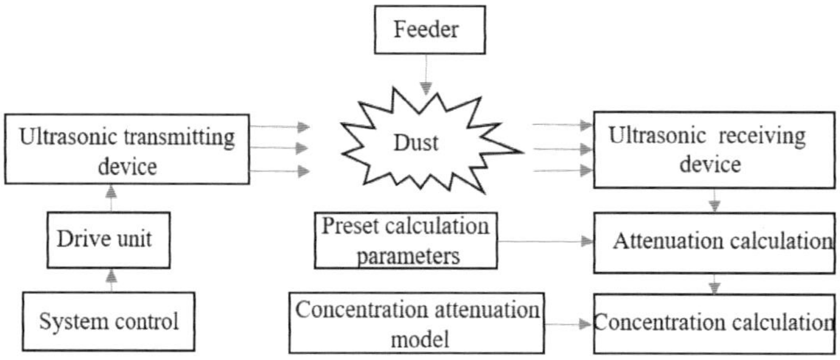

Figure 1. Concentration detection system based on acoustic attenuation

Research manuscripts reported large datasets that are deposited in a publicly available database where the data have been deposited and provided the relevant accession numbers. If the accession numbers have not been obtained at the time of submission, it will provide during review. They must be provide prior to publication.

The numerical relationship between the acoustic attenuation and the concentration of the mixture is the core technology of the acoustic attenuation method. The ECAH model is a classical model, and the mathematical expression of the model is given without derivation.

$$\left(\frac{k}{k_2}\right)^2 = 1 + \frac{3\varphi}{jk_c^3R^3}\sum_{n=0}^{\infty}(2n+1)A_n \tag{1}$$

among them:

$$k = \omega / C_s(\omega) + j\alpha_s(w) \tag{2}$$

k is the complex wave number in the mixed medium, k_2 is the wave number in the continuum, φ represents the volume concentration of the suspended particles, R is the radius of the particles, ω is the ultrasonic angular frequency, C_S and α_S are the acoustic velocity and the attenuation coefficient of the ultrasonic medium in the mixed medium, respectively.

The model has many computational parameters, such as high calculation parameters and high computational complexity, and it is not suitable for real-time calculation. Therefore, it is necessary to establish a simpler solution model. According to the test acoustic wavelength is much larger than the dust particle size, dust particle density is much larger than the air density of the two practical characteristics, ECAH model can be reduced step by step to reduce the difficulty of calculation, reduce the calculation of the task, equation (1) can be reduce to:

$$\left(\frac{k}{k_2}\right)^2 = 1 + \frac{9\varphi}{jk_c^3 R^3} A_1, \tag{3}$$

among them: $A_1 \cong i(k_c R)^3 (1 + T + is)$, substituting equation (3) can obtain:

$$\left(\frac{k}{k_2}\right)^2 = 1 + 9\varphi(1 + T + is) \tag{4}$$

among them: $\delta_s = \sqrt{\frac{2\eta_1}{\rho_1 \omega}}$, $T = \frac{1}{2} + \frac{9\delta_s}{4R}$, $S = \frac{9\delta_s}{4R}\left(1 + \frac{\delta_s}{R}\right)$

The EACH model under long wavelength and large density ratio is simplified, and the calculation parameters are reduced to seven. Compared with the EACH model, the computational complexity is reduced by 29%. The MATLAB is used to calculate the time consuming comparison test under the same conditions, using a simplified algorithm can reduce the calculation time by 24%.

2.2. MEMS piezoelectric ultrasonic transducer

The numerical relationship between the acoustic attenuation and the concentration of the mixture is the core technology of the acoustic attenuation method. The ECAH model is a classical model, and the mathematical expression of the model is given without derivation.

Acoustic field method for concentration detection, the characteristics of ultrasonic transducers on the test results have a decisive role. In this paper, the transducer is optimized from the aspects of piezoelectric material, center frequency, direction angle, ultrasonic radiation surface and acoustic matching layer. The structure as shown in figure 2.

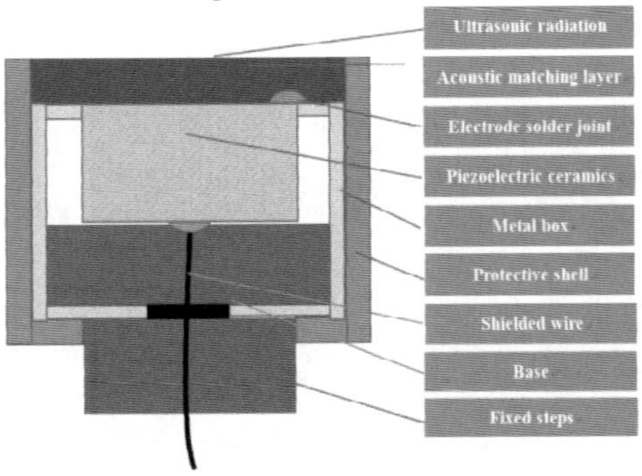

Figure 2. MEMS piezoelectric ultrasonic transducer structure

Acoustic waves in the process of propagation due to the increasing area of acoustic waves caused by the unit to receive area on the energy reduction, the attenuation of acoustic waves is mainly reflected in absorption attenuation and scattering attenuation, the acoustic intensity is inversely proportional to the propagation distance, that satisfies:

$$P = P_0 / x^2 \tag{5}$$

the attenuation intensity is used to characterize the intensity of acoustic waves with the propagation medium and attenuation characteristics. Experimental study has shown that both absorption attenuation and scattering attenuation follow the exponential form and can be express as:

$$\text{Absorption attenuation: } P = P_0 e^{-\alpha_{ist} x} \tag{6}$$

$$\text{Scattering attenuation: } P = P_0 e^{-\alpha_{sca} x} \tag{7}$$

the two forms of attenuation are consistent, and the attenuation coefficients are related to the dielectric properties. Therefore, the two declines are call "medium-dependent attenuation", the corresponding definition expression is:

$$P=P_0e^{-(\alpha_{ist}+\alpha_{sca})x}=p_0e^{-\alpha x} \tag{8}$$

in the equation (8): P_0 is acoustic intensity of the acoustic source, P is observation acoustic intensity, x is observation distance, α is correlation coefficient of the medium. The correlation coefficient of the medium is relate to the change of the acoustic intensity and the observation distance, which is a constant. In the case of observation distance and medium certain circumstances, from the equation 1.8 can obtain: dust concentration and the sensor acoustic intensity is into an exponential relationship, acoustic intensity changes can be converted into voltage signals through the signal processing circuit, and then sent to the computer, dust concentration can be calculated from the voltage measurement.

3. Measuring device design

This section may be divide by subheadings. It should provide a concising and precising description of the experimental results, their interpretation as well as the experimental conclusions is can be draw. Authors should discuss the results and how they can be interpret in perspective of previous studies and of the working hypotheses. The findings and their implications should be discuss in the broadest context possible. Future research directions may be highlight.

In order to improve the accuracy of the measurement, the change of the sensor induce voltage which is as large as possible under the change of the dust concentration. The parameters of the sensor are set as follows: Center frequency, Test interval, Pitch and outer radius. For the test, as shown in figure 3. The working principle of the test device is applied 3.3v drive signal to the sensor, through the continuous jet to form a high concentration of clouds, changing the size of the dust to adjust the size of the dust concentration, the concentration of dust in the MEMS piezoelectric ultrasonic transducer can be changed. For each dust concentration, the nozzle of the high-pressure gas is close when the display value of the measuring device is stabilize, and the measure value of the dust is record by the data processing software.

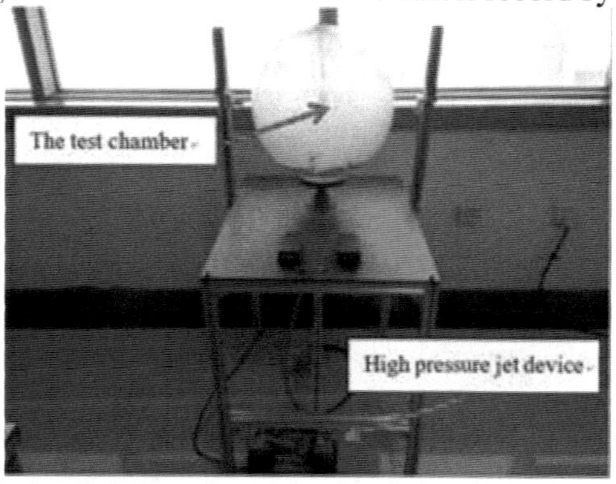

Figure 3. Micro dynamic cloud simulation device

The voltage measurement device uses STM32T8U6 microcontroller as signal processing and control core, and the principle diagram as shown in figure 4. After the MEMS piezoelectric ultrasonic transducer transforms the acoustic intensity signal into the voltage signal, the signal processing unit amplifies and filters the voltage signal, the filtered signal is sent to the single chip microcomputer for A/D conversion and calculation processing, and then by the display device output dust concentration measurement.

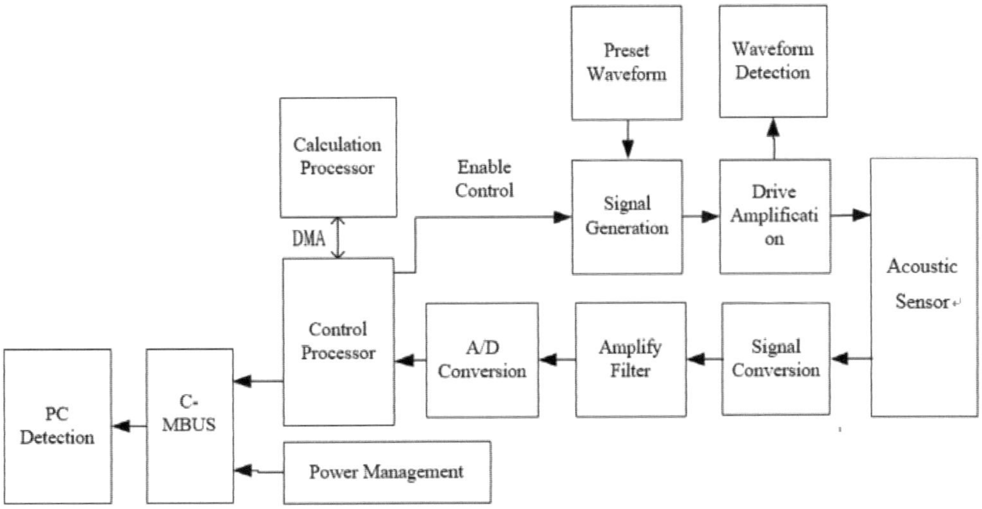

Figure 4. System principle

4. Experiment method and result analysis

4.1. Experiment method

The experiment uses iron powder to simulate dust, talc powder is the main component of Fe_2O_3, it is reddish brown, fine micron grade powder, as shown in figure 5. The experiment uses a small cloud concentration simulation device to generate dynamic clouds. The maximum average concentration of single injection is 900g/m^3, which can form high concentration cloud by continuous injection. By changing the mass of the dust to adjust the concentration, you can change the density of the MEMS piezoelectric ultrasonic transducer. For each dust concentration, the nozzle of the high-pressure gas is close when the display value of the measuring device is stabilize, and the measured value of the dust is recorded using the data processing software.

(a) Micron copper powder sample

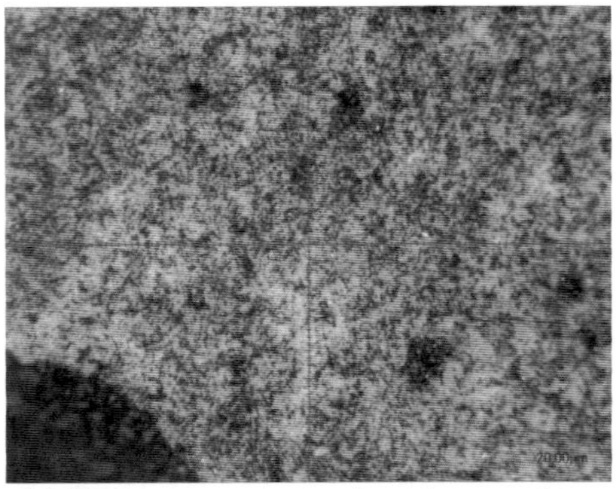

(b) Micron copper powder micrograph

Figure 5. Micron copper powder sample and micrograph

4.2. Experiment results

Ultrasonic transducer is used to test, adjusting the weight of dust, through the dust concentration, the actual value and the measured value of the curve shown in figure 6: Dust concentration between changes, the two curves are consistent, with consistency. Table 1 for the specific experiment data, the maximum measurement error is 10.4%, the minimum is 2.56%, the average error of 7.14%.

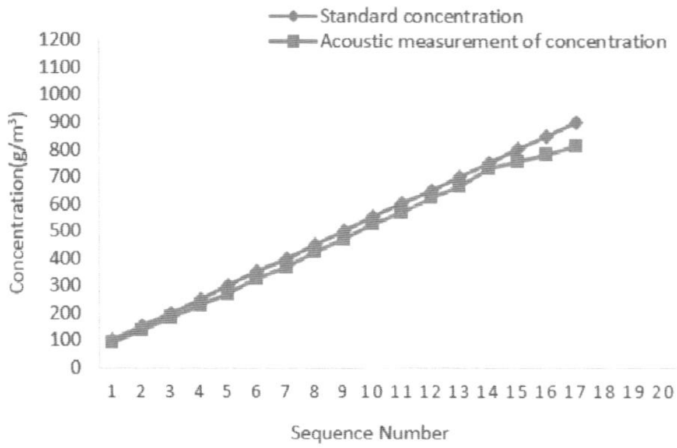

Figure 6. Experimental results

Table 1: Dust concentration test data.

	Dust concentration (g/m^3)								
Dust concentration (g/m^3)	900	850	800	750	700	650	600	550	500
	817	789	752	737	641	614	578	531	459
	826	805	759	729	673	626	559	529	471
Measurement (g/m^3)	809	781	761	741	661	631	567	513	462
	794	764	770	726	668	629	572	528	468
	821	775	735	722	677	613	569	522	475
Average (g/m^3)	813.4	782.8	755.4	730.8	664	622.6	569	524.6	467
Measurement error (%)	9.62	7.91	5.58	2.56	5.1	4.2	5.17	4.62	6.6
	Dust concentration (g/m^3)								
Dust concentration (g/m^3)	450	400	350	300	250	200	150	100	
Measurement (g/m^3)	429	379	324	281	219	182	137	88	
	421	371	319	261	228	186	141	93	
	414	364	331	273	231	169	125	91	
	426	361	328	279	222	187	133	87	
	430	353	335	258	235	179	139	89	
Average (g/m^3)	424	365.6	327.4	270.4	227	180.6	135	89.6	
Measurement error (%)	5.78	8.6	6.46	9.87	9.2	9.7	10	10.4	

MEMS piezoelectric ultrasonic transducer is used to test, adjusting the weight of dust, through the experimental device to detect the corresponding dust concentration, the actual value and the measured value of the curve are shown in figure 6: Dust concentration between changes, the two curves are consistent. Table 1 for the specific experiment data, the maximum measurement error is 10.4%, the minimum is 2.56%, the average error is 7.14%.

According to the experimental results can be conclude as follows:

1. When the dust concentration varies from 100 to 900g/m^3, the measured value of dust concentration is close to the actual value, the average deviation is 7.14%, and linearity is good and can be used for dust concentration detection.
2. When the dust concentration is greater than 750g/m^3, the measured curve tends to be gentle, because the dust concentration is high, due to electrostatic adsorption, pressure pump pressure

and other restriction, will be precipitate in the experiment device. Therefore, the ultrasonic change no longer obvious, and the measurement device shows that the value change is no obvious.

3. Dust measurement range and the actual engineering application has a certain gap, this has a greater relationship with the test platform itself, because the dust flow rate through the pipeline is not accurate, resulting in increasing bias.

5. Conclusion

The basic principle of ultrasonic measurement of dust concentration is study, and designed MEMS piezoelectric ultrasonic transducer, dust concentration experiment is carry out using a test platform. The measuring results show that when the dust concentration is between 100 -900 g/m^3, the deviation between theory and experiments is smaller than 10.4%. The results show the consistency of theoretical analysis and experiment test. It can meet the accurate detection of dust concentration by the small cloud concentration detection system. The detection system can be use for industrial dust concentration detection, industrial dust early warning, and other high concentration of workplaces dust concentration detection, and so on.

References

[1] Ziliang Wang, Development and Application of Dust Concentration Sensor [J]. *Industrial Safety and Environmental Protection,* 2006, 32 (4): 24-27.

[2] Aiqing Guo, Status and thinking of dust monitoring technology in China [J]. *Chinese Journal of Safety Science and Technology,* 2001, 21 (5): 35-38.

[3] Zonglun Li, etc. Application of β-ray Dust Measuring Instrument in Coal Mine Dust Concentration Monitoring [J]. *China Mining Industry,* 2010, 36 (3): 107-109.

[4] Litchford R J, Sun F, Few J D, et al. Optical measurement of gas turbine engine soot particle effluents [J]. *Journal of Engineering for Gas Turbines & Power*, 1998, 120(1):69-76.

[5] Geng Lu, Method for detecting two-phase flow phase concentration based on capacitance measurement and PCA method [J]. *Journal of China Jiliang University*, 2003, 14 (1): 15-18.

[6] Sibiao Zhao, etc. Study on Dust Concentration Measurement Based on External Charge Principle [J]. *Instrument Technique and Sensor*, 2010 (6): 15-16.

[7] Ni W, Ultrasonic attenuation model for measuring particle size and inverse calculation of particle size distribution in mineral slurries [J]. *Journal of Central South University (English Edition)*, 2006, 13 (4): 445-450.

[8] Weidong Li, etc. China's coal industry dust concentration monitoring technology status and development trends [J]. *Mining Safety and Environmental Protection*, 2005, 32 (s1): 66-67.

[9] Kerker M. The Light Scattering of Light and Other Electromagnetic Radiation [J]. 1969.

[10] Chun M H, Sung C K. Parametric effects on the void fraction measurement by capacitance transducers [J]. *International Journal of Multiphase Flow*, 1986, 12(4):627-640.

[11] Mingxu Su, etc. Measurement of Particle Size by Ultrasonic Attenuation [J] .*Journal of Instrumentation*, 2004, 25 (s1): 1-2.

[12] Mingxu Su, etc. Determination of Particle Size and Concentration in Suspension by Ultrasonic Attenuation [J]. *Acta acoustica Sinica*, 2004 (5): 440-444.

Printing of highly conductive solution by alternating current electrohydrodynamic direct-write

Jiaxin Jiang [1,2,3], Gaofeng Zheng [1,2,3,*], Xiang Wang [4], Jianyi Zheng [1,2,3], Juan Liu [1,2,3], Yifang Liu [1,2,3], Wenwang Li [4] and Shumin Guo [5,6]

[1] Department of Instrumental and Electrical Engineering, Xiamen University, Xiamen 361005, China
[2] Xiamen Key Laboratory of Optoelectronic Transducer Technology, Xiamen 361005, China
[3] Fujian Key Laboratory of Universities and Colleges for Transducer Technology, Xiamen 361005, China
[4] School of Mechanical and Automotive Engineering, Xiamen University of Technology, Xiamen 361024, China
[5] School of Mathematical Sciences, Xiamen University, Xiamen 361005, China
[6] Fujian Provincial Key Laboratory of Mathematical Modeling and High Performance Scientific Computing, Xiamen University, Xiamen 361005, China

E-mail: zheng_gf@xmu.edu.cn

Abstract. Electrohydrodynamic Direct-Write (EDW) is a novel technology for the printing of micro/nano structures. In this paper, Alternating Current (AC) electrical field was introduced to improve the ejection stability of jet with highly conductive solution. By alternating the electrical field, the polarity of free charges on the surface of jet was changed and the average density of charge, as well as the repulsive force, was reduced to stabilize the jet. When the frequency of AC electrical field increased, the EDW process became more stable and the shape of deposited droplets became more regular. The diameter of printed droplets decreased and the deposition frequency increased with the increase of voltage frequency. The phenomenon of corona discharge was overcome effectively as well. To further evaluate the performance of AC EDW for highly conductive solution, more NaCl was added to the solution and the conductivity was increased to 2810μs/cm. With such high conductivity, the problem of serious corona discharge could still be prevented by AC EDW, and the diameter of printed droplets decreased significantly. This work provides an effective way to accelerate industrial applications of EDW.

1. Introduction

As a typical manufacture technology of flexible electronics, electrohydrodynamic inkjet technology has been the most promising technique for the advantages of simple process, low cost and no pollution, which has attracted interests of researchers in many fields such as semiconductor transistors, organic light-emitting diodes, micro-/nano-sensing and bio-medical [1-3]. How to achieve the precision deposition of micro/nano structures has been the research hotspot in recent years [4-5].

Electrohydrodynamic Direct-Write (EDW) is a novel and simple inkjet technology in which a stable and straight jet is driven by electrical field to fabricate orderly micro/nano structure with a shortened distance between nozzle and collector and a reduced voltage, which is helpful for decreasing the size of printed structures and increasing the material compatibility to promote its applications [6-8]. However, there are still some problems that hinder the industrial applications of EDW technique. A critical one is the ejection of highly conductive solution. In conventional EDW under DC electrical field, solution with high conductivity usually results in droplet breaking or corona discharge during the ejecting process. For highly conductive solution, the large density of free charges on the jet and the repulsive force will block the stable ejection of a single jet. The highly conductive jet discharges easily, thus it is difficult to direct-write orderly micro/nano structures [9]. The stable ejection of jet with highly conductive solution has been the key to promoting the applications of EDW.

The stable jet with highly conductive solution is the research hotspot for EDW technology, which has attracted wide interests and attentions. Yu [10] et al printed PANI with highly conductive solution on patterned architectures to control the deposition accuracy of nanofiber. Wang [11] et al fabricated conductive silver patterns on demand by pulsed electrohydrodynamic printing and prepared lots of micro/nano devices such as resistor, capacitor and inductor. Watanabe [12] et al achieved aligned nanofiber deposition with two parallel electrodes and discussed the relationship between solution conductivity and polymer structure. In recent years, AC electrical field was introduced to reduce the free charges on the jet and the repulsive force during EDW process effectively, which was beneficial to achieve stable motion of charged jet [13-15]. The rheology and deposition behaviors of jet under AC electrical field are significant to applications of EDW in the integration manufacturing of micro/nano system.

In this paper, an AC electrohydrodynamic direct-write method was proposed, where an AC electrical field, instead of a DC electrical field, was generated to print highly conductive solution. The influences of process parameters were studied.

2. Materials and Methods

The schematic diagram of EDW under AC electrical field was shown in figure 1, including an AC power source, a precise syringe pump, a stainless steel nozzle, a motion platform and a computer. The AC power source (HVP-402NP1, China) was used to generate a high electrical field between the nozzle and the collector. The stainless steel nozzle with an inner diameter of 0.26mm was used as the spinneret. The precise syringe pump (Harvard 11 Pico Plus, USA) was used to supply polymer solution to the spinneret. A silicon wafer, which was cleaned in acetone and deionized water before the experiment, was placed on the motion platform (Parker Linear Motor, USA) to act as the collector. The anode of the power source was connected to the nozzle, and the cathode was connected to the grounded collector. The trajectory and velocity of the motion platform was controlled by a computer. The ejection behavior of the charged jet was captured by a CCD camera (Sony SSC-DC80, Japan) and the morphology of the micro/nano structure was recorded by the optical microscope (Mitutoyo, Japan).

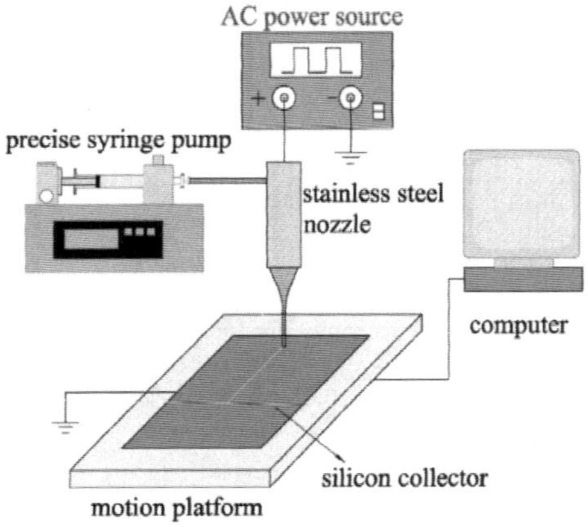

Figure 1. Schematic diagram of EDW under AC electrical field.

Polyethylene Oxide (PEO, *M*w=300kg/mol) solution with a concentration of 3wt% was used as the material of the experiments. PEO powder was dissolved in the mixture of deionized water and ethanol (*v*:*v*=2:1) for the preparation of PEO solution. Then, NaCl was added to adjust the conductivity, which increased the amount of free charges on the jet.

3. Materials and Methods

The ejection behavior of charged jet with highly conductive solution under DC electrical field was first investigated. For solution of high conductivity, the amount of free charges on the jet increased and the radius of jet decreased. The stretching length of the Tayler Cone became larger and the distance between the tip of the Taylor Cone and the collector was shorter, which brought more difficulties for the inducing of single jet with highly conductive solution under DC electrical field, as shown in figure 2. Under DC electrical field, because of the large repulsive force during the ejection process, the jet would have the problems such as whipping, shrinking and breaking.

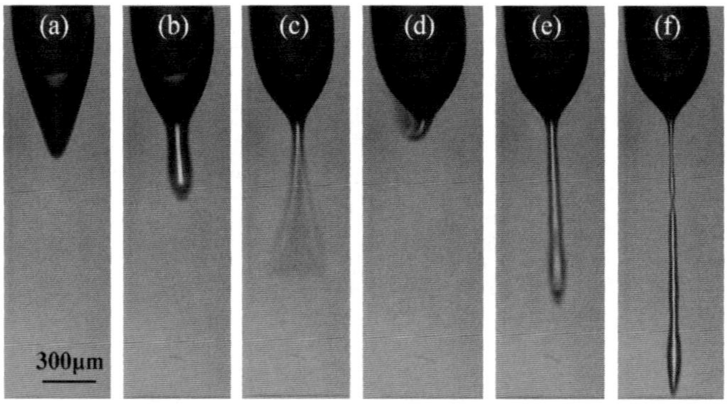

Figure 2. Ejection behavior of charged jet with highly conductive solution under DC electrical field.

However, by utilizing the AC electrical field, the polarity of free charges on the surface of the jet was changed by the alternation of electrical field, and the average density of charge, as well as the repulsive force, was reduced, which contributed to improving the stability of charged jet with highly conductive solution. In our work, a series of EDW experiments with solution of different conductivities were carried out, and the influence of process parameters on the printed patterns was investigated.

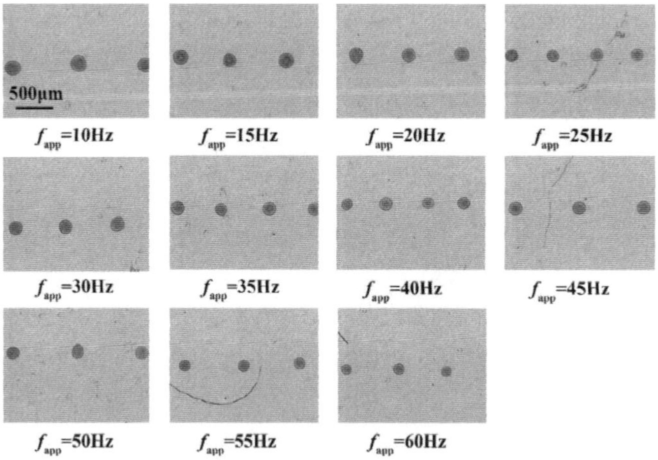

Figure 3. Printed patterns under various frequencies with a solution conductivity of 146.5μs/cm.

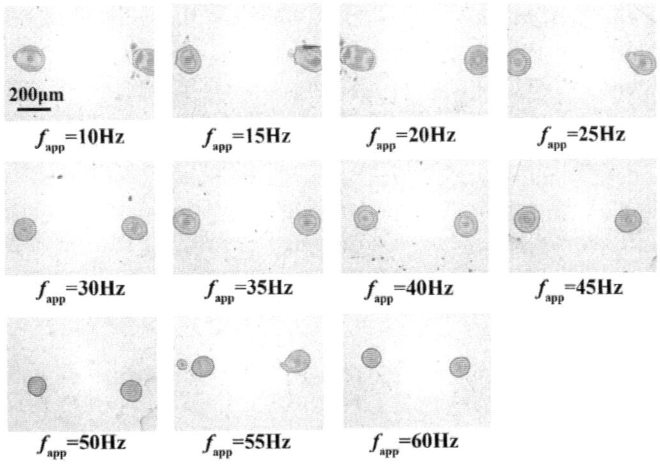

Figure 4. Printed patterns under various frequencies with a solution conductivity of 770μs/cm.

Because of high conductivity, electrostatic discharge occurred when the distance between the nozzle and the collector was short. So the distance was fixed to be 3mm in our experiments. The positive voltage was 2600V, while the negative voltage was -1000V. The duty ratio was 50%, the solution supply rate was 50μl/hr, and the velocity of collector was 10mm/s. Figure 3 and figure 4 showed the printed patterns under various voltage frequencies when the conductivity of solution were 146.5μs/cm and 770μs/cm, respectively. When the voltage frequency of AC electrical field increased, the EDW process became more stable and the shape of deposited droplets became more regular. Experimental results showed that the diameter of printed droplets decreased with the increase of voltage frequency.

The effect of voltage frequency on the deposition frequency of droplets was further discussed, as shown in figure 5(a). When the voltage frequency increased, the deposition frequency had a trend of increasing. When the conductivity was 146.5μs/cm, the deposition frequency increased by leaps and bounds with the increase of voltage frequency. When the conductivity was 770μs/cm, however, the deposition frequency increased at a lower speed. The diameter of printed droplets was investigated as well, as shown in figure 5(b). The deposition diameter decreased with the increase of voltage frequency. With the increase of solution conductivity, the deposition of droplets became more unstable. Therefore, the diameter of printed droplets for the conductivity of 770μs/cm was larger than the diameter for the conductivity of 146.5μs/cm.

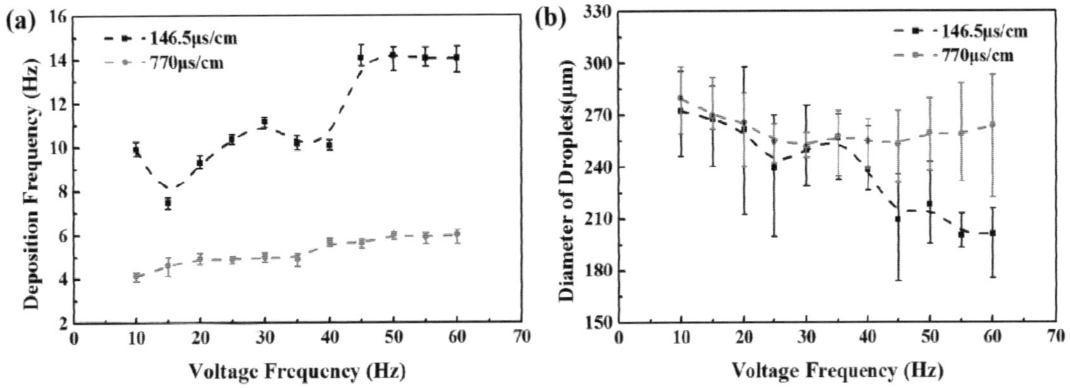

Figure 5. Effect of the voltage frequency on the diameter of printed patterns and the deposition frequency.

To further examine the performance of AC EDW with solution of high conductivity, more NaCl was added to the solution and the conductivity was increased to 2810μs/cm. With such high conductivity, there would be serious corona discharge in DC EDW. This problem could be well overcome under the AC electrical field. The printed patterns under various voltage frequencies were shown in figure 6. When the conductivity was 2810μs/cm, the diameter of printed droplets decreased and the deposition frequency increased significantly with the increase of voltage frequency, as shown in figure 7. Results showed that when the conductivity of solution was beyond a certain degree, the rheology and deposition behavior of jet was different. Because of the high density of charges, the droplets were broken up during the deposition process, thereby the deposition frequency increased and diameter of printed patterns decreased significantly compared with the results of solution with lower conductivity.

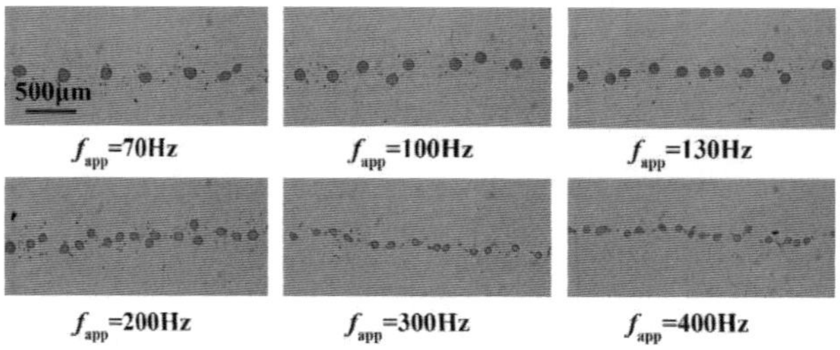

Figure 6. Printed patterns under various frequencies with a solution conductivity of 2810μs/cm.

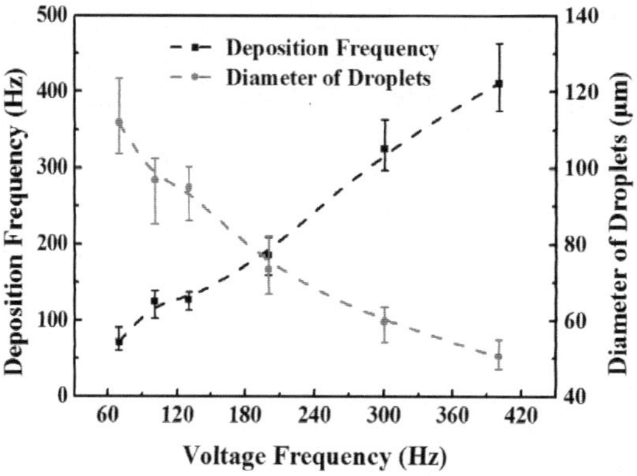

Figure 7. Effect of the voltage frequency on the diameter of printed patterns and the deposition frequency when the conductivity of solution was 2810μs/cm.

4. Conclusions

AC electrical field was introduced into EDW to improve the stability of jet ejection. Under DC electrical field, due to the large repulsive force, it is difficult to induce jet ejection with highly conductive solution. However, with the introduction of AC electrical field, the free charges on the jet and the repulsive force were reduced effectively, which promoted the improvement of the stability of charged jet. When the frequency of AC electrical field increased, the printed patterns became stable and the shape of deposited droplets became regular. The diameter of printed droplets decreased and the deposition frequency increased with the increase of voltage frequency. With a solution conductivity of 2810μs/cm, the deposition frequency increased and the diameter of printed patterns decreased significantly. This work provides an effective way to improve the stability of jet ejection especially with highly conductive solution, which is available to push the applications of electrohydrodynamic inkjet technology in the fields of micro/nano system integration.

Acknowledgments
This research was financially supported by the National Natural Science Foundation of China (No. 51575464, U1505243), Xiamen Municipal Science and Technology Projects (No. 3502Z20163005), Fundamental Research Funds for the Central Universities (No. 20720160086), Education and Scientific Research Projects for Middle-aged and Young Teachers of Fujian Province of China (No. JAT160002).

References
[1] Kim K, Kim G, Lee B R, Ji S, Kim S Y, An B W, Song M H and Park J U 2015 High-resolution electrohydrodynamic jet printing of small-molecule organic light-emitting diodes *Nanoscale* **7** 13410
[2] Laurila M M, Khorramdel B and Mäntysalo M 2017 Combination of E-Jet and Inkjet Printing for Additive Fabrication of Multilayer High-Density RDL of Silicon Interposer *IEEE Trans.* Electron Devices 64 1217-24
[3] Kim S Y, et al. 2016 High-resolution electrohydrodynamic inkjet printing of stretchable metal oxide semiconductor transistors with high performance *Nanoscale* **8** 17113-21
[4] Hashimdeen S H, Miodownik M and Edirisinghe M J 2013 Print head design and control for electrohydrodynamic printing of silk fibroin *Mater Sci Eng C Mater Biol Appl* **33** 3309-18
[5] Huang Y, Bu N, Duan Y, Pan Y, Liu H, Yin Z and Xiong Y 2013 Electrohydrodynamic direct-writing *Nanoscale* **5** 12007-17
[6] Sun D, Chang C, Li S and Lin L 2006 Near-Field Electrospinning *Nano Lett.* **6** 839-42
[7] Fang F, Chen X, Du Z, Zhu Z, Chen X, Wang H and Wu P 2015 Controllable Direct-Writing of Serpentine Micro/Nano Structures via Low Voltage Electrospinning *Polymers* **7** 1577-86
[8] Zheng G, Sun L, Wang X, Wei J, Xu L, Liu Y, Zheng J and Liu J 2016 Electrohydrodynamic direct-writing microfiber patterns under stretching *Appl. Phys. A* **122** 112
[9] Datta R S, et al. 2015 Ionic liquid entrapment by an electrospun polymer nanofiber matrix as a high conductivity polymer electrolyte *RSC Adv.* **5** 48217-23
[10] Yu G, et al. 2016 Patterned, highly stretchable and conductive nanofibrous PANI/PVDF strain sensors based on electrospinning and in situ polymerization *Nanoscale* **8** 2944-50
[11] Wang X, Xu L, Zheng G, Cheng W and Sun D 2012 Pulsed electrohydrodynamic printing of conductive silver patterns on demand *Sci. China Technol. Sc.* **55** 1603-7
[12] Watanabe T, Tanaka M and Kawakami H 2016 Fabrication and electrolyte characterization of uniaxially-aligned anion conductive polymer nanofibers *Nanoscale* **8** 19614-9
[13] Ghashghaie S, Bazargan A M, Ganji M E, Marzbanrad E, Zamani C, Raissi B, Keyanpour-Rad M and Bahrevar M A 2011 An investigation on the behavior of electrospun ZnO nanofibers under the application of low frequency AC electric fields *J. Mater. Sci.-Mater. El.* **22** 1303-7
[14] Liu J, Lin Y, Jiang J, Liu H, Zhao Y and Zheng G 2016 Bead-on-string structure printed by electrohydrodynamic jet under alternating current electric field *Appl. Phys. A* **122** 809
[15] Wei C, Qin H, Chiu C P, Lee Y S and Dong J 2014 Drop-on-demand E-jet printing of continuous interconnects with AC-pulse modulation on highly insulating substrates *Journal of Manufacturing Systems* **37** 505-10

Cu Pillar Low Temperature Bonding and Interconnection Technology of for 3D RF Microsystem

G X Shi [1,2,3], K Q Qian [1], M Huang [1,2], Y W Yu[1,2], J Zhu [1,2]

[1]Nanjing Electronic Devices Institute, Nanjing 210016 China
[2]Science and Technology on Monolithic Integrated Circuits and Modules Laboratory Nanjing 210016 China
[3]E-mail:xiong525625@163.com

Abstract. In this paper 3D interconnects technologies used Cu pillars are discussed with respect to RF microsystem. While 2.5D Si interposer and 3D packaging seem to rely to cu pillars for the coming years, RF microsystem used the heterogeneous chip such as GaAs integration with Si interposers should be at low temperature. The pillars were constituted by Cu (2 micron) -Ni (2 micron) -Cu (3 micron) -Sn (1 micron) multilayer metal and total height is 8 micron on the front-side of the wafer by using electroplating. The wafer backside Cu pillar is obtained by temporary bonding, thinning and silicon surface etching. The RF interposers are stacked by Cu-Sn eutectic bonding at 260 ℃. Analyzed the reliability of different pillar bonding structure.
Keywords: Three dimension integration, Temporary bonding, Interposer.

1. Introduction

Today, the large array of vertical 3D interconnections technologies has been developed on silicon interposers to the IC chips together formed complex system. But interconnects seem to be even more diverse and complex than TSV. In 1970s the flip-chip technology has already be used for chip stacking[1], For today's applications and several supply chains are coexisting and competing to each other. The initial main driver was to increase the interconnect density to follow the integration trend. But factors like dimensions constraints, process compatibility, reliability or costs have led to the development of different technologies[2], solder Cu pillars is a major technology for many applications, it seems that Cu pillars could meet the specifications of all emerging 3D RF microsystem. We propose in this paper to discuss Cu pillars interconnects and solder bonding for 3D RF microsystem.

2. Materials and Methods

In this wafer to wafer stacking, the RF application's bandwidth limitation is directly dependent on interconnects. Copper pillars capped with tin-based solder with a pitch of 20μ m is mass production (Figure. 1). But with the coming active interposers and 3D heterogeneous stacking like GaAs packaging, the interconnect requested pillars will rapidly to shrink down to 20μ m in production this year. Strong efforts are made to reduce even more the pitch of cu pillars, but many issues remain to be solved.

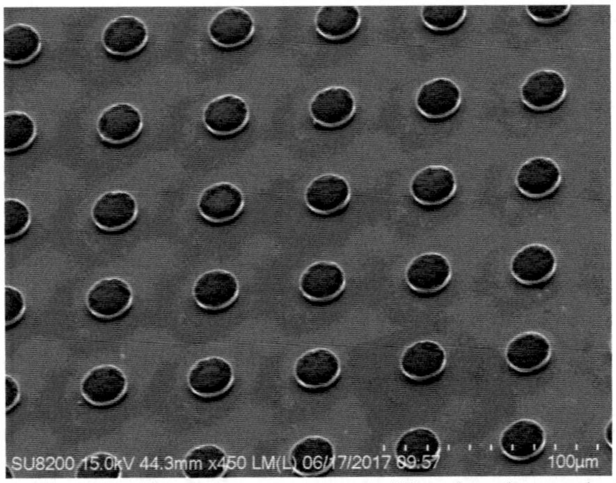

Figure 1 . Front side 20μm copper pillars by electroplating

The benefit of Cu– Sn SLID bonding for interconnects and hermetic sealing, and especially for 3D RF microsystem, is quite apparent. With interconnects or bond lines stable for large temperature variations, subsequent bonding steps, processing at elevated temperatures, or thermal annealing, for instance to activate getter materials, will not compromise the initial bonds and integrity of the package. Figure 2 shows Cu–Sn SLID bonding where Sn is placed between two Cu pillars, a pressure is applied, and the increased temperature will first cause a diffusion process between Sn and Cu to form the η -phase Cu6Sn5. Then, given enough time, the final intermetallic compound will be the stable ε -phase Cu3Sn. Since the metal layers used typically 8μ m, electroplating is used to deposit both Cu and Sn, as this offers the most economical processing[3].

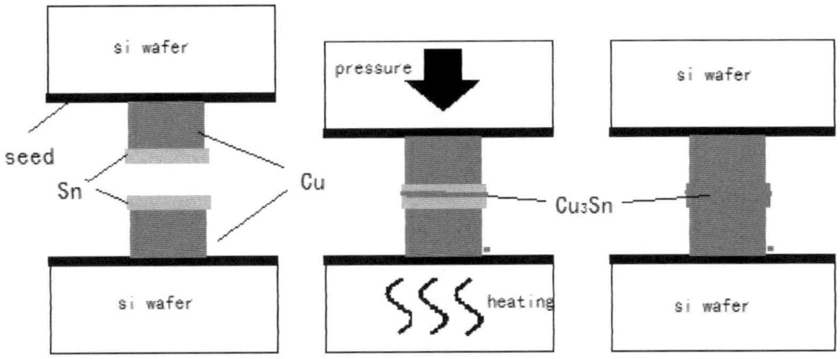

Figure 2. Cu– Sn SLID bonding

The minimum Cu/Sn thickness ratio for obtaining a steady - state Cu3Sn SLID bond, meaning that the diffusion process has come to completion, can be calculated from Eq.1 and the properties given in Table 1. Thus, the required layer thickness ratio for Cu to Sn is ⩾ 1.3. This means that for a Sn layer of 2.5 μ m, a Cu layer of a minimum thickness of 3.25 μ m is required. This amount of Cu could theoretically be the sum of the amount of Cu on both sides of the joint; however, the diffusion process may not be fully symmetric due to different temperatures of the top and bottom wafer, and because inter-diffusion will occur faster where the layers are deposited on top of each other than at the bonding interface. Thus, to ensure a sufficient amount of Cu is present in the joint, if one side as a source should be somewhat inhibited due to oxides or similar, a Cu thickness of 5μ m on each bonding partner, with a total thickness of 3μ m for Sn, Fig3 show the 8 μ m height pillars has been found to work well.

$$\frac{h_A}{h_B} = \frac{v_A}{v_B} = \frac{x}{y}\frac{M_A}{\rho_A}\frac{\rho_B}{M_B} \qquad (1)$$

Where h_A and h_B denote the layer thicknesses of the pure metals A and B, V their respective volumes, M their respective molar masses, and ρ their respective densities[4].

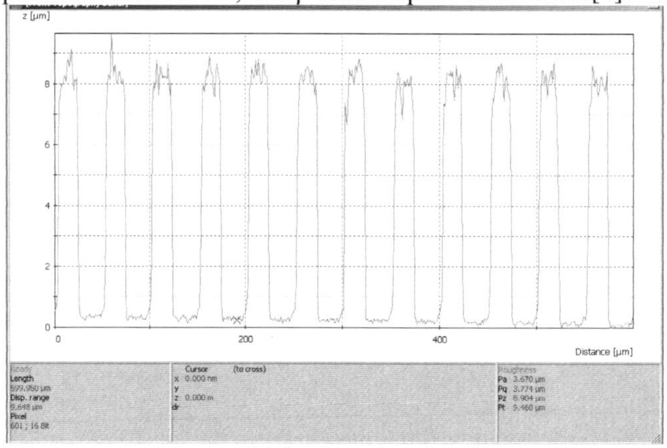

Figure 3. The 8µm height pillars

Table 1. Relevant material properties for Cu, Sn, and the corresponding intermetallic phases present in SLID bonding[4]

material	*M*(g)	*ρ* (gcm^{-3})	*M/ ρ*	*E* (GPa)	R(µΩcm)	*Thermal Expansion (10^{-6})*	*Melting Temperatur e(℃)*
Cu	63.55	8.92	7.12	110	1.7	17	1084
Sn	118.71	7.28	16.31	41	11.5	23	232
Cu_3Sn	309.35	8.9	34.76	133	8.93	19	676
Cu_6Sn_5	974.83	8.27	117.86	117	17.5	16.3	415

3. Bonding Processes

As an alternative to using a continuous bond force, a more detailed bonding profile can take advantage of the ductile Sn layers to conform better to each other, given any nonuniformity across the wafer, compared to the significantly harder Cu surface, and the inter-diffusion process between Cu and Sn would occur with a slow temperature ramp past the melting point of Sn. Combining this with reducing the bond force at 240°C, a little above the melting temperature for Sn (232°C) [5]to account for any thermal gradients between the wafers, the reduction in bonding pressure can reduce squeeze - out of any molten Sn. From bonding data profile where a greater temperature ramp rate to 150°C (7°Cmin^{-1}) is followed by a smaller temperature ramp rate (3°Cmin^{-1}) until reaching the melting temperature of Sn and the final soak temperature. This was done so as to reduce the amount of Sn left at the interface when reaching 240°C, and to allow for the diffusion process to occur in the bond line early in the process. The wafers were brought into contact with a bond force of 7 kN (20 MPa) after the temperature had reached 150°C. Following making of contact, the force was either reduced to 1 kN (3 MPa) at a temperature right above the melting point of Sn (240 °C)[6], or maintained high throughout the soak time. Using this two - step temperature profile, combined with a soak time of at 10 min at 260 °C, the bond line will be fully converted to IMC, that all available Sn has been converted to Cu_3Sn (versus stoichiometric 25 at% Sn). The bonding result show in figure 4 .

Figure 4. Pillar bonding and wafers stacking results.

The key requirements for successful SLID bonding is to remove, or convert, the oxides that are formed on the bonding surface. Both Cu and Sn oxidize, but Cu oxidation is known to be faster than Sn oxidation. Thin SnO layers can be tolerated and incorporated into the molten Sn when bonding. Therefore, the most critical oxide to remove is that of Cu prior to, and during, bonding, which can be accomplished by formic acid vapor usually requires temperatures above 100°C to be effective[7]. Note that both etching and formic acid vapor treatment are followed by a rinse step in deionized water to remove any residual compounds. To avoid a rinse step, a plasma process may be used; however, one must ensure that the temperature is kept low so as not to unintentionally melt the Sn layer. A high bonding yield is also ensured by bonding in an inert atmosphere, use of a reactive gas, or bonding in vacuum, to avoid any re-oxidation of the metal surfaces at elevated temperatures.

4. Conclusions

Since the first demonstrations of Cu–Sn SLID bonding as an excellent interconnection method for 3D RF microsystem applications, it has over recent years been demonstrated for wafer-level bonding with TSVs together with hermetic sealing and packaging at the wafer-level.

References

[1] Munding, A. et al. Cu/Sn solid–liquid inter-diffusion bonding, in Wafer Level 3D-ICs Process Technology, 2008 *New York. Springer Science*, 131–170.
[2] Iino,Y. Partial transient liquid phase metals layer technique of ceramic–metal bonding. 1991 *J. Mater. Sci. Lett. New York.*10 p104 –106.
[3] Duvall, D.S. et al. TLP bonding: a new method for joining heat resistant alloys. 1974 *Weld. J.* 53 203 –214.
[4] Peter Ramme 2012 *Handbook of Wafer Bonding* (New York: WILEY-VCH)
[5] Reza Ghodssi 2011 *MEMS Materials and Processes Handbook* (New York: Springer)
[6] Rao R. Tummala 2008 *Introduction to System-on-package* (New York: Mc Graw Hill)
[7] John H. Lau 2010*Advanced MEMS Packaging* (New York: WILEY-VCH)

Kinematics and Mechanics analysis of trap-jaw ant Odontomachus monticola

Wenteng Hao, Guang Yao, Xiangyu Zhang and Deyuan Zhang *

School of Mechanical Engineering and Automation, Beihang University, Xueyuan Road No.37, Haidian District, Beijing 100191, P.R. China

*Correspondence: zhangdy@buaa.edu.cn

Abstract. Trap-jaw ants of the genus Odontomachus exhibit spectacularly rapid predatory and fugitive strikes. In order to reveal the extraordinary impact resistance of the apical teeth material, we analyzed the kinematics and mechanics of the closing mandibles. Odontomachus monticola is an Odontomachus species and extensive in China. We video-recorded jaw-strikes to measure the closing velocity and acceleration. The experimental results showed that O. monticola's mandibles closed at a highest velocity of 35.42 m/s and a highest acceleration of 750,000 m/s^2 within an average duration of 0.16 ms. In addition, in order to measure the strike force, we developed an extraordinary measuring method with poly (vinylidene fluoride) (PVDF) piezoelectric film. First, the dynamic calibration of the PVDF piezoelectric film was conducted, then the calibrated piezoelectric film was struck by O.monticola. Finally, the mandible strike force was calculated according to the calibration result and the output signal. The measurements results demonstrated that the strike force ranges from 102.2 N to 235.2 N, which is impressive contrast with O.monticola's body weight.

1. Introduction

Arthropods are celebrated for their morphological variation, and many species have developed extraordinary characteristics for a diversity of functions such as the exceptional jumping ability of fleas [1], the extreme rapid takeoffs of locusts and flies [2], and the crushing strikes of stomatopods [3]. These exceptional speeds and accelerations are always accomplished under the help of specific innovations such as latches, lever arms and spring mechanisms which assist in storing and releasing huge amounts of energy [4]. One surprising example is the extremely fast mandible strikes of trap-jaw ants, which have the rapidest speed in the animal kingdom [5].

The morphological and neurobiological characterization of the mandible strikes of trap-jaw ants were carried out by Gronenberg and his groups in a list of papers [6-10], and the species Odontomachus bauri's mandible strikes can achieve impressively high speed of more than 60 m·s^{-1} [5]. In addition to equipping trap-jaw ants with the capability to disable prey, the mandible snaps have been evolutionarily co-opted for ballistic movement. These movements are shown as 'bouncer defense' jumps [11], where the ants are impelled horizontally away from a threat, and 'escape jumps', where the mandibles are placed against or aimed at the substrate then fired, launching the ants into the air [5]. The mandible was modeled as a thin rod of uniform density, which rotates around a fixed axis, by Patek et al [5], whereas actually the interior of mandibles consists mainly of a large tracheal sac filled with air [12]. The mass of mandibles is very light, however, O.monticola's escape jumps surprisingly yielded heights ranging from 6.1 to 8.3 cm and the mean horizontal distance of defense jumps reached 22.3 cm (5.3–39.6cm) [5], which are both much longer than its mean body length of 1.31 cm. Furthermore, there is always obvious sound when the jaw strikes. Therefore, we presumed that the extremely light mandibles generate significantly instantaneous strike force to propel the ants to jump. In addition, the actual strike force is still unexplored, no researchers have measured trap-jaw ants' instantaneous mandible strike force.

Therefore, investigations were firstly conducted on visualizing the complete strike process of Odontomachus monticola by using available imaging technology. And then, the mandible strike kinematics was analyzed simultaneously. In addition, the actual strike force was measured by an extraordinary measuring method with poly (vinylidene fluoride) (PVDF) piezoelectric film. The objectives of our work were as follows: (i) to characterize the kinematics of trap-jaw strikes using high-resolution video images; (ii) to measure the strike force using PVDF piezoelectric film and the force sensor.

2. Materials and Methods

O.monticola workers were collected in Yunnan province in May, 2016. Colonies were maintained in an ecological nest with porous stones and fed a diet of mealworms, every two days. The body mass of O.monticola ranges from 16.6 mg to 22.5 mg (Sartorius BS 224 S Max=220g, d=0.0001g electronic balance). Their mandibles' lengths range from 1.61mm to 1.89mm (OLYMPUS SZX16 stereomicroscope).

2.1. Kinematics analysis

In order to film of trap-jaw strikes, a high-speed camera (50,000 frames per second, 20 μs shutter speed; Olympus I-Speed 3) was used. Ants were fixed using a drop of dental wax (applied to the top of head) to the end of a thin rod that could rotate to keep the jaws perpendicular to the high-speed camera's axis. And the ants can strike freely. Then the ants were stimulated to strike by touching their 'trigger hairs' with a thin metal probe. The kinematic data was used to calculate velocity, acceleration and the lag time (if any) between the first mandible to close and the second.

2.2. Strike force measurement

The instantaneous mandible strike force is difficult to be obtained with common miniature force sensors because of the micro-scale of the mandible and the mandible's apical teeth. Therefore, we developed an extraordinary measuring method to acquire the strike force.

PVDF piezoelectric film is a new type of polymer piezoelectric material. Because of its light weight, thin thickness, high sensitivity, extremely fast dynamic response and wide frequency respond range, it has broad application prospect in the force impact measurement field [13]. PVDF piezoelectric film will come to electrical polarization and produce electric charges when exerted by stress. And the charges are proportional to the mechanical stress. When the mechanical stress is removed, the piezoelectric film will return to electrically neutral [13]. We selected SDT1-028K piezoelectric sensor (Made by MEAS Cooperation, USA), which can be used for impact monitoring. This kind of piezoelectric film (28 mm × 13.2 mm) is only 75 μm thick, and the piezoelectric constant is -33 pc/N. The SDT1 piezoelectric sensor consists of a rectangle silvered piezoelectric film and a concentric cable of the length of 18 inches. And the piezoelectric film is doubled back to form a self-shielded area allowing it to work in high-EMI environment.

First, the dynamic calibration is essential to measure the strike force using the PVDF piezoelectric film. Therefore, a particular and effective dynamic calibrating method was adopted to calibrate the piezoelectric film. Through the observation and analysis of the apical teeth's shape, as shown in figure 1, the apical teeth are made up with three different teeth including two canine teeth (1,2) and one cheek tooth (3). It is obvious that the two canine teeth are on the outside of the mandible, while the cheek tooth is closer to the root of the mandible than the other two teeth.

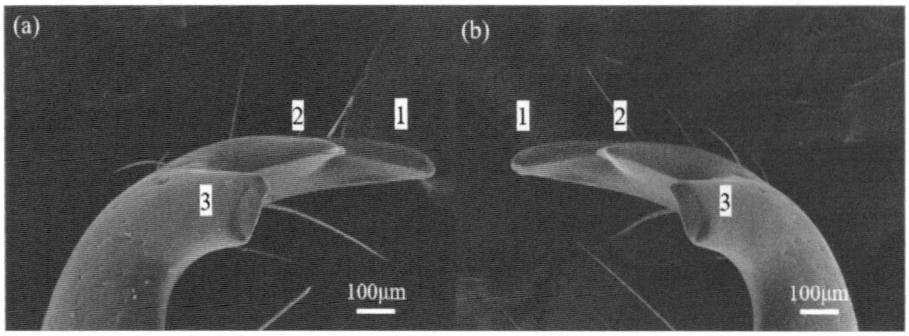

Figure 1. Scanning electron micrographs of the apical teeth. (a) a dorsal view of left mandible's apical teeth. (b) a dorsal view of right mandible's apical teeth.

In order to calibrate the piezoelectric film accurately, a method of free fall was adopted. An impact hammer shown in figure 2 was made by solidifying a small steel ball (Diameter: 8mm) with the cut-off apical teeth by using epoxy glue at the temperature of 60°C for 6 hours. Then a string for holding in hand is glued at the end of the impact hammer. The non-uniform mass distribution in the impact hammer can guarantee the apical teeth pointing to the force sensor during the free fall process.

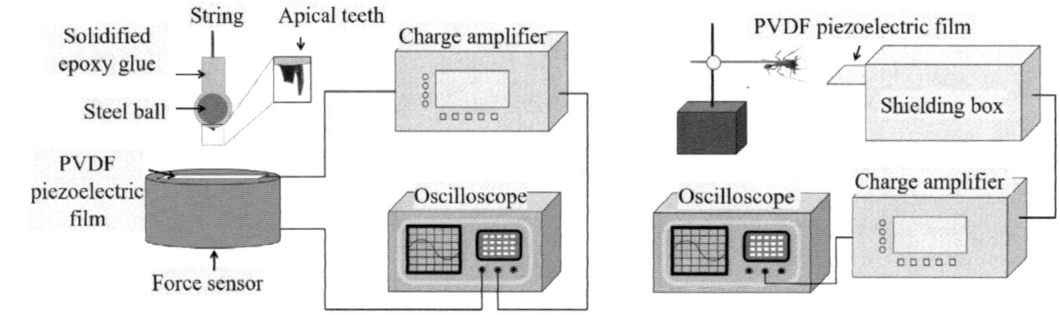

Figure 2. Strike force measurement experiment setup. (a)The dynamic calibration setup of PVDF piezoelectric film. (b)The mandibles striking experiment setup.

The dynamic calibration setup is shown in figure 2 (a). The PVDF piezoelectric film was glued tightly on the surface of the impact cap of the PCB force sensor (Made by PCB Cooperation, USA) by cyanoacrylate glue. The impact cap was installed at a heavy pedestal and connected with the oscilloscope (Tektronix TDS 2012C). The positive and negative electrode of the PVDF piezoelectric sensor were connected with the charge amplifier (B&K 2692) to amplify the charge signals, which were ultimately delivered to the oscilloscope as well. At first, the force sensor on which the calibrated PVDF piezoelectric film was glued was struck by the free falling apical teeth of the mandible. The heights of free fall range from 5 cm to 21 cm with the increment of 2 cm. Different heights of free falling apical teeth produced different impact forces and different output charges. Then through the linear fitting of the symbols, the calibration result can be obtained.

Furthermore, a platform (shown in figure 2 (b)) where the calibrated PVDF piezoelectric film was struck by the O.monticola was set up. Likelihood, the ants were fixed using a drop of dental wax (applied to the top of head) to the end of a thin rod that could rotate to keep the jaws perpendicular to the plane of the piezoelectric film. And the ant can strike freely, meanwhile the PVDF piezoelectric film was kept horizontal. The positive and negative electrode of the PVDF piezoelectric film were connected with the charge amplifier to amplify the charge signal, which was ultimately delivered to the oscilloscope. Then the ants were stimulated to strike by touching their 'trigger hairs' with a thin metal probe. The output signals were recorded every time when the ant struck the film. Finally, the strike force was calculated according to the output signals and the calibration result.

3. Results and Discussion

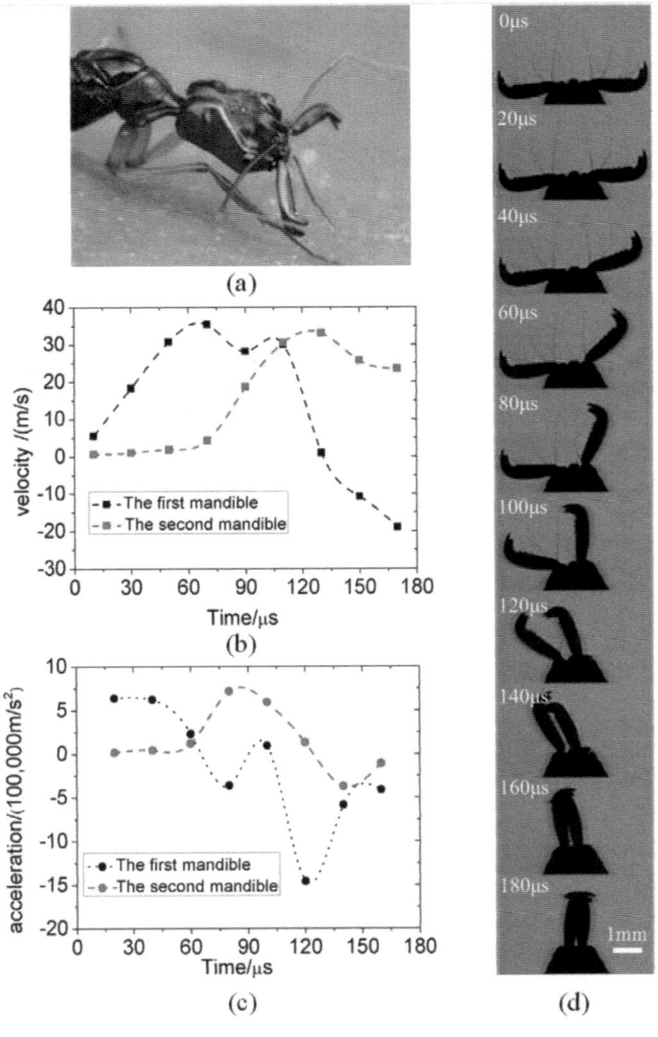

Figure 3. The high-speed kinematics of trap-jaw strikes. (Scale bars: 1 mm) (a) A side view of an O. monticola worker with mandibles cocked in preparation for a strike. (b) The first mandible to fire attains a lightly higher velocity than the second mandible. (c) The second mandible to fire attains a higher acceleration than the first mandible. (d) High-speed video images present a typical unobstructed strike (20 μs between each frame). The duration time of the strike is 0.18 ms.

The high-speed video results are shown in figure 3. As clear from the results, O.monticola's mandibles closed at a highest velocity of 35.42 m/s and a highest acceleration of 717,299 m/s^2 within an average duration of 0.18 ms. From figure 3 (b), before the first mandible scissored past the second one, the first mandible accelerated to 35.42 m/s, and then a mild fluctuation was witnessed; however, the second mandible's velocity gently increased during the first 70 μs, and then experienced a considerable boost from 4.29 m/s to 30.49 m/s. From the figure 3 (c), before the first mandible scissored past the second one, its acceleration continued to decrease, and even became negative during the first 90 μs, while the second mandible's acceleration continuously raised and enjoyed the highest acceleration approximately at 90 μs.

With high-resolution video images, asynchronous locomotion of the mandibles was observed. In all strikes, the left and right mandibles closed sequentially, with an obvious average interval of 0.08 ms. Both mandibles braked before reaching the midline, demonstrating that maximal force generates before the mandibles meet.

Table 1. The calibrated data of the free fall calibration experiment.

Height/cm	Impact force/N	The output voltage of the film/mv	The output charge of the film/pc
5	7.12	140	70
7	8.01	172	86
9	8.90	184	92
11	10.68	210	105
13	12.19	220	110
15	13.34	256	128
17	16.81	320	160
19	22.68	442	221
21	31.14	618	309

The calibrated data is presented in table 1. The height of the free falling apical teeth was changed to obtain different strike forces. And according to the parameter settings of the charge amplifier, the value of the charge signal is two times of the value of the voltage signal.

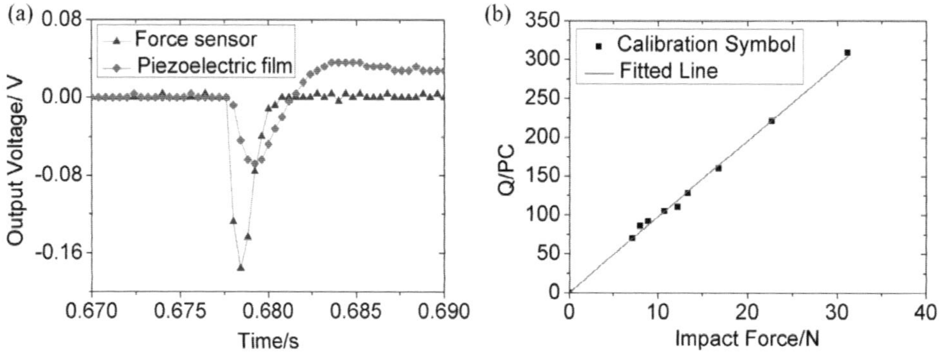

Figure 4. The calibration results of the PVDF piezoelectric film. (**a**) The two different output signal of one struck. (**b**) The calibration line of the PVDF piezoelectric film.

The calibration results are shown in figure 4. The output signals of PVDF piezoelectric film and the force sensor of one struck are illustrated in figure 4 (a). From the graph, the output signals of these two were generated simultaneously. The amplitude of the voltage of the force sensor reached 176 mV, while the amplitude of the voltage of the PVDF piezoelectric film was 68 mV. And it can be seen that response time of the force sensor was much shorter than that of the PVDF piezoelectric film, only taking 3.2 ms. Figure 4 (b) presents the calibration line of the PVDF piezoelectric film. Through the linear fitting of the symbols, the fitted equation was obtained:

$$Q(pc) = 9.78F(N) + 0.09 \tag{1}$$

where Q is the output charge and F is the strike force.

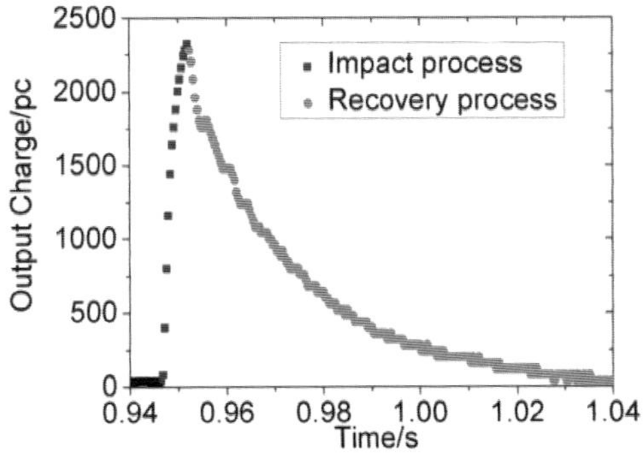

Figure 5. The output charge when an O.monticola worker struck the PVDF piezoelectric film.

The output signal of one struck is indicated in figure 5. The duration of strike was 6 ms (the blue symbols) and after that the PVDF piezoelectric film gradually returned to electrically neutral (the magenta symbols). As clear from the graph, the maximum of the output charge is 2300 pc. All the output charge signals ranged from 1,000 to 2,300 pc during the repetitive impact experiments of 30 times.

Finally, the strike force ranging from 102.2 N to 235.2 N was calculated according to the output charge signals and the calibrated result.

4. Conclusions

The strike process of O.monticola's mandibles was characterized. The highest closing speed of the mandibles surprisingly exceeding 35 m/s occurred on the first mandible, while the highest acceleration was occupied by the second mandible. And the asynchronous movements of the mandibles were demonstrated with an average interval of 0.08 ms. In addition, a new method was provided to obtain the accurate micro-scale strike force, which can effectively overcome the drawback of qualitative measurement by the piezoelectric films in traditional measuring methods. The result ranges from102.2 N to 235.2 N which is astonishingly impressive contrast with O.monticola's body weight. The extreme instantaneous strike force may imply the extraordinary impact resistance of the apical teeth material.

Acknowledgments

The authors would like to thank the financial support from the National Natural Science Foundation of China (No.51475029).

References

[1] Bennet-Clark H C and Lucey E C 1967 The jump of the flea: a study of the energetics and a model of the mechanism *J. Exp. Biol* **47** 59-67

[2] Bennet-Clark H C 1976 *The Insect Integument*, ed Hepburn H R (Amsterdam: Elsevier) pp 421-43

[3] Patek S N, Korff W L and Caldwell R L 2004 Biomechanics: deadly strike mechanism of a mantis shrimp-this shrimp packs a punch powerful enough to smash its prey's shell underwater *Nature* **428** 819-20

[4] Gronenberg W 1996 Fast actions in small animals: springs and click mechanisms *J. Comp. Physiol.* A **178** 727-34

[5] Patek S N, Baio J E, Fisher B L and Suarez A V 2006 Multifunctionality and mechanical origins: ballistic jaw propulsion in trap-jaw ants *PNAS* **103**12787-92

[6] Gronenberg W 1995 The fast mandible strike in the trap-jaw ant Odontomachus.1.Temporal properties and morphological characteristics *J. Comp. Physiol.* A **176** 391-8

[7] Gronenberg W 1995 The fast mandible strike in the trap-jaw ant Odontomachus.2.Motor control *J. Comp. Physiol.* A **176** 399-408
[8] Gronenberg W 1996 The trap-jaw mechanism in the dacetine ants Daceton armigerum and Strumigenys sp *J. Exp. Biol* **199** 2021-33
[9] Gronenberg W and Tautz J 1994 The sensory basis for the trap-jaw mechanism in the ant Odontomachus bauri *J. Comp. Physiol.* A **174** 49-60
[10] Just S and Gronenberg W 1999 The control of mandible movements in the ant Odontomachus *J. Insect Physiol* **45** 231-40
[11] Carlin N F and Gladstein D S 1989 The 'bouncer' defense of Odontomachus ruginodis and other odontomachine ants (Hymenoptera: Formicidae) *Psyche* **96** 1-19
[12] Gronenberg W, Tautz J and Hoildobler B 1993 Fast trap jaws and giant neurons in the ant odontomachus *Science* **262** 561-3
[13] Khan A, Abas Z, Kim H S and Oh I K 2016 Piezoelectric thin film:an integrated review of transducers and energy harvesting *Smart Materials and Structures* **25**

IOP Conf. Series: Journal of Physics: Conf. Series **986** (2018) 012030 doi:10.1088/1742-6596/986/1/012030

Data fusion algorithm for rapid multi-mode dust concentration measurement system based on MEMS

Maohao Liao[1,*], Wenzhong Lou[2], Jinkui Wang[3], Yan Zhang[4]

[1,2,3,4] School of Mechatronical Engineering, Beijing Institute of Technology, Beijing, China

[*]Corresponding author's E-mail: liaomaohao@163.com

Abstract. As single measurement method cannot fully meet the technical requirements of dust concentration measurement, the multi-mode detection method is put forward, as well as the new requirements for data processing. This paper presents a new dust concentration measurement system which contains MEMS ultrasonic sensor and MEMS capacitance sensor, and presents a new data fusion algorithm for this multi-mode dust concentration measurement system. After analyzing the relation between the data of the composite measurement method, the data fusion algorithm based on Kalman filtering is established, which effectively improve the measurement accuracy, and ultimately forms a rapid data fusion model of dust concentration measurement. Test results show that the data fusion algorithm is able to realize the rapid and exact concentration detection.

1. Introduction

Industrial dust explosion accidents occur frequently and respiratory diseases caused by dust particles emitted from industrial production have been more and more common in recent years, thus it is of great practical significance for the control of environmental pollution and human health to limit the concentration of dust [1]. How to accurately measure the dust concentration is a key step in dust control. Previous studies have shown that the detection effect is difficult to meet the practical needs of dust detection projects when using a single measurements method for dust concentration measurement [2]. To solve this problem, we have developed a multi-mode dust concentration measurement system based on multiple MEMS sensors suitable for use in detection of industrial dust.

To get the accurate results when simultaneously measuring the same target with several different measurement methods, it needs to perform deep level fusion in sensing units, measure circuits or solution algorithms, thus the measurement performance or the data of detection can be complementary to each other to compensate for the defects of the single measure method and complete the rapid, accurate and efficient detection of the complex measure object [3].

Multi-mode detection method is a kind of organic combination of different detection methods. Different methods need to be fused at different levels in sensing unit, measure circuit or solution algorithm, and data fusion is the core of multi-mode detection [4]. Data fusion algorithm can improve the detection accuracy and resolution, but at the same time it will bring a certain amount of calculation, which results in increasing computing time and computing resources. How to choose the optimal multi-source data fusion algorithm based on the calculation model characteristics and detection characteristics of the basic detection method, and form the optimal match between the detection speed and the detection precision is the core problem of the fusion algorithm [5].

2. Measuring system

Typically, according to the difference of system architecture, multi-sensor data fusion detection system can be divided into the centralized, the distributed and the mixed types. The structure of the proposed multi-mode detection system in this paper is relatively simple, but the relationship of data is more complex, because both of the ultrasound field measure method and the electric field measure method are required to be compensated by environmental parameters, so the mixed system structure has been chosen, that not only can guarantee the system to achieve good integration, but also to make full use of the advantage of the simple structure system, to avoid a huge calculation and lots of communication tasks. The final design of the system is shown as follows:

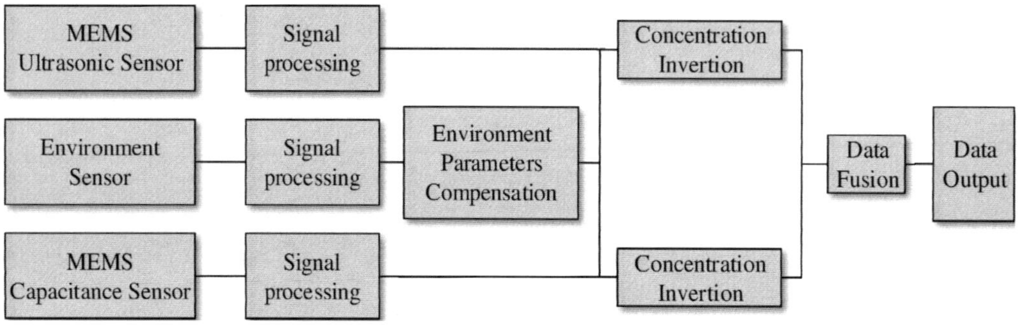

Figure 1. Multi-mode measuring system schematic diagram

The method of ultrasound field detection is mainly based on the theory of ultrasound attenuation: absorption attenuation and scattering attenuation are produced in the propagation of ultrasound in the medium, and attenuation level is represented by attenuation coefficient α . Absorption, attenuation and scattering attenuation are the processes of energy conversion and dissipation in a propagating medium. The attenuation coefficients α contains characteristic information of the propagating medium. Therefore, the concentration of the mixture can be detected by measuring the acoustic attenuation coefficient [6].

The electric field detection method is mainly based on the equivalent dielectric constant theory: different substances have different dielectric constants, and the mixture also has an equivalent dielectric constant, which is determined by the composition of the mixture and its mixing ratio, that is, when the composition is known, the equivalent dielectric constant contains the proportionality information of the different components. Therefore, the density of the mixture can be calculated by measuring the dielectric constant of the mixture using a capacitive sensor with fixed structural parameters [7].

According to the measuring system principle, the concentration measurement system which is designed in this paper includes at least the MEMS capacitance sensor, MEMS ultrasonic sensor, the signal driver module, the signal receiving and amplifying module and the control & calculation module, in addition it should have the power supply and the certain peripheral circuits.

3. Algorithm

Kalman filter algorithm is a data processing method based on recursive and error estimation. The method optimizes the statistical characteristics of the current data according to the error estimation and the data detection value of the previous state, thus providing the filter (corrected) calculated value of the current detection value [8]. The data fusion method based on Kalman filter algorithm is to calculate the weighted average of multiple sensor data. The weight of each sensor data is inversely proportional to its detection variance. Because of the recursive computing idea, the method has good computational characteristics, which is easy to implement in the programming language, and has high computational real-time and sufficient computational precision. Therefore, Kalman filter has been widely studied and applied [9]. The mathematical principle of the Kalman filter algorithm is given as follows:

Assume that the state equation and the detection equation of the dynamic measuring system are respectively:

$$X_k = \Phi_{k,k-1} X_{k-1} + \Gamma_{k,k-1} W_k \quad (1)$$

$$Z_k = H_k X_k + V_k \quad (2)$$

In the above equation (1) and (2), X_k is the system state at time k, $\Phi_{K,K-1}$ and $\Gamma_{K,K-1}$ are the state transition matrix at time k-1 to time k, Z_k is the detection value at time k, H_k is the parameter of the detection system, W_k and V_k represent the Gaussian white noise of the system process and the observation.

If the estimated state and observations satisfy equation (1), the system process noise and the observation noise satisfy equation (2), then the estimated value $\hat{X}_k$ of the observed X_k at time k can be solved according to the following equation.

Further predicted:

$$\hat{X}_{k,k-1} = \Phi_{k,k-1} \hat{X}_{k-1} \quad (3)$$

State estimated:

$$\hat{X}_k = \hat{X}_{k,k-1} + K_k [Z_k - H_k \hat{X}_{k,k-1}] \quad (4)$$

Filter gain matrix:

$$K_k = P_{k,k-1} H_k^T R_k^{-1} \quad (5)$$

Predicted estimate covariance:

$$P_{k,k-1} = \Phi_{k,k-1} P_{k-1,k-1} \Phi_{k,k-1}^T + Q_k \quad (6)$$

Updated estimate covariance:

$$P_k = [I - K_k H_k] P_{k,k-1} \quad (7)$$

The above is the five basic formulas of the Kalman filter [10]. From the above reasoning, once given the initial values X_0 and P_0, according to the observed value Z_k at time k, we can recursively calculate the state estimation $\hat{X}_k$ at time k(k =1,2,…,N).

According to the Kalman filter model, when there is a dust concentration detection using the ultrasound field method and the electric field method respectively, there is a system state equation (8) and two detection equation (9) and (10):

$$X_k = \Phi_{k,k-1} X_{k-1} + \Gamma_{k,k-1} W_{k-1} \quad (8)$$

$$Y_k{}^1 = H_k{}^1 X_k + V_k{}^1 \quad (9)$$

$$Y_k{}^2 = H_k{}^2 X_k + V_k{}^2 \quad (10)$$

If the detection noise variance matrices R^1 and R^2 guarantee that $H_k{}^{1T} R^1 H_k{}^1 + H_k{}^{2T} R^2 H_k{}^2$ is a nonsingular matrix, the state value X has a linear minimum variance estimation:

$$\hat{X} = R(H_k{}^1 R^{1^{-1}} Y^1 + H_k{}^2 R^{2^{-1}} Y^2) \quad (11)$$

The generalized detection vector can be defined as:

$$Y_K = [Y_k{}^1 \ Y_k{}^2] \quad (12)$$

Then the generalized detection equation as follows:

$$Y_K = H_K X_K + V_K \quad (13)$$

$$H_K = [H_k{}^1 \ H_k{}^2]^T \quad (14)$$

$$V_K = [V_k{}^1 \ V_k{}^2]^T \quad (15)$$

According to the classical Kalman filter algorithm and the optimal fusion estimation theorem [11], the iterative algorithm of multi-sensor data fusion algorithm can be obtained as follows:

$$\hat{X}_k = \hat{X}_{k,k-1} + K_k\left[Y_k{}^1 - H_k{}^1 \hat{X}_{k,k-1}\right] + K_k{}^2\left[Y_k{}^2 - H_k{}^2 \hat{X}_{k,k-1}\right] \quad (16)$$

$$P_k = [P_{k,k-1}{}^{-1} + (H_k{}^{1T} R_k{}^1 H_k{}^1)^{-1} + (H_k{}^{2T} R_k{}^2 H_k{}^2)^{-1}]^{-1} \quad (17)$$

4. Sample Test

In order to evaluate the effect of the above algorithm, the effect evaluation and verification of the data fusion algorithm are further studied. The research process is as follows:

Step 1. The initial data acquisition

To generate the standard environment of even dust concentration, we use customized dust chamber, and the dust concentration was measured by the ultrasound field method and the electric field method with MEMS sensors respectively. The data of the dust stabilization stage were selected as the analysis object, and the data length was 150 points. The concentration curves obtained from the two test methods are as follows:

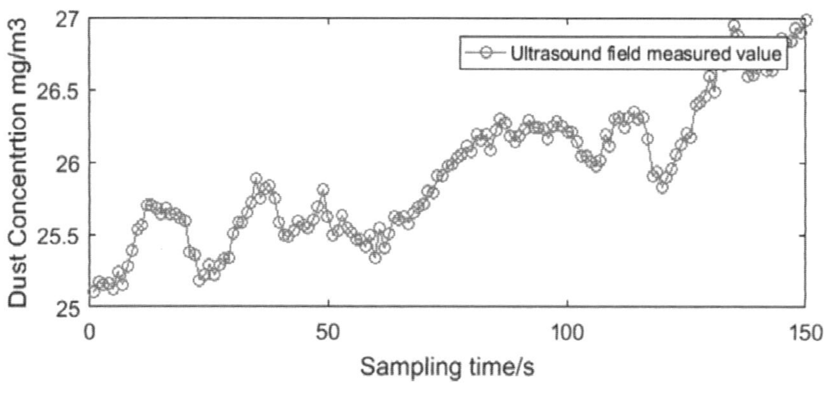

Figure 2. Dust concentration value measured by ultrasound field method

It can be seen from the graph that the system noise of the dust concentration value obtained by the electric field method is larger than that of the sound field method, but the trend of the concentration detection value of them is basically the same, which indicates that the dust concentration cannot be

IOP Conf. Series: Journal of Physics: Conf. Series **986** (2018) 012030 doi:10.1088/1742-6596/986/1/012030

accurately detected by a single measure method.

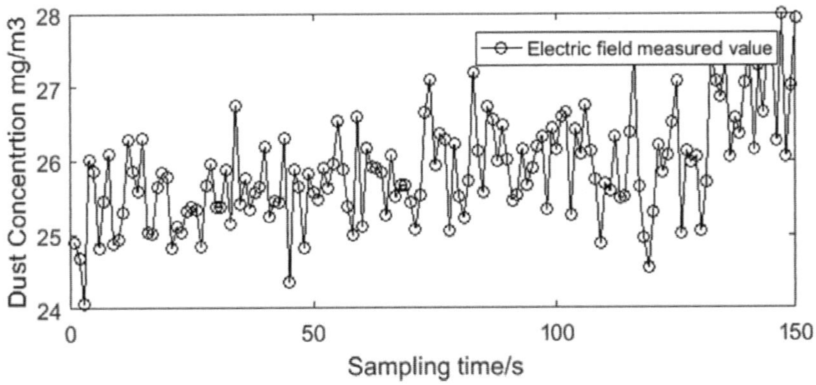

Figure 3. Dust concentration value measured by electric field method

Step 2. Extraction of state equations and measure equations
Using the MATLAB software to analyze the two sets of data, we can get the corresponding linear fitting curve and the corresponding noise variance, as follows:

$$\widehat{X}_{k} = \widehat{X}_{k,k-1} + K_k\left[Y_k^{\ 1} - H_k^{\ 1}\widehat{X}_{k,k-1}\right] + K_k^{\ 2}[Y_k^{\ 2} - H_k^{\ 2}\widehat{X}_{k,k-1}] \tag{18}$$

$$P_k = [P_{k,k-1}^{\ \ -1} + (H_k^{\ 1T}R_k^{\ 1}H_k^{\ 1})^{-1} + (H_k^{\ 2T}R_k^{\ 2}H_k^{\ 2})^{-1}]^{-1} \tag{19}$$

The above two formulas are used as the true dust concentration state equation, the detection equation of ultrasound field method and the electric field method, and the corresponding data variance is used as the noise variance of the detection equation.

Step 3. Data fusion processing
According to the data fusion model, we use MATLAB software to program for acoustic-electrical data fusion, and compare with the concentration value obtained by the single detection method, the contrast curves was shown as follows:

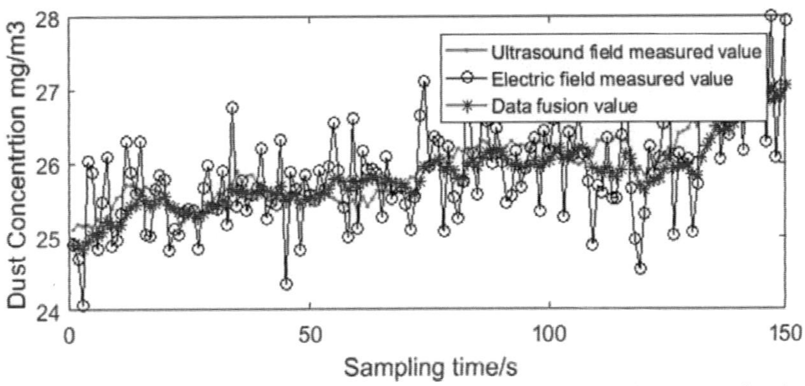

Figure 4. Comparison curves between measured values and processed values

As can be seen from the figure, with the use of data fusion algorithm, the detection noise of the dust concentration data is significantly reduced, effectively improve the accuracy of concentration data.

5. Conclusion

According to the comparison of the curves shown above, it can be seen that the acoustic-electrical data fusion model based on Kalman filter effectively reduce the detection error and the detection noise, so as to effectively improve the measuring system's measure precision and obtain more reliable dust concentration data. Therefore, the multi-sensor data fusion model and algorithm based on Kalman filter we present in this paper are reasonable and effective.

References

[1] Sibiao Zhao. Study on Dust Concentration Measurement Based on External Charge Principle. *Instrument Technique and Sensor*, 2010 (6): 15-16.

[2] S. S. Park, J. Kim, J. Lee, Combined dust detection algorithm by using MODIS infrared channels over East Asia, *Remote Sensing of Environment*, 2014, 141: 24-39.

[3] F. Liang, M. Gao, Q. Xiao, Evaluation of a data fusion approach to estimate daily PM2.5 levels in North China, *Environmental Research*, 2017, 158:54-60.

[4] Hall D. L., Linas J. *Handbook of multi-sensor data fusion*. New York: CRC Press, 2001.

[5] White F. A model for data fusion. *Proc. 1. St. National Symposium on Sensor Fusion*, 1988.

[6] Y. Lu, W. Lou, M. Guo, Rapid Temperature Measurement of Meteorological Detection System Based on MEMS, *Key Engineering Materials*, 2014, 609: 1185-1188.

[7] Geng Lu, Method for detecting two-phase flow phase concentration based on capacitance measurement and PCA method. *Journal of China Jiliang University*, 2003, 14 (1): 15-18.

[8] J. B. Reley, Y. C. Agrawal. Sampling and Inversion of Data in Diffraction Particle Sizing, *APPLIED OPTICS*, 1991, 30(33):480-4817

[9] Kahn Henry D, Rosati Jacky A, Bray Andrew P, Statistical evaluation of data from multi laboratory testing of a measurement method intended to indicate the presence of dust resulting from the collapse of the World Trade Center, *Environmental Monitoring and Assessment*, 2012, 18410.

[10] Y.K Chen, X.C Si, Z.G Li, Research on Kalman-filter based multisensory data fusion. *Journal of Systems Engineering and Electronics*, Vol. 18, No. 3, 2007, pp.497-502

[11] Gao J.B, Harris C J. Some remarks on Kalman filters for the multi-sensors fusion. *Information Fusion*, 2002, 3: 191-201

A fractal-like texture of multicrystalline silicon for enhancing optical performance

Y J Guo, L Q Wu and W B Li

School of Mechanical Engineering, Hangzhou Dianzi University, Hangzhou, 310018, P.R. China

E-mail: wuliqun@hdu.edu.cn

Abstract. Appropriate texturing is one major topic for multicrystalline silicon wafers to improve the cost effectiveness. In this paper, a new fractal-like texture being applied to multicrystalline wafers is introduced. The focus of this paper is the study of the light-trapping ability of the novel fractal-like texture for multicrystalline silicon solar cell. With a simple model of the fractal-like texture, the conditions needed to achieve lower reflectance are discussed first. It is shown how the effect of the micro structure of multicrystalline silicon with the fractal-like characteristics. Finally, in our present work, the finite element method(FEM) is used to simulate the propagation of electromagnetic field in the structure of multicrystalline silicon. We obtain an average reflectivity of 3% in the wavelength range 300–800 nm on the surface of multicrystalline silicon wafer, which has the fractal-like structure feature.

1. Introduction

Multicrystalline silicon has been hailed as the cornerstone of the microelectronics industry and photovoltaic industry, which is widely used in solar cells. However, the photoelectric conversion efficiency of multicrystalline silicon solar cells has been no breakthrough. It is an effective method to improve the absorption rate of the surface by texturing of multicrystalline silicon. Appropriate texturing is one effective way to decrease the reflectance of multicrystalline silicon wafers [1].

In the past few years, in order to optimize the optical properties of multicrystalline silicon wafers, many kinds of texture structure of multicrystalline silicon wafers have been proposed, such as inverted-pyramid texture [2], honeycomb texture [3-5], or random pyramid texture [6,7] on the top surfaces. However, a severe drawback has been observed, these textured surfaces have not obtained a sufficiently low reflectivity. Therefore, how to optimize structure for obtaining lower-reflectance surface is still an important research direction in the field of multicrystalline silicon solar cells.

The aim of this work is to introduce a novel fractal-like texture micro structure which may lead to the development of the PV technology. In this paper, to achieve sufficiently high prediction capabilities, three models of multicrystalline silicon including semi-ellipse texture, inverted-pyramid texture, and fractal-like texture were proposed [8]. Simulations are usually used to analyze a series of parameter variations and to predict effects due to structure changes [9]. The study of surface morphology and reflectance of different textured silicon based surfaces have been analyzed by COMSOL Multiphysics in this paper. Firstly, we give a schematic diagram of the light trapping of three models. Further more, specific discussions are given of reflectivity through the fractal-like texture microstructuring. Finally we derive that a remarkable lowering of the reflectance of fractal-like texture compared to other textures is achieved.

2. Methods

One way to acquire lower reflectivity is to introduce specific structure into silicon base surface. Texturing of multicrystalline silicon wafer means creating an especial surface which may be allowing more of the incident light being absorbed into the solar cell [10].

By previous research, it is not difficult to get only a small part of the study is used to simulate light trapping of texture structure. In order to study the light trapping effect of different surface structures of multicrystalline silicon and the corresponding silicon surface reflectivity. We carried out a ray tracing simulation[11-13] by COMSOL Multiphysics to estimate the light trapping influence in the multicrystalline silicon solar cells with different texture structures.

It is not necessary to research details of the different texture structures in order to understand some basic optical performance parameters. Three different structure models with periodic array were designed: semi-ellipse texture, inverted-pyramid texture, and fractal-like texture as shown in Figure 1. As is known to all, when the incident light from a medium incident to another medium, in the interface of two media, the propagation direction of the light will change, part of the light is reflected back to the original medium, which is named as the reflection ray, another part of the light incident another medium, which is called for the refraction ray. So we can know that the light from air into silicon arriving at the interface is divided ,the refraction light is absorbed, at the same time the reflection light is reflected back into the air.

Figure 1 shows that how the incident light trapping in the fractal-like texture compares with the other two kinds of texture structures. From figure 1(a)-(b),we can know that light can be reflected and reflected on the silicon surface by two times. But if we create a fractal-like texture structure, which has many nanoscale structures on both sides, as shown in figure 1(c). In the ideal case, when the light is irradiated on the surface of the multicrystalline silicon, the light in this fractal-like texture features allowing multiple reflection and refraction to make more photons enter into the multicrystalline, which will greatly reduce the surface reflectivity.

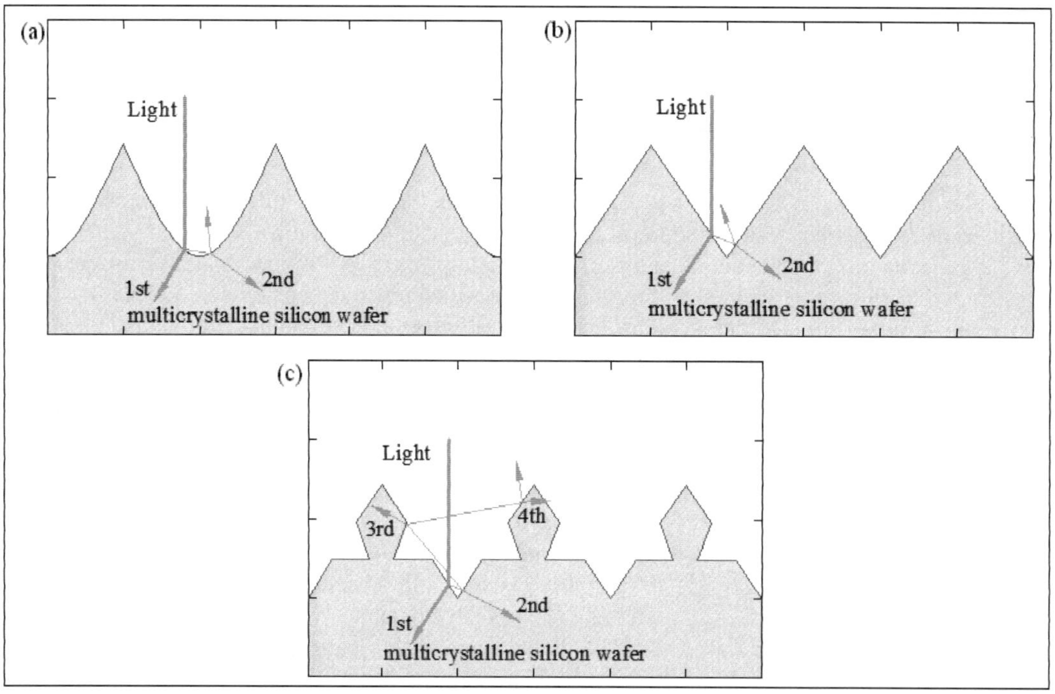

Figure 1. three structures of light reflection and refraction in the trap including semi-ellipse texture(a), inverted-pyramid texture(b) and fractal-like texture(c).

After that, we investigated these texturing structures to determine which produced the lowest reflectance using the RF Module by COMSOL Multiphysics simulation software. Under normal circumstances, the way to use COMSOL is to design the model structure firstly. A 3-D model, as shown in figure 2, is constructed with periodic array.

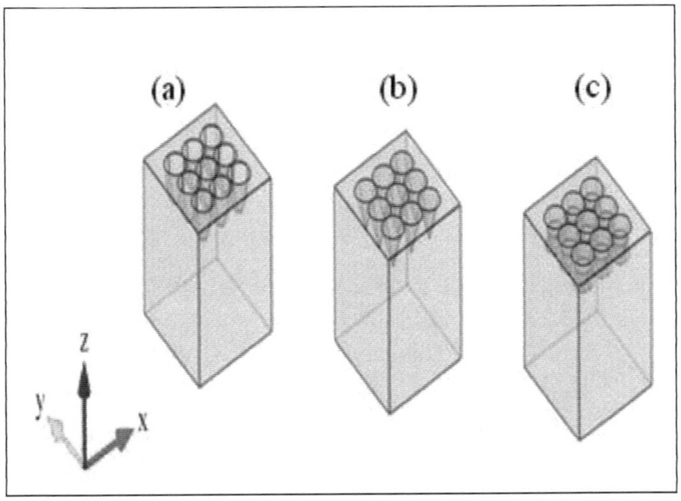

Figure 2. three different models of light trapping geometry of silicon are constructed with periodic array: semi-ellipse texture(a), inverted-pyramid texture(b) and fractal-like texture(c).

However, in order to reduce the computational difficulty, in this paper we use the floquet periodicity as a boundary condition to replace the periodic array structure model. In addition, Perfectly Matched Layers (PML) are set in the incidence and output port, which can effectively absorb into the perfect match Layer area of the incident light. So we can see that a 2-D geometry with creating the cross section is described which is shown in figure 3.

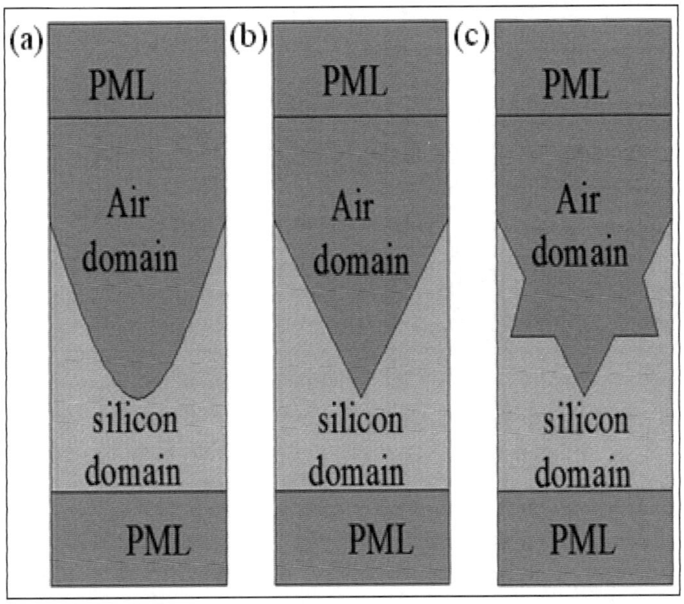

Figure 3. 2-D profile geometry depicts the various domains of the different texture models: semi-ellipse texture(a), inverted-pyramid texture(b) and fractal-like texture(c).

In the geometry which is described above, each region needs to be configured for the corresponding materials. The refractive index of the air domain is assumed to 1. But base material is multicrystalline silicon ,which refractive index is different in different wavelengths. Therefore, it is necessary to use interpolation function set silicon refractive index of real and imaginary parts. Then the propagation of the electromagnetic field on the multicrystalline surface is simulated by the finite element method

(FEM), which is known that the solution converges to a given boundary value problem. Therefore, the finite element method is a good method to verify the solution. The constitutive equation is

$$\nabla \times \mu_r^{-1}(\nabla \times E) - k_0^2 \varepsilon_r E = 0 , \quad (1)$$

where $\mu_r = 1 , \varepsilon_r = n^2 , k_0 = 2\pi / \lambda$; λ is the wavelength of the incident light, n is the refractive index of the material.

Finally, to evaluate the effect of the light trapping, we carried on the simulation of the light trapping effect in the wavelength range from 0.3 to 0.8 μm using the model above as shown in figure 3. The reflectance and surface electrical field intensity are compared in the same size with the ideal case, with the case of the inverted-pyramid texture, with the case of the semi-ellipse texture, and with that of the fractal-like texture.

3. Results and Discussion

In order to test and verify the effect of the fractal-like texture surface compared with the other two structures, the results of surface electrical field intensity and reflectivity for three structures previously mentioned are measured. And it should be noted that the results being simulated from a finite element model with the same size, materials, boundary conditions except for shape.

Figure 4 shows the distribution of surface electrical field intensity of the semi-ellipse texture, inverted-pyramid texture and fractal-like texture at the wavelength of 600 nm from a beam incident at the vertical angle. It can be seen from the figure 4, when the multicrystalline silicon texture morphology is different, the electric field intensity z component distribution is also different. And also it shows that the semi-elliptical textured surface electric field intensity z component shows stripe evenly distributed.

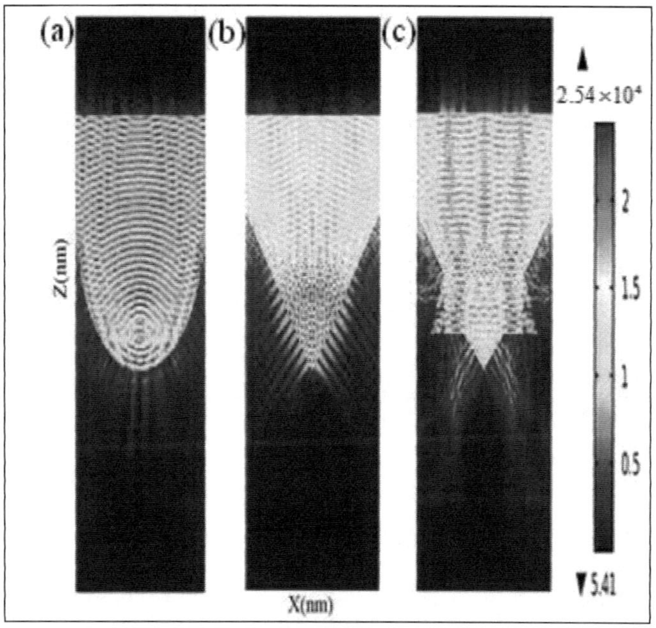

Figure 4. Distribution of surface electrical field intensity of the semi-ellipse texture(a), inverted-pyramid texture(b) and fractal-like texture(c) at the wavelength of 600nm.

The reflectivity of the fractal-like texture multicrystalline silicon is compared with the other texturings in figure 5, which has significantly reduced reflectivity. This is attributed to this fractal-like texture effects which can make a large number of photons into the inside of the multicrystalline silicon far above the other texturing. Simulation of reflection based on the fractal-like texture assumption show that it can obtain lower surface reflectance ,is superior to the other two textured surfaces. The reflectivity may be further improved by optimizing the structure.

The result states clearly that a obvious light trapping effect can be obtained using the fractal-like texture, which is better than the other texturings.

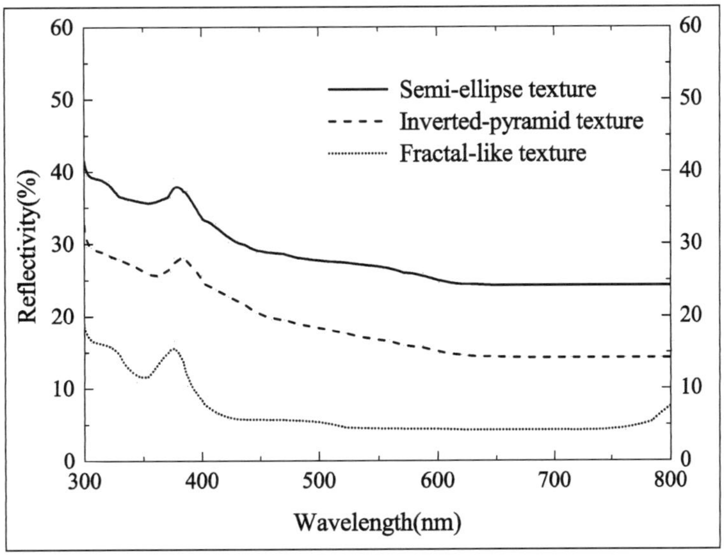

Figure 5. The reflectivity distribution of the semi-ellipse texture, inverted-pyramid texture and fractal-like texture at the wavelength range of 300-800 nm.

4. Conclusions

In summary, the fractal-like texture provides better optical properties than the other texture structures. Figure 1 shows the different light trapping effect between the fractal-like texture and the other two structures under ideal conditions. In accordance with the above simulation results, the fractal-like texture designed shows extremely good reflection results. The resultant multicrystalline silicon with a fractal-like structure as a solar cell can absorb more solar photon energy, and consequently has higher efficiency. The conclusion is that the fractal-like texture used in this work is very effective for enhancing optical performance. The novel "fractal-like texture" structure is introduced in this letter that may continue to increase the cost effectiveness of solar cells and extend the application of multicrystalline silicon.

Our further work will relate to model and optimize the other textures that could be manufactured through new processing technology.

Acknowledgments

This work is supported by the National Natural Science Foundation of China under Grant No. 51775154, the Key project of Natural Science Foundation of Zhejiang Province under Grant No. LZ15E050004, the Open Program of Jiangsu Key Laboratory of 3D Printing Equipment and Manufacturing under Grant No. BM2013006, Zhejiang Provincial Natural Science Foundation of China under Grant No. LQ17E050012 and Hangzhou Hui Shijia Electronics Technology Co. ,Ltd. development projects under Grant No. JDC201701.

References

[1] D. Zhang, C. Sun, and W. Yue, 2017, *Research & Progress of Sse*, **37**, 144-48.
[2] L. Yang, Y. Liu, Y. Wang, W. Chen, Q. Chen, J. Wu, A. Kuznetsov, and X. Du, 2017, *Sol. Energ. Mat. Sol. C*, **166**, 121-26.
[3] B. Yang, and M. Lee, 2013, *Appl. Surf. Sci.*, **284**, 565-68.
[4] I. M. Peters, H. Hauser, N. Tucher, and B. Bläsi, 2016, *IEEE J. Photovolt.*, **6**, 1480-87.

[5] S. Wang, Z. Wang, R. Futamura, M. Endo, and K. Kaneko, 2017, *Chem. Phys. Lett.*, **673**, 38-43.
[6] A. Khanna, P. K. Basu, A. Filipovic, V. Shanmugam, C. Schmiga, A. G. Aberle, and T. Mueller, 2015, *Sol. Energ. Mat. Sol. C*, **132**, 589-96.
[7] Q. Zhang, and Y. Xu, 2016, *Neural Comput. Appl.*, **27**, 593-602.
[8] Z. Xu, H. Qiao, H. Huangfu, X. Li, J. Guo, and H. Wang, 2015, *Opt. Commun.*, **356**, 526-29.
[9] P. P. Altermatt, 2011, *J. Comput. Electron.*, **10**, 314-30.
[10] L. A. Dobrzański, A. Drygała, K. Gołombek, P. Panek, E. Bielańska, and P. Zięba, 2008, *J. Mater. Process. Tech.*, **201**, 291-96.
[11] T. YAGI, Y. URAOKA, and T. FUYUKI, 2006, *Sol. Energ. Mat. Sol. C.*, **90**, 2647-56.
[12] S. Lien, C. Yang, C. Hsu, Y. Lin, C. Wang, and D. Wuu, 2012, *Mater. Chem. Phys.*, **133**, 63-68.
[13] W. J. Helms, and A. D. Thompson, 2016, *Radio Sci.*, **8**, 1125-32.

Temperature Effect and Parameter Optimization of Silicon Resonant Accelerometer Based on Electrostatic Rigidity

W B Xiao[1], L Li[1*] and R Hao[2]

[1]School of Mechanical Engineering, Southwest Jiaotong University, Chengdu 610031, China
[2]School of Mechatronics Engineering, University of Electronic Science and Technology of China, Chengdu 611731, China

[*]Correspondence: lili@home.swjtu.edu.cn; Tel: +86-028-66366430

Abstract. Temperature stability, one of the most important parameters of highly accurate MEMS accelerometers, is always calibrated by the temperature coefficients of performance, thus reducing the temperature coefficients has significance to acquire a higher stability of micro devices. This paper proposed an in-depth analytical study on the temperature coefficients of silicon resonant micro-accelerometer based on electrostatic rigidity. And an corresponding optimization method was introduced to reduce the influence of the temperature on the scale factor and improve the device stability over temperature. The results showed the temperature coefficient and the temperature sensitivity of the scale factor were 9Hz/g/°C and 1280ppm/°C, respectively, which are equal to 1.6% and 49% of the original values respectively.

1. Introduction

Resonant principle has been widely used for measuring physical parameters such as mass, acceleration, force, flow or pressure due to its good resistance to noise or electrical interference and simple interface to a digital system[1]. Micro accelerometer based on the resonance exhibited many advantages over other type ones because they converted the input acceleration into frequency, a quasi-digital signal with a higher resistance to disturbance[2-7]. As a mechanical sensor, however, the resonant accelerometers were inevitably subject to the effects of temperature change, which is characterized by the temperature coefficients of performance usually including scale factor and zero bias. In some applications, such as the inertial navigation system(INS), the acceleration measurement error caused by the temperature stability of scale factor after the second integral would seriously affect the positioning precision of the navigation system. Therefore, scale factor stability needed to be carefully considered in the design of accelerometers.

This paper focused on the frequency thermal stability of the silicon resonant accelerometer based on electrostatic rigidity. Current investigations indicated that the performance drift of micro devices could result from the temperature effects of residual stress, structure, materials and circuitry[8-10]. The temperature change could cause the change of Young's modulus and structural dimensions of resonator[11]. The residual stress induced by the mismatch of thermal expansion coefficient between silicon and substrate would experience a redistribution during the temperature change[12-14]. The stability issue could be solved by constant temperature control[15], temperature compensation[16] and isolated encapsulation stress, etc[17-18]. However, those techniques always came with a more complicated system and larger volume of micro devices. In this paper, a novel measure was proposed to improve the temperature stability of silicon resonant accelerometer based on parameter optimization. It was found that the temperature stability of scale factor could be improved dramatically by adjusting the stiffness ratio.

2. Modeling and Theory Analysis

2.1. Principle of the Silicon Resonant Accelerometer

The simplified structure of silicon resonant accelerometer based on electrostatic rigidity is show in figure 1[1]. It consists of the double-ended tuning fork (DETF), driving capacitors, sensing capacitors, proof mass and supporting springs.

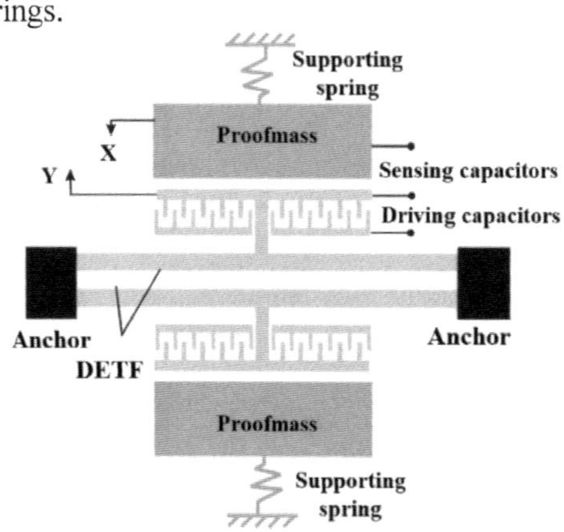

Figure 1. A schematic diagram of the resonant accelerometer.

The DETF resonant state was sustained by the oscillating circuit connecting to driving and sensing electrodes. The sensitive proof mass supported elastically by folded beams and was bonded to a glass substrate, which moves near or away from the electrode of DETF under inertial force and changed the gap distance of sensing capacitor. Then, the stiffness of the DETF were reduced or increased by the electrostatic force form between mass and DETF and resulted in the change of the output frequency. The dynamic equation of the vibrating beam of DETF can be expressed as[1] :

$$m\ddot{Y} + c\dot{Y} + \kappa Y = F_{elec}(x,Y,t) \tag{1}$$

where x,Y are the vertical displacement of the vibrating beam and proof mass, respectively, m, c and κ are effective mass of the vibrating beam, damping coefficient, effective mechanical stiffness of vibrating beam, respectively. The electrostatic force $F_{elec}(x,Y,t)$ can be expressed as:

$$F_{elec}(x,Y,t) = \frac{\varepsilon A V_{d2}^2}{2(g_0 - x - Y)^2} - \frac{N_1 \varepsilon h (V_{d1} + V_c \sin(\Omega t))^2}{2g_0} \tag{2}$$

where, the first part on the right is sensing electrostatic forces, while the second part is driving electrostatic forces. g_0 is the gap distance of driving capacitor. The V_{d2} is DC voltage of the proof mass, the DETF is connected to ground. The driving voltage is $V_c \sin(\Omega t)$ with DC bias voltage V_{d1}, which is added in the fixed comb electrodes, Ω is frequency of AC voltage. The other parameters shown in equation (2) were described in [1].

2.2. Residual stress of the resonant beam

The residual stress was mainly caused by the thermal expansion difference between silicon chip structure and glass substrate layer during processing. When the temperature of silicon accelerometer dropped to room temperature from processing condition, the thermal stresses including normal stresses and shearing and transverse normal (peeling) stresses, arose in the interface of the attachment material. In this paper, the residual stress(or thermal stress) was predicted by the theory of Tri-material assembly by E.Suhir [19-21]. The assembly structure could be expressed by figure 2(a). Note that it's not a typical adhesively assembled structure. Actually the upper component #1 and #2 are the silicon mirco beam structure, component #1 represents the resonant beam and component #2 represents the anchor points. They have the same material, so the interface between them is fictitious for ease of analysis. The lower component #3 is the glass substrate. In this picture, l is a half-length of the resonant beam, L is the length of one anchor point.

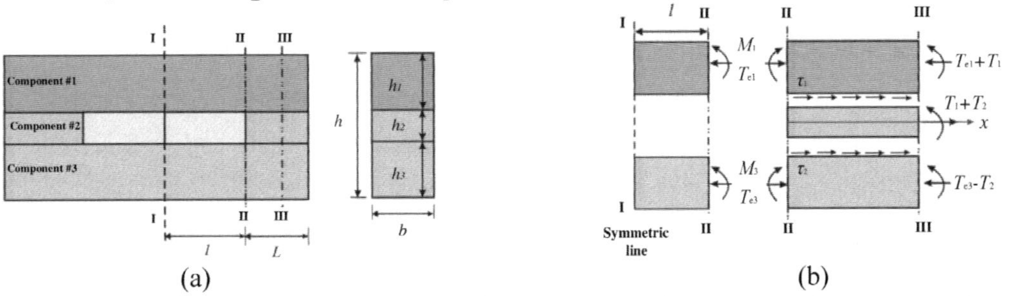

Figure 2[20]. (a)The assembly structure model; (b)Forces and moments at the x cross section.

The coefficients of linear thermal expansion of the assembly components are related as follows: $\alpha_1 = \alpha_2 < \alpha_3$, Shearing stresses in the interfaces, when the assembly is cooled down by the temperature ΔT, are schematically shown in figure 2(b). Section I-II is the half structure of resonant beam, There is no direct contact between #1 and #3 layer structure, so it is only subject to the force from the dividing plane I, including axial forces(T_{e1} and T_{e3}) and he moments (M_1 and M_3) at the x direction. And the section II-III is a complete three-layer structure, and the first and third layers are subjected to the opposite forces in the structure I(see figure 2(b)).

According to [20], the axial force T_{e1} and stress σ acting on the resonant beam can been expressed as:

$$T_{e1} = \frac{(\alpha_3 - \alpha_1)\Delta T}{\frac{(h_1 + 2h_2 + h_3)^2}{4(D_1 + D_3)} + \lambda_1 + \lambda_3} + \frac{(\kappa_2 + \kappa_3)\tau_2(0) - (\kappa_1 + \kappa_2)\tau_1(0)}{(\frac{(h_1 + 2h_2 + h_3)^2}{4(D_1 + D_3)} + \lambda_1 + \lambda_3)l} \tag{3}$$

$$\sigma = \frac{T_{e1}}{h_1} + \frac{h_1\theta_0 E_1}{2l} \tag{4}$$

In the equation (3), the second term on the right is a smaller amplitude than the first term, this term is omitted. In the equation (4), the first term on the right represents the axial stress introduced by the shearing force, and the second term represents the axial stress introduced by the moment. Rewrite equation (4):

$$\sigma = \frac{n_E h_3 (n_E h_3^3 + h_1^2(4h_1 + 6h_2 + 3h_3))}{h_1^4 + n_E^2 h_3^4 + 2n_E h_1 h_3 (2h_1^2 + 6h_2^2 + 2h_3^2 + 6h_1h_2 + 3h_1h_3 + 6h_2h_3)} E_1(\alpha_{si} - \alpha_{glass})(T_0 - T) \tag{5}$$

where, $n_E = {E_3}/{E_1}$,define E_{1eff} as the equivalent Young's modulus for component 1#. $E_{1eff} = \frac{n_E h_3 (n_E h_3^3 + h_1^2(4h_1 + 6h_2 + 3h_3))}{h_1^4 + n_E^2 h_3^4 + 2n_E h_1 h_3 (2h_1^2 + 6h_2^2 + 2h_3^2 + 6h_1h_2 + 3h_1h_3 + 6h_2h_3)} E_1 = \eta E_1$, where, the parameter η is only related to the thickness and the ratio of the Young's modulus between the structural layers. Then, the residual axial load of the resonant beam can be expressed as:

$$N = \sigma A_s = (\alpha_{si} - \alpha_{glass})(T_0 - T) E_{1eff} A_s \tag{6}$$

where, A_s is the cross-sectional area of the resonant beam, T_0 is the initial temperature on the die packaging process, and T is the working ambient temperature.

2.3. *Improvement of Stability of Scale Factor*

The resonant frequency of DETFs can be expressed as:

$$f_i = \frac{1}{2\pi}\sqrt{\frac{\kappa - \kappa_{ei}}{m}}, i = 1,2 \tag{7}$$

where, κ_{e1}, κ_{e2} represent the electrostatic stiffness of two resonant beams in DETFS, respectively, and their values are obtained based on the mechanical dynamics equation (1). κ is the effective mechanical stiffness of vibrating beams. The value of κ with residual stress can be expressed as:

$$\kappa = \frac{E_1 I}{l^3} a_1 - \frac{N}{l} a_2, \kappa_0 = \frac{E_1 I}{l^3} a_1 \tag{8}$$

where, I is the moment of inertia of the beam cross-section. a_1 and a_2 are constants determined by vibration modes of vibrating beam. κ_0 is the effective mechanical stiffness of vibrating beams regardless of the residual stress effect. Combining equation (6) and equation (8), the temperature coefficient of mechanical stiffness is:

$$\frac{\partial \kappa}{\partial T} = \frac{a_2 E_{1eff} A_s}{l}[(\alpha_{si} - \alpha_{glass}) - \frac{\partial \alpha_{si}}{\partial T}(T_0 - T)] \tag{9}$$

here, the thermal expansion coefficient of the glass (α_{glass})also has good thermal stability as a constant. Here, we ignored the situation that Young's modulus and the geometric dimensions were affected by the temperature and treated it as a constant. In fact, the effect of temperature on them is much less than that on the thermal expansion coefficient of Si(α_{si}) [3].

$$\alpha_{si}(t) = \left\{3.725 \times [1 - e^{-5.88\times10^{-3}(T+273-124)}] + 5.548 \times 10^{-4}(T + 273)\right\} \times 10^{-6} \tag{10}$$

The scale factor of the resonant accelerometer can be expressed as:

$$S_f = \Delta_f \big|_{a=g} = f_0 \frac{6\gamma a(M\kappa - mk_s)[g_0\kappa k_s - 2\gamma(\kappa + k_s)]^3}{g_0\kappa[g_0\kappa k_s - \gamma(3\kappa + 2k_s)]^4}\big|_{a=g}, f_0 = \frac{1}{2\pi}\sqrt{\frac{\kappa}{m}} \tag{11}$$

where, f_0 is the mechanical natural frequency of DETFs. $\gamma = \frac{\varepsilon A V_{d2}^2}{2g_0^2} = \frac{N_1 \varepsilon h_1 V_{d1}^2}{2g_0}$ Remaining parameters shown in above are described in table 1. As the working ambient temperature T changed, the residual thermal stress would change, causing a drift in the scale factor.

$$\begin{aligned} CT_{S_f} &= \frac{\partial S_f}{\partial T} = f_0 \frac{\partial \kappa}{\kappa \partial T} \frac{3g\gamma[g_0\kappa k_s - 2\gamma(\kappa + k_s)]^2}{g_0\kappa[\gamma(3\kappa + 2k_s) - g_0\kappa k_s]^5}\{(g_0\kappa k_s)^2(\kappa M - 3k_s m) \\ &+ g_0\gamma\kappa k_s[8k_s^2 m + 15\kappa k_s m - 5\kappa^2 M] + \\ &2\gamma^2[3\kappa^3 M + \kappa^2 k_s(7M - 9m) - \kappa k_s^2(17m + 2M) - 2k_s^3 m]\} \end{aligned} \tag{12}$$

is the temperature coefficient of scale factor.

$$CET_{S_f} = \frac{CT_{S_f}}{S_f} \tag{13}$$

is defined as the temperature sensitivity of the scale factor. The larger the value is, the more significant the temperature effect and the worse the thermal stability of the scale factor of the resonant accelerometer would be. equation (11) and equation (12) are substituted into equation (13) so as to obtain the expression of CET_{S_f} .

$$CET_{S_f} = \{\frac{1}{2} + \frac{4\gamma^2 k_s^3 m + 2\gamma m k_s^2 (11\gamma - 3g_0 k_s)\kappa}{(M\kappa - mk_s)[g_0\kappa k_s - \gamma(3\kappa + 2k_s)][g_0\kappa k_s - 2\gamma(\kappa + k_s)]} + \frac{2k_s[g_0^2 k_s^2 m + 6\gamma^2 (m - M) + g_0 \gamma k_s (M - 5m)]\kappa^2}{(M\kappa - mk_s)[g_0\kappa k_s - \gamma(3\kappa + 2k_s)][g_0\kappa k_s - 2\gamma(\kappa + k_s)]} - \frac{M[6\gamma^2 - 5g_0\gamma k_s + g_0^2 k_s^2]\kappa^3}{(M\kappa - mk_s)[g_0\kappa k_s - \gamma(3\kappa + 2k_s)][g_0\kappa k_s - 2\gamma(\kappa + k_s)]}\} \frac{\partial\kappa}{\kappa\partial T} \quad (14)$$

Previous studies had shown that the thermal stability of the resonant accelerometer was improved in virtue of designing the isolating structures[17-18], temperature compensation[16] or other methods[15],which accordingly reduced the residual thermal force. equation (14) showed that by optimizing the configuration of the structural parameters (such as: $M, m, k_s, g_0 ....$), the absolute values in curly braces{}were infinitesimal, and the thermal stability of the resonant accelerometer could be effectively improved even if the value of the residual stress is not changed.

3. Result and Discussion

Example was calculated by MTALAB-based theory, and the simulation parameters were expressed in table 1. According to equation (14), the stiffness of the folded beam k_s was adjusted depending on the stiffness ratio K($K = \kappa_0 / k_s$). The optimal stiffness ratio K=60.5 was obtained after finding the minimum absolute value of braces {} in equation (14) at a room temperature T=25°C. After optimization, both the temperature coefficient and the temperature sensitivity of the scale factor were significantly improved, as shown in figure 3. Under initial conditions, of K=55, the temperature coefficient and the temperature sensitivity of the scale factor were 9Hz/g/°C and 1280ppm/°C, respectively, and values of the two parameters were reduced to 1.6% and 49% of the original values respectively after optimization.

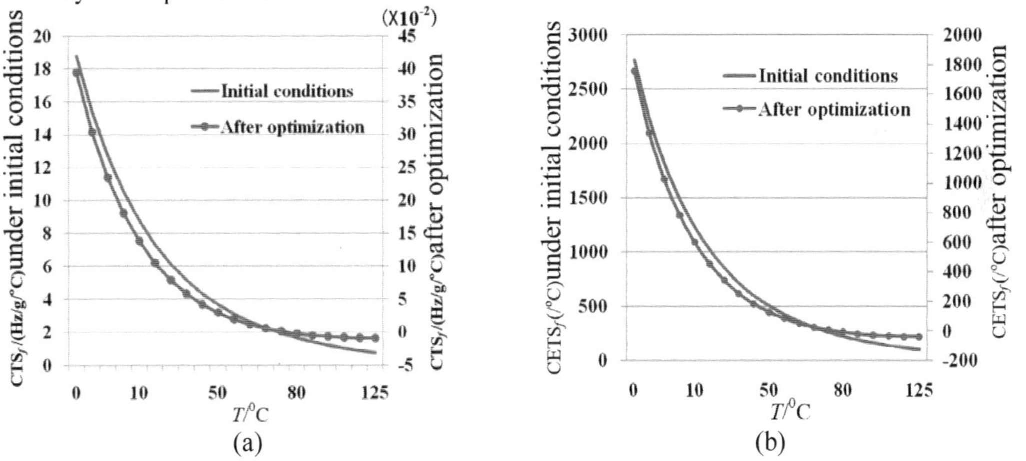

Figure 3. Compare the values of CT_{Sf} and CET_{Sf} under the initial conditions(K=55) and after optimization(K=60.5). (a) Simulated the temperature coefficient of scale factor CT_{Sf} versus temperature T(°C); (b) Simulated the temperature sensitivity of the scale factor CET_{Sf}. versus temperature T(°C).

Table 1. The simulation parameters.

Parameter	E_1 ,E_2(Gpa)	E_3(Gpa)	$h_1(um)$	$h_2(um)$	$h_3(um)$	$l(um)$
Value	160	62.7	40	4	50	400
Parameter	v_1 ,v_2	v_3	ε (F/m)	$m(kg)$	$M(kg)$	$A(m^2)$
Value	0.22	0.2	8.85e-12	2.86e-9	5.66e-8	5.12e-8
Parameter	$V_{d1}, V_{d2}\,(V)$	$V_c\,(V)$	α_{glass} (/°C)	$g_0(um)$	$b(um)$	$L(um)$
Value	8	0.1	3.25e-6	2	8	200
Parameter	$T(^oC)$	$T_0(^oC)$	a_1	a_2	η	$g(m/s^2)$
Value	25	355	500	-12.3	0.39	9.8
Parameter	C_1, C_2, C_3, C_4=-0.03,-9.37,0,42.89 ; C_5, C_6, C_7, C_8=0.04,12.41,0,10.76					

4. Conclusions

This paper mainly studies the temperature stability of the silicon resonant accelerometer based on electrostatic rigidity under residual thermal stress. In the study, the analytical formula of scale factor about residual thermal stress has been deduced, based on which the relevant parameters have been optimized. Simulation results show that the temperature stability of the scale factor has been improved significantly.

References

[1] Zhang F, Xiaoping H E, Shi Z, et al. Structure Design and Fabrication of Silicon Resonant Micro-accelerometer Based on Electrostatic Rigidity[J]. Lecture Notes in Engineering & Computer Science, 2009, 2176(1):37-45.

[2] B. L. Lee, C. H. Oh, Y. S. Oh, and K. Chun, "A Novel Resonant Accelerometer: Electrostatic Stiffness Type", The 10th International Conference on Solid State Sensors and Actuators (Transducer '99),Sendai, Japan, June 7-10, 1999, pp.1546-1549.

[3] Seonho Seok, Hak Kim, and Kukjin Chun. An Inertial-grade laterally-driven MEMS Differential Resonant Accelerometer.IEEE 2004:654-657.M. Young, The Techincal Writers Handbook. Mill Valley, CA: University Science,1989.

[4] Su S X P, Yang H S, Agogino A M. A resonant accelerometer with two-stage microleverage mechanisms fabricated by SOI-MEMS technology[J]. IEEE Sensors Journal, 2005, 5(6):1214-1223.

[5] Comi C, Corigliano A, Langfelder G, et al. A high sensitivity uniaxial resonant accelerometer[C]// IEEE, International Conference on MICRO Electro Mechanical Systems. IEEE, 2010:260-263.

[6] Huang L, Yang H, Gao Y, et al. Design and implementation of a micromechanical silicon resonant accelerometer[J]. Sensors, 2013, 13(11):15785-804.

[7] Jing Z, Yan S, Qin S, et al. Microelectromechanical Resonant Accelerometer Designed with a High Sensitivity[J]. Sensors, 2015, 15(12):30293-310.

[8] Wu Zhou, Yu Chen, Bei Peng, Hui Yang, Huijun Yu, Heng Liu, and Xiaoping He. Air Damping Analysis in Comb Microaccelerometer. Adv. Mech. Eng., Volume 2014, Article ID 373172, 6 pages.

[9] Wu Zhou, Jiangbo He, Xiaoping He, Huijun Yu, Bei Peng. Dielectric charging induced drift in micro device reliability-a review. Microelectronics Reliability 66 (2016) 1–9

[10] Wu Zhou, Peng Peng, Huijun Yu, Bei Peng, and Xiaoping He. Material Viscoelasticity-Induced Drift of Micro-Accelerometers. Materials 2017, 10, 1077.

[11] Melamud R, Hopcroft M, Jha C, et al. Effects of stress on the temperature coefficient of frequency in double clamped resonators[C]// The, International Conference on Solid-State Sensors, Actuators and Microsystems, 2005. Digest of Technical Papers. Transducers. IEEE, 2005:392-395 Vol. 1.

[12] Gianchandani Y B, Najafi K. A bulk silicon dissolved wafer process for microelectromechanical

devices[J]. Journal of Microelectromechanical Systems, 1992, 1(2):77-85.
[13] Walwadkar S S, Cho J, Farrell P W, et al. Tailoring of Stress Development in MEMS Packaging Systems[J]. Mrs Proceedings, 2002, 741.
[14] He L, Xu Y P, Qiu A. Folded silicon resonant accelerometer with temperature compensation[C]// Sensors, 2004. Proceedings of IEEE. IEEE, 2004:512-515 vol.1.
[15] Li T, Liu Y, Dong J, et al. Temperature characteristics of micro accelerometers with constant temperature chip sensor[J]. Journal of Tsinghua University, 2010, 50(7):1013-1017.
[16] Zheng C, Chen J. Study of a Novel Method of Temperature Compensation for MEMS Accelerometer[J]. Chinese Journal of Sensors & Actuators, 2015.
[17] Zwahlen P, Dong Y, Nguyen A M, et al. Breakthrough in high performance inertial navigation grade Sigma-Delta MEMS accelerometer[C]// Position Location and Navigation Symposium. IEEE, 2012:15-19.
[18] Dong J H, Qiu A P, Shi R. Temperature influence mechanism of micromechanical silicon oscillating accelerometer[C]// Power Engineering and Automation Conference. IEEE, 2011:385-389.
[19] E. Suhir. Calculated thermally induced stresses in adhesively bonded and soldered assemblies[C]. ISHM International Symposium on Microelectronics, Atlanta, USA, 1986
[20] Lai Liushan. The Effect of the Adhesive Void in Packaging Structure on the Thermal Drift of Micro-accelerometer[D].University of Electronic Science and Technology of China, 2016.
[21] E. Suhir, “Analysis of Interfacial Thermal Stresses in a Trimaterial Assembly,” J. Appl. Phys., vol. 89, no. 7, pp. 3685–3694, 2001.

IOP Conf. Series: Journal of Physics: Conf. Series **986** (2018) 012033 doi:10.1088/1742-6596/986/1/012033

The design method and research status of vehicle detection system based on geomagnetic detection principle

Y H Lin [1], R Bai [1] and Z H Qian [1]

[1]Center for Integrated Spintronic Devices, Hangzhou Dianzi University, Hangzhou, Zhejiang, China

E-mail: 540140435@qq.com

Abstract. Vehicle detection systems are applied to obtain real-time information of vehicles, realize traffic control and reduce traffic pressure. This paper reviews geomagnetic sensors as well as the research status of the vehicle detection system. Presented in the paper are also our work on the vehicle detection system, including detection algorithms and experimental results. It is found that the GMR based vehicle detection system has a detection accuracy up to 98% with a high potential for application in the road traffic control area.

1. Introduction

Control and management play an important role in improving and optimizing the road traffic. Vehicle detection systems are used to obtain traffic flow information. The research on vehicle detection dates back to the 1980s [1]. The commonly used vehicle detection technology include ring coil, ultrasonic, infrared, geomagnetic and video. However, these detection techniques generally have a varied of disadvantages, including low detection accuracy, complicated installation, high cost and high power consumption [2]. In recent years, due to the rapid development of solid state magnetic sensors with low power and high sensitivity, it becomes feasible to develop vehicle detection with high precision, easy installation, low power consumption and low cost, that can meet practical applications.

In this paper, reviewed first are the magnetic sensors and vehicle detection systems. Followed are the design and results by our research group, demonstrating that a detection accuracy could be more than 98%.

2. Geomagnetic Sensors and Detection Principle

Geomagnetic sensors, which can detect changes in the geomagnetic field, can be used for road vehicle detection and this technology is widely used in the field of road traffic detection. The geomagnetic sensors used for vehicle detection in the current market mainly include AMR sensor, GMR sensor and TMR sensor. These three kinds of geomagnetic sensors have different principles, and also exhibit different characteristics.

The AMR sensor is short for anisotropic magneto resistive sensor based on anisotropic magneto resistive effect, which is defined as the change in the resistivity of ferromagnetic material with the change of its magnetization and the angle of the direction of the current [2]. The AMR bridge structure under the action of external magnetic field is shown in Figure 1-(a).

$$R_1 = R_2 = R_3 = R_4 = R \tag{1}$$

$$V_{out} = V_{CC} \times \frac{\Delta R}{R} \tag{2}$$

In the equation (1), R is the resistance of four anisotropic magnetic resistors in the AMR bridge. In the equation (2), V_{out} is the output voltage of AMR sensor under the action of external magnetic field and ΔR stands the change in resistance values of four anisotropic magnetic resistors under external magnetic field. V_{CC} is the supply voltage.

The GMR sensor is short for giant magneto resistive sensor based on giant magneto resistive effect, which is defined as that the resistivity of the magnetic material has great changes in the external magnetic field compared with no external magnetic field. The GMR bridge structure under the action of external magnetic field is shown in Figure 1-(b).

$$V_{out} = V_{CC} \times \frac{\Delta R}{2R + \Delta R} \tag{3}$$

In the equation (3), V_{out} is the output voltage of GMR sensor under the action of external magnetic field and R is the resistance of four giant magnetic resistors in the GMR bridge. ΔR stands the change in resistance values of four giant magnetic resistors under external magnetic field. V_{CC} is the supply voltage.

The TMR sensor is short for tunnel magneto resistive sensor based on tunnel magneto resistive effect, which is that the magnetization direction of the ferromagnetic slice can be independently switched under the action of an external magnetic field. The TMR bridge structure is shown in Figure 1-(c).

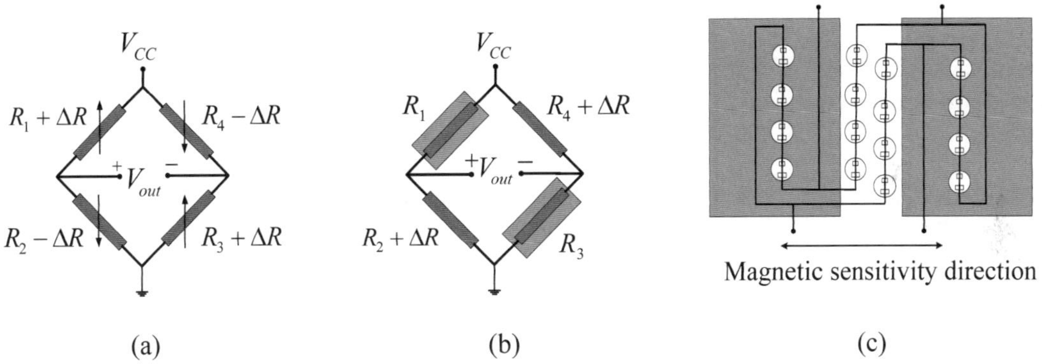

Figure 1. Bridge structure of geomagnetic sensor.

In the vehicle detection, the key parameter is the geomagnetic field intensity, which locates in the range of 0.5Oe-0.6Oe. From Table 1, it can be seen that the three kinds of geomagnetic sensors can meet the application requirements of geomagnetic vehicle detection. The AMR sensor is the most widely used geomagnetic sensor, which has the advantages of good linearity and low hysteresis. However, due to the fact that the linear region is narrow and the detection field is easy to saturate, the bias voltage of the sensor needs to be removed by simply switching the magnetic properties of the iron nickel film. The GMR sensor is simple to use and can work normally only with a stable power supply. GMR sensors are fabricated by semiconductor integrated process, and they are characterized by small size, high sensitivity and good linearity. They are widely used in electronic compass, gear sensors, current sensors, and magnetic switches having a very broad market prospects. The TMR sensor is the fourth generation of magneto resistive sensors in the world. Compared with GMR sensors and AMR sensors, it has great advantages in size, power consumption, sensitivity, linearity, range of operation and temperature stability. However, the technology of TMR sensor is still not mature enough, and products in this field are rarely to be seen.

Table 1. Parameters of geomagnetic sensor.

Sensor	Size (*mm*)	Temperature (℃)	*Consumption* (*mA*)	Sensitivity (*mv/V/Oe*)	*Working range* (*Oe*)
AMR	1*1	<150	1~10	1	0.001~10
GMR	2*2	<150	1~10	3	0.1~30
TMR	0.5*0.5	<200	0.001~0.01	20	0.001~200

Currently, there are many researches on geomagnetic detection vehicle detection system around the word. Kang of Korea designed a vehicle detector based on AMR sensor. The detector consists of a single axis AMR sensor and a corresponding peripheral circuit. The experimental results from three stages show that the vehicle detector has good detection performance in low speed traffic measurement with an average error of 0.3% [3].

Honeywell uses an AMR sensor of three axes for vehicle detection test, and analyses the output characteristics on three axes. The experimental results show that when the vehicle is farther from the sensor, the sensitivity of the sensor is lower. The single axis AMR sensor can be used to detect the direction and the appearance of the vehicle. The dual axis AMR sensor can be used to determine the speed and analyze the direction of the vehicle. The three axis AMR sensor can be used in vehicle classification [4].

The MIT Knaian uses HMC1021S, a single axis AMR sensor produced by Honeywell, to conduct vehicle detection test. The system receives and processes the data through the TMS430 microprocessor, and transmits the data to the base station through the wireless communication module. For parking space detection, the system uses the state machine algorithm. The vehicle detection system is tested on a street, and the results show that the system can accurately detect the traffic flow at different times [5].

The Harbin Institute of Technology has designed a hardware platform which uses single axis TMR sensor as detection unit, ZigBee as wireless communication protocol. The system can monitor vehicle information and upload data in real time. The system adopts the baseline adaptive state machine detection algorithm. Through the comparison and analysis of the actual test results, using BP neural network as the system vehicle classification algorithm, the recognition accuracy can reach 88.92%.

3. Design and Detection Algorithm

In 2012, our research team designed a vehicle detection system based on GMR sensors. The system uses the single axis SAS03-1 sensor produced by SpinIC as the geomagnetic field information acquisition module. The test result shows that the system can meet the practical application requirements. In 2014, based on the previous work, our research team developed a wireless parking space detection system based on GMR sensor of three axes. The system adopts the improved fixed threshold algorithm. The baseline value is updated by collecting the baseline value of the parking space in the condition of no car regularly. In 2015, our research group improved the algorithm of parking detection system, and developed a two level baseline tracking detection algorithm. The algorithm sets two thresholds, controlling the baseline tracking by small threshold, and judging the parking space status through the large threshold. By setting two thresholds, the algorithm avoids the system errors caused by the drift of the baseline, and also compensates for the system errors due to vehicle disturbances.

In 2017, our research team designed a parking area detection system integrating the ZigBee network into the Ethernet. In the hardware aspect, the detection node of the system uses dual axis SAS022-1 and single axis VA100F3 produced by SpinIC as the geomagnetic field information acquisition module, CC2530 as the main controller and RF transceiver, ZigBee protocol for wireless communication, TPS63001 as the master chip for power management module. The detection nodes object is shown in Figure 2 and Figure 3 shows the hardware framework of the detection node.

Figure 2. Detection nodes object.

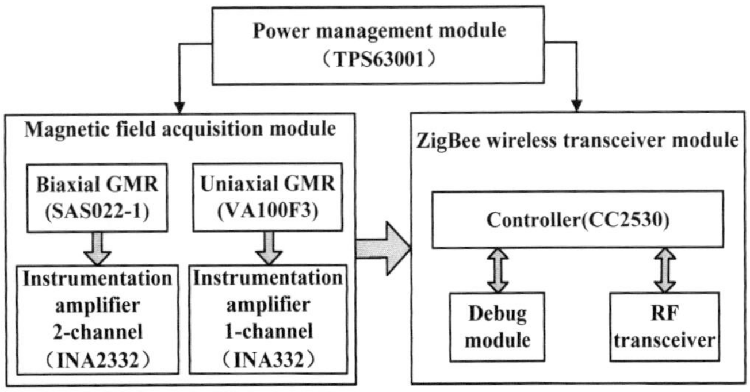

Figure 3. Hardware framework of detection node.

In the software aspect, the system first uses the sliding average filter algorithm to filter the signals collected by the GMR sensors, and this method can effectively remove noise interference. The equation (4) shows the principle of the sliding filter algorithm.

$$T(k)=\begin{cases}\dfrac{A(1)+A(2)+\cdots+A(k)}{k} & ,k<N\\[2ex] \dfrac{A(k-N+1)+A(k-N+2)+\cdots+A(k)}{N} & ,k\geq N\end{cases} \tag{4}$$

In the equation (4), $A(k)$ stands the original signals collected by GMR sensors and $T(k)$ is the filtered signals. N is the window length of the sliding average filter defined as 5 in the system.

The system uses the baseline adaptive state machine detection algorithm to determine the parking space status. The flow chart of the baseline adaptive state machine detection algorithm is shown in Figure 4. In the flow chart, T_{min} stands for small thresholds, and T_{max} means large thresholds. Firstly, the baseline value is determined in the vehicle detection region without vehicle and the value is defined as T_v. After the baseline value has been successfully initialized, the magnetic time sequence of the sensor is input to the state machine defined as T. If $|T-T_v|$ is less than T_{min}, it is non-triggering state at this time and the state can be judged as the interference of the external environment. Then the weighted function should be used to update the baseline value and continue to collect the magnetic field data of the detection area as shown in equation (5); if $|T-T_v|$ is greater than T_{min} and less than T_{max}, it is Semi trigger state at this time and the state can be judged as the interference of the vehicle. Then the baseline updates should be stopped at this point; if $|T-T_v|$ is greater than T_{max}, it is fully triggered state at this time and the vehicle has completely entered the detection area at this point. Also the baseline remains unchanged.

$$T_v(k)=\begin{cases}T_v(k-1)\times(1-a)+T(k)\times a\\ T_v(k-1)\end{cases},(0\leq a\leq 1) \tag{5}$$

In the equation (5), $T_v(k)$ is the current updated baseline value, and $T_v(k-1)$ is the baseline value of the previous state. $T(k)$ is the current measurements, and a is the weighting coefficient.

The detection algorithm combines the baseline adaptive algorithm with the state machine algorithm and avoids the shortcomings of the two algorithms, the accuracy of the system is therefore improved.

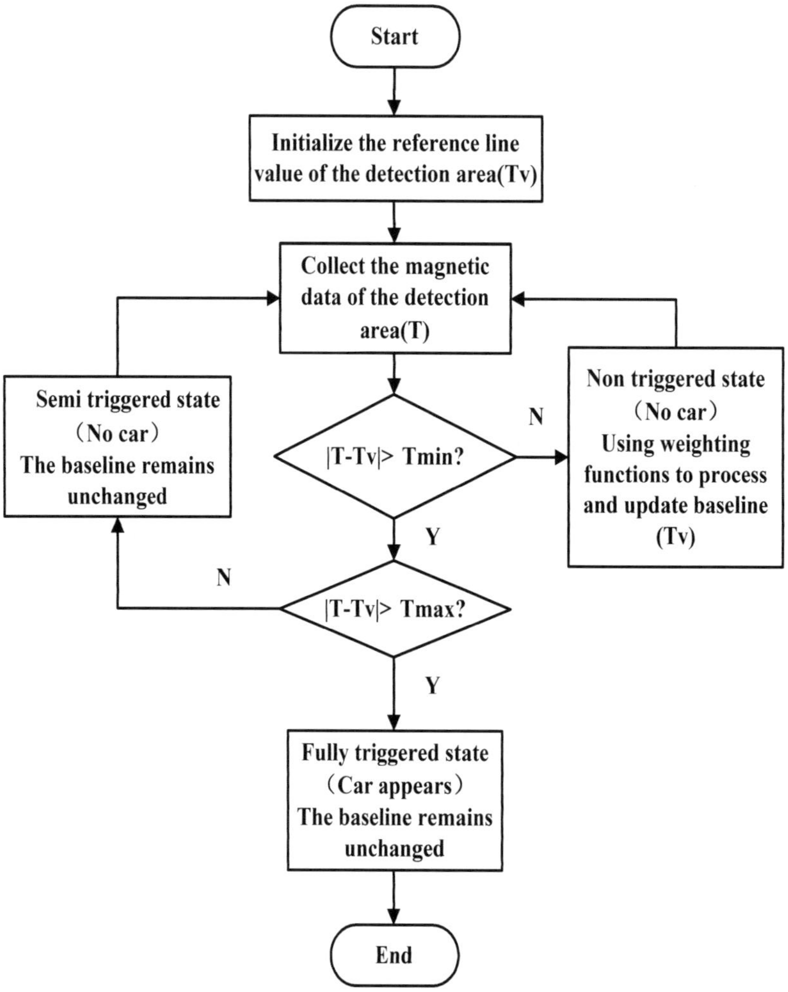

Figure 4. Baseline adaptive state machine algorithm.

4. Experimental Results

In 2017, our research team tested the improved vehicle detection system. Considering the actual application of parking spaces, the detection nodes were placed in the center of the parking space to minimize the false detection rate and improve the stability of the system. For the No.1 parking space, a household car was made to pass from the south to the north with a low speed, while for the No.2 parking space the same car was made to pass from the east to the west. And the nodes collected the data of magnetic field disturbance generated by vehicle.

Figure 5-(a), (b) and (c) show the curves of the output voltage of X axis, Y axis and Z axis respectively as the vehicle passes through the GMR sensor in the parking spaces. In the figures, X1, Y1, Z1 represent the voltage output curves of the vehicle passing from the east to the west. X2, Y2, Z2 represent the voltage output curves of the vehicle passing from the south to the north. A1 and A2 indicate the time when the headstock passes through the sensor. B1 and B2 indicate the time when the tailstock passes through the sensor.

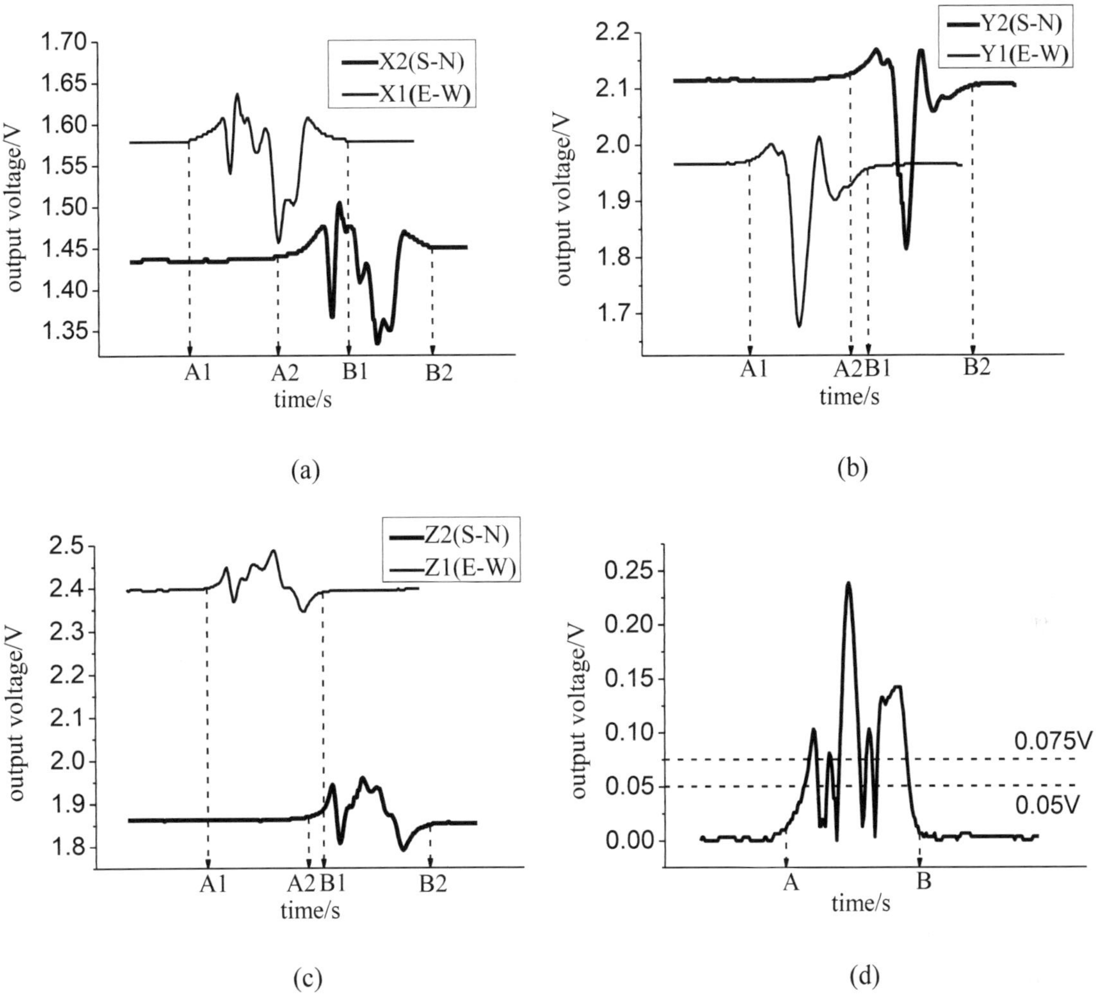

Figure 5. Three axis curve of output.

According to the voltage output, the following conclusions can be obtained: Firstly, the law of magnetic disturbance of parking space is not related to the geographical direction of parking space. It is only determined by the direction of sensitive axis. Secondly, the same sensitive axis shows the same output voltage under the same vehicle interference. Thirdly, when the vehicle is far away from the sensor, the output voltage of the three axes remains unchanged. When the vehicle passes through the sensor, the output voltage changes significantly, especially when the wheel and the engine pass through the sensor. This is mainly because the wheel and engine are composed of ferromagnetic materials, which can cause a great disturbance to the earth's magnetic field, and leads to a great change of the output voltage.

Judging from the variation of the single axis, the disturbance of the vehicle may be caused by certain error. In order to improve the reliability of the system, the output voltage of the three axes should be analyzed comprehensively [6]. In this study, the output voltage is analyzed in accordance with equation (6).

$$H(k)=|X(k)-x|+|Y(k)-y|+|Z(k)-z| \tag{6}$$

In the equation (6), $X(k)$, $Y(k)$ and $Z(k)$ respectively represent the output voltages. x, y and z respectively indicate the stable baseline voltage of the three axes of the sensor without interference from the vehicle. The $H(k)$ represents the result of the combined processing of the three axes output voltage as shown in Figure5-(d), where the point A indicates that the headstock passes the sensor, and the point B indicates that the end of the vehicle passes the sensor. Through this method, the overall variation of the geomagnetic field under the influence of vehicle interference can be analyzed more clearly and intuitively.

According to Figure5-(d), when the vehicle is far away from the sensor, the output of $H(k)$ is 0 and remains stable. From the A moment to the B moment, $H(k)$ has a more obvious output change. When the vehicle engine and wheel close to the sensor, $H(k)$ has the most obvious change. The changing of $H(k)$ are basically above 0.05V, and at this time the vehicle is in the parking space. Therefore, $H(k)$ parameter can provide a reliable reference for the system to set the appropriate threshold in practical applications. Finally, the system is installed in the parking lot for real test, and test results show that the detection accuracy can reach 98%.

5. Conclusions

The vehicle detection system designed by our research team has a detection accuracy up to 98% with the advantages of easy installation, low power consumption and low cost. The system with three-axial GMR sensor can avoid the certain error by single axis. Moreover, the detection algorithm of the baseline adaptive state machine avoids the drawbacks of the baseline adaptive algorithm and the state machine algorithm.

Although the research with respect to the vehicle detection system has made a great progress, the vehicle detection system has some limitations when it is put into practical use in the increasingly complicated traffic environment. Further studies in the future is still needed for the improvement of the vehicle detection system in hardware characteristics and detection algorithms. The present study provides valuable reference for the development of geomagnetic vehicle detection systems in the future.

References

[1] Mimbela L E Y and Klein L A 2000 Summary of vehicle detection and surveillance technologies used in intelligent transportation systems
[2] Caruso M J, Bratland T, Smith C H and Schneider R 1998 A new perspective on magnetic field sensing *Sensors* **15** 34-47
[3] Kang M H, Choi B W, Koh K C, Lee J H and Park G T 2005 Experimental study of a vehicle detector with an AMR sensor *Sensors and Actuators A Physical* **118** 278-284
[4] Caruso M J and Withanawasam L S 1999 Vehicle detection and compass applications using AMR magnetic sensors *Sensors Expo Proceedings* **477** 39
[5] Knaian A N 2000 A wireless sensor network for smart roadbeds and intelligent transportation systems *Massachusetts Institute of Technology*
[6] Ding J, Cheung S Y, Tan C W and Varaiya P 2004 Signal processing of sensor node data for vehicle detection *Intelligent Transportation Systems* 70-75

IOP Conf. Series: Journal of Physics: Conf. Series **986** (2018) 012034 doi:10.1088/1742-6596/986/1/012034

Design, Simulation and Characteristics Research of the Interface Circuit based on nano-polysilicon thin films pressure sensor

Xiaosong Zhao[1, 2], Xiaofeng Zhao[1*], Liang Yin[2]

[1]The Key Laboratory of Electronics Engineering, College of Heilongjiang Province, Heilongjiang University, Harbin, China.
[2]MEMS center, Harbin institute of technology , Harbin, China

*zhaoxiaofeng@hlju.edu.cn

Abstract. This paper presents a interface circuit for nano-polysilicon thin films pressure sensor. The interface circuit includes consist of instrument amplifier and Analog-to-Digital converter (ADC). The instrumentation amplifier with a high common mode rejection ratio (CMRR) is implemented by three stages current feedback structure. At the same time, in order to satisfy the high precision requirements of pressure sensor measure system, the $1/f$ noise corner of 26.5 mHz can be achieved through chopping technology at a noise density of 38.2 nV/sqrt(Hz).Ripple introduced by chopping technology adopt continuous ripple reduce circuit (RRL), which achieves the output ripple level is lower than noise. The ADC achieves 16 bits significant digit by adopting sigma-delta modulator with fourth-order single-bit structure and digital decimation filter, and finally achieves high precision integrated pressure sensor interface circuit.

1. Introduction

With the development of microelectronics mechanical system (MEMS) technology and the improvement of integrated circuit technology, integrated sensor with signal processing circuit has been applied in car electronics, aeronautics and astronautics, etc. Due to the output of pressure sensor is about a few tens of millivolt, amplifier often uses to amplify the output then utilizing the ADC to convert the analog signal into digital signal. However, the output signal of pressure sensor is always effected by noise and offset of interface circuit, which seriously reduced the pressure sensor performance. To maintain interface circuit of pressure sensor accuracy, the instrumentation amplifier should be low noise and high CMRR. The input noise and offset of typical complementary metal oxide semiconductor (CMOS) amplifiers is at the millivolt level, limiting their accuracy severely. The need for precise pressure measurement is the driving force behind a continuous effort to reduce the offset and noise of CMOS amplifier. The technology can be taken use of includes chopping and auto-zero technology. Because auto-zero technology adopts sampling, so noise aliasing will be induced to signal band, which would reduce circuit performance. Therefore, chopping technology is more suitable for low-noise instrumentation amplifier. In addition, in order to facilitate subsequent digital circuit processing, a ADC with high signal to noise ratio (SNR) for high precision measurement system is needed. Among various ADC architectures, the Sigma-Delta ADC can realize high SNR by using oversampling and noise shaping technologies [1], which trade speed for accuracy and low requirement to integrated circuit manufacturing process as well as reduce the difficulty of the analog integrated circuit design. Those properties exactly fit for pressure sensor interface circuit.

Based on characteristics of nano-polysilicon thin films pressure sensor, a high CMRR and high precision interface circuit presents in this paper. Chopping technology was chosen in this design to achieve low-noise instrumentation amplifier. For the complete realization of digital output interface circuit, a Sigma-delta ADC was designed in the system.

2. Principle of pressure sensor interface circuit

2.1. Characteristics of pressure sensor

In order to measure the environment pressure, a block diagram of pressure sensor interface circuit was present in figure 1(a). The sensitive element of pressure sensor is four nano-polysilicon thin films piezoresistors (*R*1, *R*2, *R*3 and *R*4) connected by the Wheatstone bridge structure, which has been represented in the dashed block (as shown figure 1(a)). The characteristics of the pressure sensor is shown in figure 1(b).The sensitivity of pressure sensor is 0.146 mV/kPa, when the supply is 3 V [2]. The pressure sensor interface circuit is generally divided into two parts: instrumentation amplifier and ADC. The instrumentation amplifier can amplify the output signal of pressure sensor and the ADC can digitize the output of amplifier.

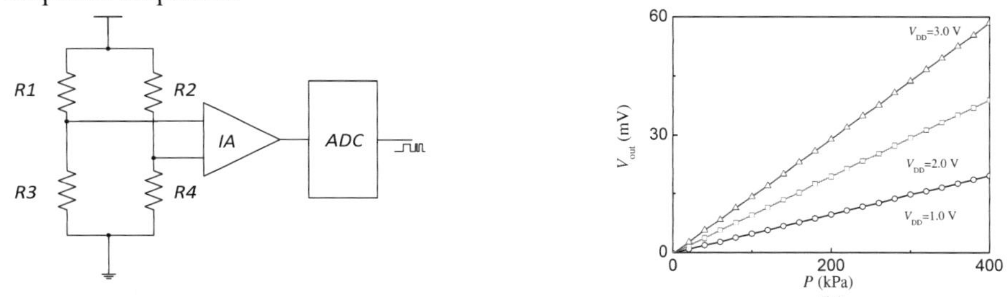

(a) (b)

Figure 1. An interface circuit block diagram and sensitive characteristics of pressure sensor based on nano-polysilicon thin films
(a) the overall block diagram of interface circuit (b)the characteristics of the pressure sensor

2.2. Principle of instrumentation amplifier

The instrumentation amplifier adopts current feedback instrumentation amplifier (CIFA) structure is shown in figure 2. This structure has high CMRR, meanwhile output common-mode voltage has no relationship with input common-mode voltage, which is fit for sensor interface circuit with the advantage of low power consumption. In order to satisfy the characteristic of low noise required by the system, chopping technology is used to reduce low frequency noise and offset. By using a three stage nested-Miller topology, and chopping both the input stage and intermediate stage as well as the third stage noise is too small to neglect, which can satisfied the requirements of the system. The ripple induced by chopping technology adopts continuous time ripple-reduce-loop (RRL) to suppress the influence of ripple on the system [3-4], achieving the demands of system on noise and offset.

The principle of the whole instrumentation amplifier is as follows: the input voltage is corresponding to the current through G_{m1}, and the feedback voltage is converted to corresponding current through Gm4, then, the current difference between the output current of G_{m1} and G_{m4} is nulled by G_{m2} and G_{m3}.

According to principle of instrumentation amplifier, the gain (A_v) of proposed instrumentation amplifier can be indicated as eq. (1),

$$A_V \simeq \frac{G_{m1}}{G_{m4}} \cdot \frac{R_G + R_{21} + R_{22}}{R_1} \tag{1}$$

where G_{m1} and G_{m4} is amplifier transconductance of input stage, R_G, R_{21} and R_{22} are resistances which use to modify the gain.

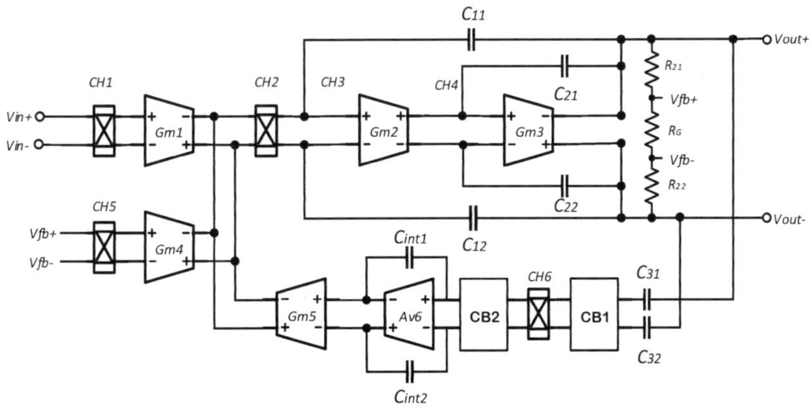

Figure 2. The overall block diagram of instrumentation amplifier circuit

As shown in the figure 2, CH1, CH5, CH2 and CH6 is chopper, the noise and offset of the input stage and the intermediate stage are modulated to high frequency by chopper CH2, transmitted to the output end through capacitance ($C_{11,}$ C_{12}). Without RRL circuit, the ripple in output end will be large, which will influence the performance of measuring circuit. The principle of RRL is: output high frequency ripple current is converted into high frequency voltage through two capacitance ($C_{31,}$ C_{32}), and then converted into direct voltage through chopper CH6, and take use of G_{m5} to convert voltage into current to make the feedback of the output end of G_{m1} and G_{m4} so as to achieve the goal of negative feedback. Signal transfer function of the whole loop can be presented as:

$$H(s)=\frac{1}{1+L(s)} \tag{2}$$

$$L(s)=\frac{A_{v6}G_{m5}}{C_2 f_{ch1}} \tag{3}$$

where $H(s)$ is the signal transfer function of RRL and $L(s)$ is the transfer function from input end of chopper CH3 to output end of chopper CH2, A_{v6} is the gain of integration amplifier, C_2 is high frequency path capacitance, f_{ch1} is chopping frequency.

Through feedback loop, the output voltage of ripple is:

$$V_{out,ripple}=\frac{V_{OS}G_{m3}}{2A_{v6}G_{m}} \tag{4}$$

where the V_{os} is the input offset.

The unity-gain bandwidth f_{notch} of the loop can be derived from eq. (2) by setting L(s) =1;

$$f_{notch}=\frac{G_{m5}C_3}{2\pi C_2 C_{in}} \tag{5}$$

where C_{int} is integration capacitance, C_3 is capacitance that using to transfer current to voltage.

Therefore, RRL achieves notch-filtering function with the $2f_{notch}$ bond width of chopping frequency. So, the ripple can be filter out within frequency at chopping frequency f_{ch} around $2f_{notch}$. The magnitude of output ripple would be determined by G_{m3}, G_{m5} and $A_{v6.}$

2.3. *Details implementation of the instrumentation amplifier*

Because output noise equivalent to the input needs to be divided by the first-level gain, so the first-level gain must be very large. This design employed a folded cascode gain-boosted topology as shown in the figure 3(a), G_{m1} and G_{m4} use the same output stage, which lowers the influence of device mismatching on system, at the same time omits the difficulty of designing feedback transconductance and saves power dissipation. The auxiliary amplifiers also adopt full differential folded-cascode operational amplifier structures. The gain of such structure can implement over 140 dB, and the 1/*f* noise is about 10 μV/sqrt(Hz).

Then, the second stage is implemented using a folded-cascode topology. The schematic diagram of the second and output stages is shown in figure 3(b). In order to increasing the range of output voltage, a class-AB output stage is implemented to achieve rail-to-rail output. Because the noise of output stage is suppressed by the gain of the preceding stages, the output stage does not use chopping technology.

The detailed implementation of the cascade buffers (CB) is shown in figure 3(c), CB isolates from C_3 to enhance the ripple reduction. To increase the output impedance of CB, a gain-boosting topology is employed.

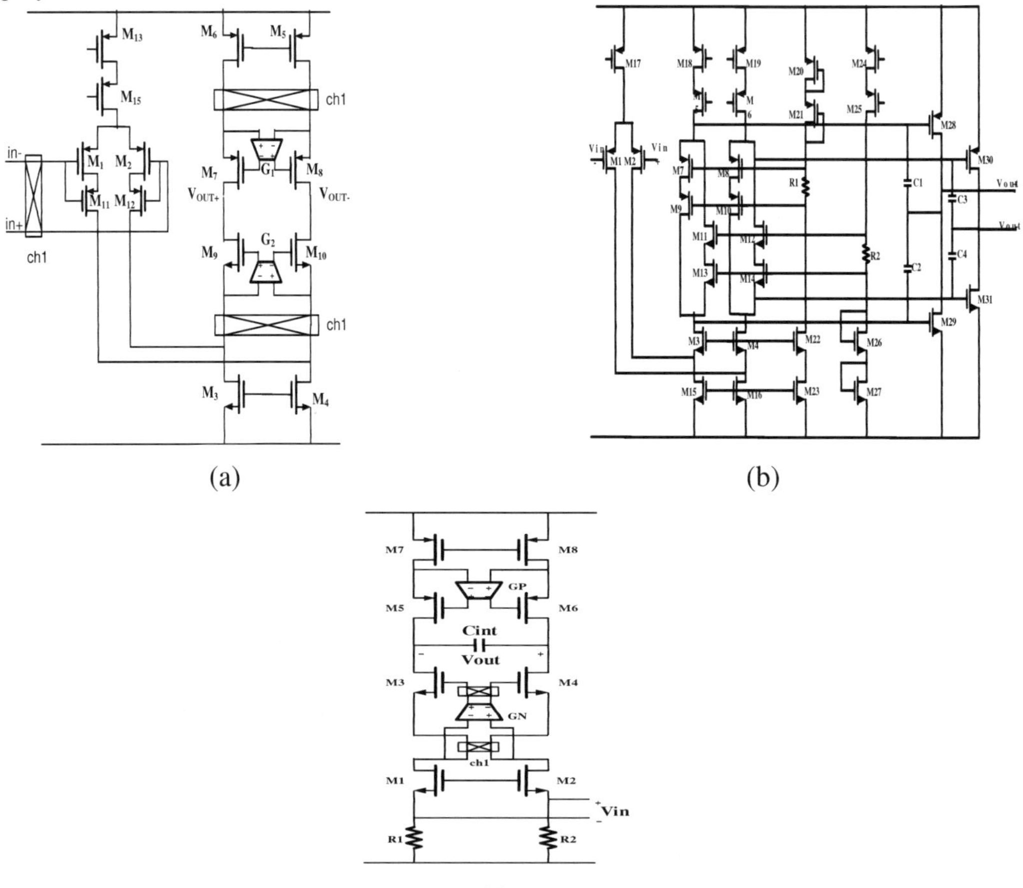

(a) (b) (c)

Figure 3. The schematic diagram details implementation of the instrumentation amplifier (a) schematic diagram of the input stage (b) schematic diagram of second and output stages circuit (c)Implementation of the gain-boosted Cascades buffer

2.4. *Principle of Sigma-Delta ADC*

For implementing high precision analog-to-digital converter, Sigma-delta modulator is chosen. Four-order single-bit modulator is selected to achieve signal bandwidth with 10kHz and ADC with 16 bits resolution. Sigma-delta modulator mainly adopts over-sampling technology and noise-shaping technology to achieve high resolution. The SNR can be indicated as eq. (6)

$$SNR = 6.02N + (2L+1)10\log_{10} OSR - 10\log_{10}\left(\frac{\pi^{2L}}{2L+1}\right) \tag{6}$$

where L is the modulator order, N is the number of quantizer bit and OSR is oversampling ratio [5].

Quantized noise within signal bandwidth can be modulated to the high frequency through these two technologies, then acquire the interest signal by using digital decimation filter to remove the useless noise, and finally accomplish the design of pressure sensor interface circuit.

2.5. The details implementation of the Sigma-Delta ADC

2.5.1. Sigma-Delta Modulator

The structure of Sigma-delta Modulator is shown in figure 4. The design of Sigma-delta Modulator is mainly the design of NTF, which determines the upper limit of system and minimizes the influence of non-ideal factors on system as much as possible. The CIFF structural fourth-order single bit is adopted in the design, with Simulink ideal model simulation, the preliminary structure and structural coefficients are obtained, as shown in table 1. Considering non-ideal factors (e.g. the influence of operational amplifier, switch, capacitance, clock and other modules in Sigma-delta system), the circuit structural parameters are designed based on non-ideal model. In addition, according to the requirements of system for every module parameters, the specific module is designed.

The integrator adopts switch capacitance structure which can reduce the affect of clock jitter. As show in the figure 4, the amplifier of the first stage integrator employed by gain-boosting structure, which can reduce the effect that amplifier gain changed by output voltage. Due to the noise floor of Sigma-delta modulator mainly determined by the first stage amplifier, the chopping technology also fit for reduce the noise of the first stage amplifier. The amplifier of rest of integrator can be shaping by noise-shaping technology, so the basic structure of fully differential folded-cascode amplifier can realize the system requirements. By appropriate designing the timming, the circuit that satisfy the system requirements is designed finally.

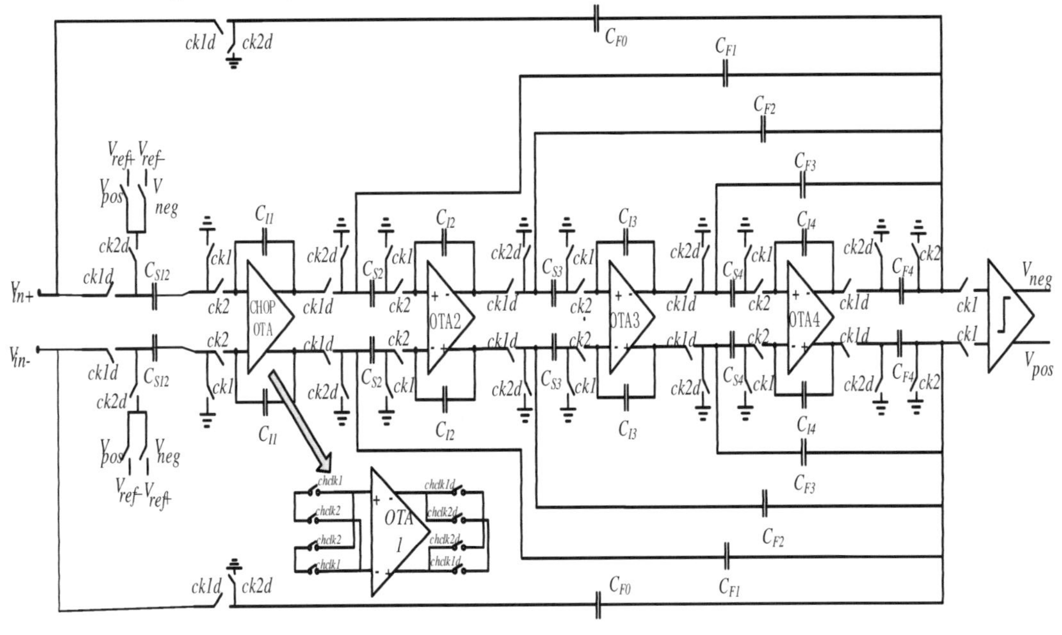

Figure 4. Implementation of CIFF fourth-order single bit Sigma-delta Modulator

Table 1. Structural coefficients of Sigma-Delta Modulator

C1	C2	C3	C4	A1	A2	A3	A4
C_{S1}/C_{I1}	C_{S2}/C_{I2}	C_{S3}/C_{I3}	C_{S4}/C_{I4}	C_{F1}/C_{F0}	C_{F2}/C_{F0}	C_{F3}/C_{F0}	C_{F4}/C_{F0}
1/3	1/3	1/3	1/5	1.8	2.8	1	1.2

2.5.2. Digital decimation filter

Considering to reduce the cost and power consumption of digital circuit, the cascaded integrator comb (CIC) filter was use as decimation filter which can be implement by integrator and differentiators. In order to filter the out-band-noise, five order comb filter was selected. For downsampling ration is N, the transfer function of comb filter in Z domain can be indicated as eq. (7)

$$H(z)=\left[\frac{(1-z^{-N})}{N(1-Z^{-1})}\right]^5=\frac{1}{1-Z^{-1}}\frac{1}{1-Z^{-1}}\frac{1}{1-Z^{-1}}\frac{1}{1-Z^{-1}}\frac{1}{1-Z^{-1}}\cdot\frac{1-Z^{-N}}{N}\frac{1-Z^{-N}}{N}\frac{1-Z^{-N}}{N}\frac{1-Z^{-N}}{N}\frac{1-Z^{-N}}{N} \quad (7)$$

The expression can be transfer to block structure as shown in figure 5.

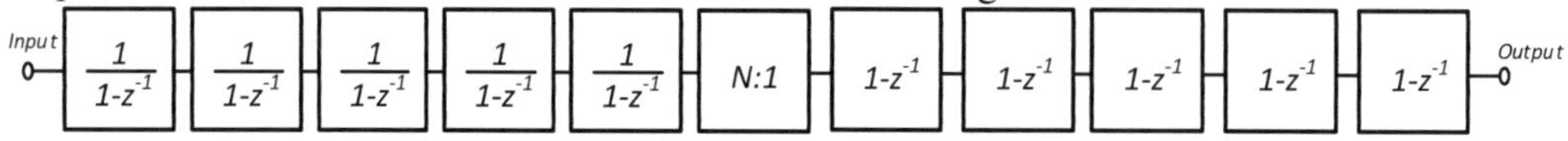

Figure 5. Block structure of CIC filter

However, after the CIC filter processing, the passband damping is serious, a Finite Impulse Response (FIR) trimming filter is needed. Then through Verilog language describes the behavior then transfer into digital circuit.

3. Results and Discussion

3.1. simulation results of AC response and noise

The pressure sensor interface circuit was realized in a 0.35μ m CMOS process. By scanning the input signal frequency range from 1 Hz to 1 GHz, simulation result presents that the unity-gain bandwidth of CIFA is 336 kHz and the gain is 40 dB. A notch around chopping frequency can be seen in the figure6, which is 6.1 kHz notch width . The outcomes satisfy the designed requirement.

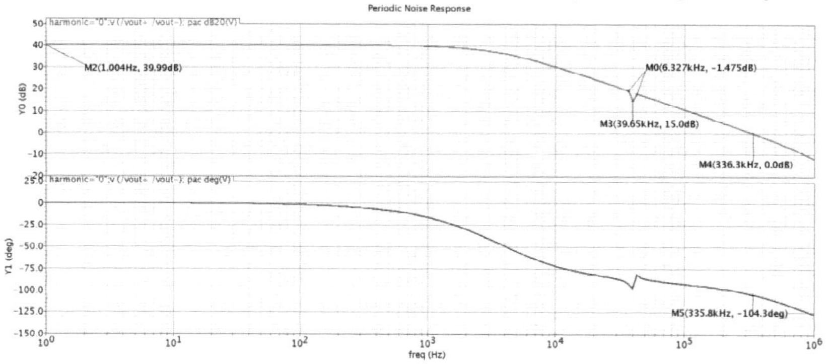

Figure 6. The periodic AC response of instrumentation amplifier

The noise simulation result is shown in figure 7, under 20 μ V offset, the 1/f noise corner is 26.5 mHz at a noise density of 38.2 nV/sqrt(Hz) of the amplifier can be achieved through chopping technology. So the noise effect induced by interface circuit can be neglected.

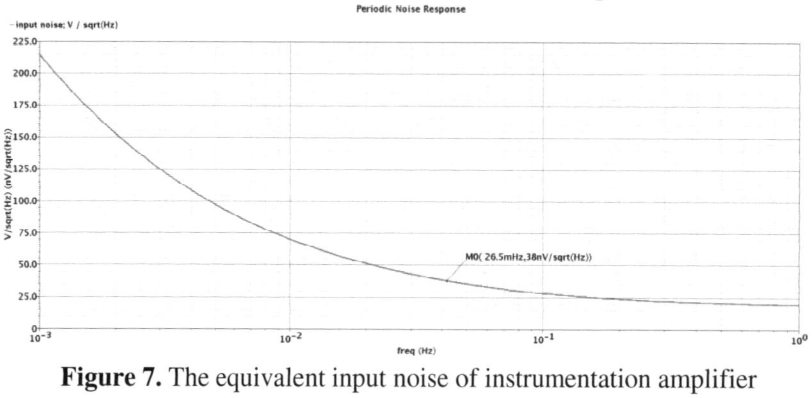

Figure 7. The equivalent input noise of instrumentation amplifier

3.2. *Simulation results of Sigma-Delta ADC*

The voltage and frequency of input signal (amplified output signal of pressure sensor) in simulation condition are 1 V and 488.28125 Hz, respectively. By computing 65536 points simulation data, the simulation result of Sigma-delta modulator is shown in figure 8. The power spectral density figure shows that the design goal of 16 bits (effective number of bits) was realized.

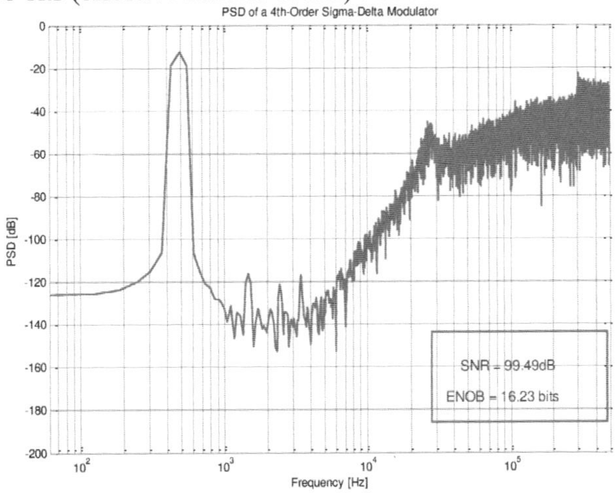

Figure 8. The Sigma-delta modulator output power spectral density

The simulation result of digital decimation filter is shown in figure 9, realizing the 16 bits significant digit output.

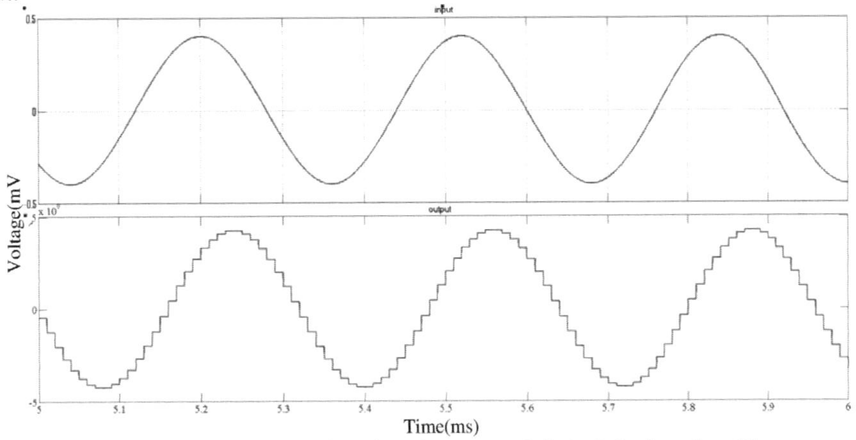

Figure 9. The input signal and output of digital decimation filter

4. Conclusions

Based on nano-polysilicon thin films pressure sensor, an integrated interface circuit was designed in the paper. After discussing the principle of instrumentation amplifier and Sigma-delta ADC, the details of design is given in the paper.The simulation result shows that the instrumentation amplifier with low noise 100 times of magnification and the ADC integrated with 16 bits significant digit is implemented. The instrumentation amplifier only consumes supply current of 620 μ A. The 1/f noise density is 38.2 nV/sqrt(Hz) at 26.5 mHz. While, the Sigma-delta ADC consumes 55 mW supply, realizing 99.49 dB SNR. The output of pressure sensor was fully high precision converted into digital signal, after the digital decimation filter. The design of high precision integrated pressure sensor interface circuit is completed.

References

[1] B.E. Boser, S.EJ. Wooley. The design of Sigma-Delta Modulator analogy-to-digital converters[J].

IEEE Journal of Solid-State Circuits, 1988, 23:1298-1308.
[2] Zhao Xiaofeng, Wen Dianzhong. Fabrication and characterization of nano-polysilicon thin film pressure sensor[J]. Journal of Semiconductors, 2008,29(10):2038-2042.
[3] Rong Wu, Kofi A. A. Makinwa, Johan H. Huijsing. A chopper current-feedback instrumentation amplifier with a 1 mHz 1/f noise corner and an AC-coupled ripple reduction loop[J]. IEEE Journal of Solid-State Circuits, 2009,44(12).
[4] Rong Wu, Johan H. Huijsing and Kofi A. A. Makinwa. A current-feedback instrumentation amplifier with a gain wrror reduction loop and 0.06% untrimmed gain error[J]. IEEE Journal of Solid-State Circuits, 2011, 46(12).
[5] Richard Schreier, Cabor C.Temes. Understanding delta-sigma data converters[M]. IEEE Press. 2005:91-121.

Acknowledgement

This work is supported by the National Natural Science Foundation of China (Grant No. 61471159, 61006057), the Special Funds for Science and Technology Innovation Talents of Harbin in China under Grant 2016RAXXJ016.

IOP Conf. Series: Journal of Physics: Conf. Series **986** (2018) 012035 doi:10.1088/1742-6596/986/1/012035

Design and Simulation of Scanner Wrapped by Flexible Microcoil Embedded in Polymer Film for Single-Optical Endoscope Application

Mengyuan Zhao[1], Zhuoqing Yang[1], Xiaojian Xiang[1], Bin Sun[2], Guifu Ding[1] and Xiaolin Zhao[1]

[1]National Key Laboratory of Science and Technology on Micro/Nano Fabrication, Shanghai Jiao Tong University, No.800 Dongchuan Road, Shanghai 200240, CHINA
[2]Institute of Biomedical and Health Engineering, Shenzhen Institute of Advanced Technology, Chinese Academy of Science, Shenzhen, 518055, CHINA

E-mail: zhaomy@sjtu.edu.cn, yzhuoqing@sjtu.edu.cn

Abstract. A single optic fiber scanner with large scanning angle, based on novel electromagnetic driven, is presented. The cylinder magnet and weight are fixed on the fiber, and vibrate under its second-order frequency by driving racetrack coils on the tube. The flexible driving coil is fabricated by uncomplicated planar MEMS technology on polyimide film, and wrapped on the tube. The electromagnetic and mechanical properties of the endoscope system are studied. Experimental results show that the maximum of the second resonant scanning angle is 9.47°.

1. Introduction

There are two categories of image acquisition of human internal organs and ducts without damaging human body, one is structure-based medical image acquisition, like computed tomography (CT).[1, 2] The CT machine generates tomographic image of the structure of the scanned tissue by using rays or ultrasonic wave, but it provides the image with low spatiotemporal resolution (millimeters, seconds); another is surface-based medical endoscopic technology with high spatiotemporal resolution (micrometers, milliseconds), the medical endoscope captures the imaging by visualizing the surface of tissue and organ of internal human body directly. Thus, the medical endoscopic technology can easily detect pathological changes of tissue and organ. The range of application for internal organs is obviously decided by the flexibility and diameter of medical endoscope, in addition, a fine and flexible endoscope can effectively reduce pain sensation and tissue trauma.[3]

There are two kinds of endoscopes in the modern medical application area, fiber bundle endoscope and image sensor endoscope. The diameter of image sensor endoscope is limited by the size of sensor (CMOS and CCD), the small size of imaging sensor will reduce imaging quality, miniaturization is extremely difficult on the premise of keeping high resolution. The fiberscope can be miniaturized by reducing the number of fiber in the tube, but it also will impact the imaging quality. To get the high resolution and smaller size, a new fiber endoscope with a rapid scanning fiber has been researched, the endoscope with a single fiber can be divided into electromagnetic and piezoelectric type depending on different driving mode. The research of piezoelectric driven fiber scanner is more mature, the theory of scanning is that PZT actuators drive fiber vibration under alternating current with a certain frequency,[4] but the applied high voltage and toxicity of PZT material are not suitable as medical endoscope. The fiber optic scanner with electromagnetic driven has much advantage that piezoelectric driven lacks, like low power, low toxicity. The electromagnetic actuator fiber scope consist of the driven part of generating driving magnetic field , like electromagnetic coil or permanent magnet, and a vibration part, it contains a fiber and the device that is fixed on the fiber and interacts with magnetic field. Some researchers use 3D MEMS process to fabricate a tilt coil as a driven part on the tube surface,[5] but it does not provide a large force output, beyond that, the manufacture is complex and expensive. Therefore we need a new design with easy processing and good driven method for providing a large device.

In this paper we report an electromagnetic single fiber endoscope with wrapped driven flexible coil, the fiber scanner is based on an optical fiber with magnet. The electromagnetic feature of the microcoil was simulated in order to theoretically analyze the interaction between magnet and coil, and the dynamic property of fiber vibration system was calculated and simulated. The device was made and properties were characterized dynamically.

2. Design and characters analysis

2.1. Design of the fiber scanner system

The design can be seen from figure 1(a), the scanning fiber system is composed of fiber weight and magnet, the shape of magnet and weight is cylinder with a round hole, and the fiber move through the hole of them and fixed by UV curing glue, the magnet and mass are placed on the middle and front part of the fiber, respectively. The fiber system is placed at the longitudinal axis of a polyimide tube. The driven device is a flexible circle microcoil that be wrapped on the outside surface of the tube. the driven coil is fabricated on flexible polyimide film, then, it is wrapped on the tube. Two of the same helix coils that are symmetrically wrapped on the surface of the polyimide tube can provide applied magnetic field for magnet. Its outer and inner diameters are 2.0mm and 1.8mm, which is fabricated and embedded in a flexible polyimide film, as shown in figure 1(b). The soft material CoNi as the magnet, because of low coercive field strength , the soft magnet is easy to be magnetized by along axis, and change the direction of magnetization by external field and when the driven coil is applied a periodic voltage, the magnet will generate a magnetized torque and impel the fiber to vibrate with the same frequency as external current. When the fiber system vibrates on its own resonant frequency, the fiber will resonate and a remarkable scan procedure can be obtained. In addition, the weight can change the resonant mode of the scanning system and can increase the resonant amplitude at high-order resonant modal; the fiber scope will achieve the goal of large scanning angle in the narrow space.

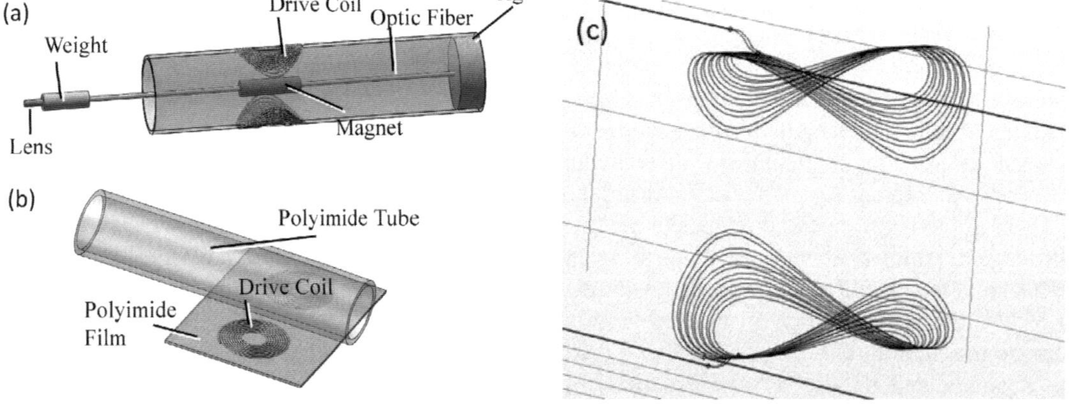

Figure 1. (a) 3D sketch of the fiber optic scanner; **(b)** 3D stretch of the drive coil and tube; **(c)** 3D model of the wrapped coil.

2.2. Electromagnetic properties analysis

Electromagnetic characters of the wrapped microcoil are the important for the fiber endoscope, in order to analyze these, finite element emulation software CMOSOL Multiphysics was used. The model of curved coil has been built in the Solidworks, and is import in COMSOL Multiphysics, as shown in figure 1(c). The geometry parameter is 8 of turns, 50of line width and pitch, and the maximum outside diameter is 3mm; the input current is 0.1A, The generating magnetic field simulation of the drive coil as shown in figure 2(a). The magnetic field is symmetric about the central axis of the coil. The magnetic field is closer to the microcoil and the strength is stronger, and only a little area can keep the relatively higher magnetic field strength. Therefore, to increase the area of high magnetic field, a coil with track type is simulation, as shown in the figure 2(b), the straightway is 2mm and the current is 0.1A. the simulation shows that the magnetic field with the same direction occurs in the straight-line area, and the magnetic field strength is high enough.

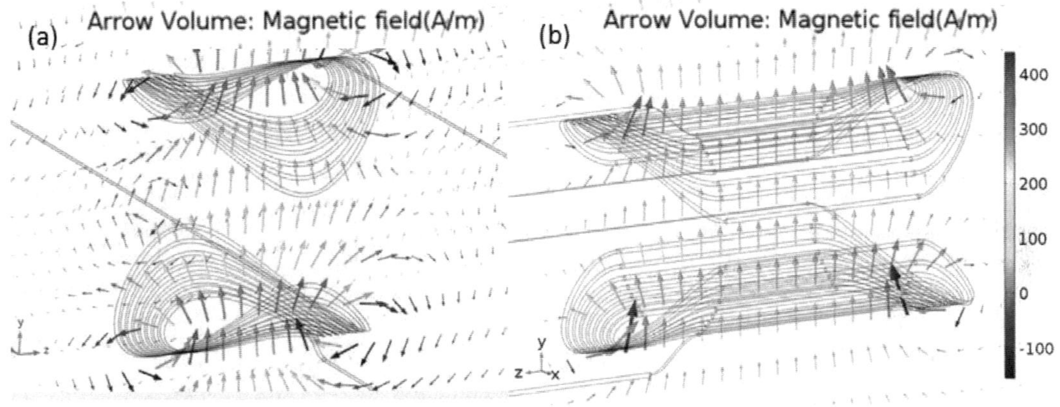

Figure 2. The simulation of the magnetic field distribution: (a) circle coil; (b) track coil.

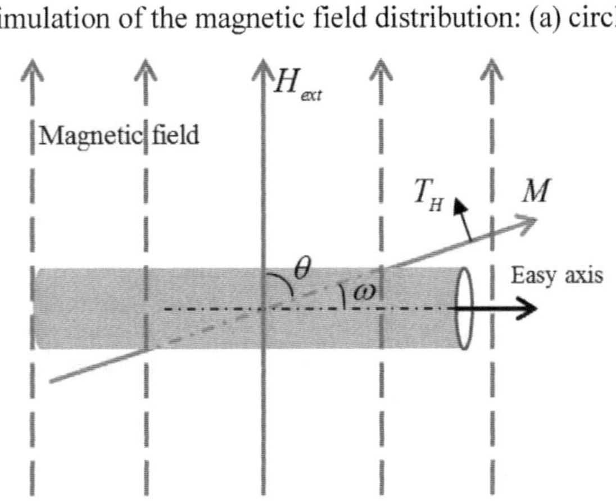

Figure 3. Schematic of torque T calculation.

In order to get a larger driven force, the theory of magnetized soft magnet should be studied, to simplify the analysis, a CoNi cylinder is placed inside a uniform magnetic field that is perpendicular to the axis of magnet as shown in figure 3. The soft magnet generates a torque T_H, it can be expressed as[6]:

$$T_H = V \times \left| \overrightarrow{M} \times \overrightarrow{H_{ext}} \right| = VMH_{ext} \sin\theta \tag{1}$$

Where V is the volume of the magnet, H is the magnetic field and M is the magnetization vector, and the angle θ is between M and H. The magnetization vector of the soft magnet varies with the changes external magnetic field along the direction of magnetization vector: $H_a = H_{ext}\cos\theta$. When the magnet starts to be magnetized by H_{ext}, a demagnetizing field H_d will be generated inside the magnet: $H_d = -N_M M / \mu_0$, where N_M is the shape-anisotropy coefficient and μ_0 is the permeability of vacuum; and the H_d is equal to the H_a at the steady state. The maximum is the saturated magnetization M_s .Taken together, the value of magnetization vector can be expressed as:

$$M = \left(\frac{\mu_0 H_{ext} \cos\theta}{N_M}, M_s \right) \tag{2}$$

When M rotates away the easy axis, a magnetic-anisotropy torque T_a is generated to stop the rotation of the magnetization vector:

$$T_a = -K_a \sin 2\omega \tag{3}$$

Where K_a is magnetic anisotropy constant, and ω is the angle between the easy axis and M. A opposite torque $-T_a$ will be exerted on the magnet, and drive the magnet (the easy axis) to coincide with the magnetization vector.

2.3. Dynamic Response Analysis

We use the simulation software CMOSOL to evaluate the dynamic properties of the fiber system, the system has its own muti-stage resonant frequencies, and corresponding different resonant modes, figure 4 (a) shows the first two order resonant mode shape, and resonant mode shape of the fiber vibration system. In the first order resonant mode, weight and magnet arrive their amplitude with same phase at same time, and the second order resonant mode is opposite phase under the same condition. The harmonic analysis is used to calculate the displacement amplitude of each resonant mode in figure 5, the harmonic force is 5e-6N, and excitation frequency is set from 10Hz to 200Hz. The results indicates the displacement amplitude of the first model is much more than the second model, and the displacement will increase with increasing the harmonic force. But the inner space of the tube has been limited, and the maximal resonant amplitude of the fiber scanner system is the inner diameter of the tube.

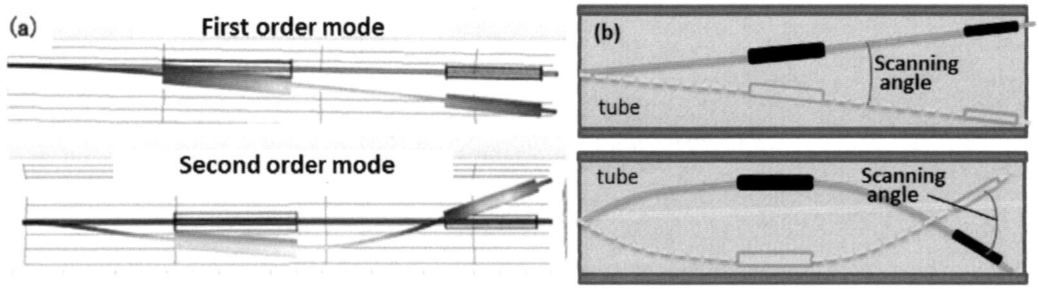

Figure 4. (a) the simulation of the first two order resonant mode; (b) the corresponding scanning angle in the tube.

When the tip of fiber touches the inner wall with different driven force under the first and second resonant frequency, the system has the same maximum vibration amplitude respectively, as shown in figure 4 (b). Compared with the vibration under the first order resonant frequency, the vertex of the scanning angle is more adjacent to the head of scanner tube, which is the static point between weight and magnet, and it can provide a larger visual field.

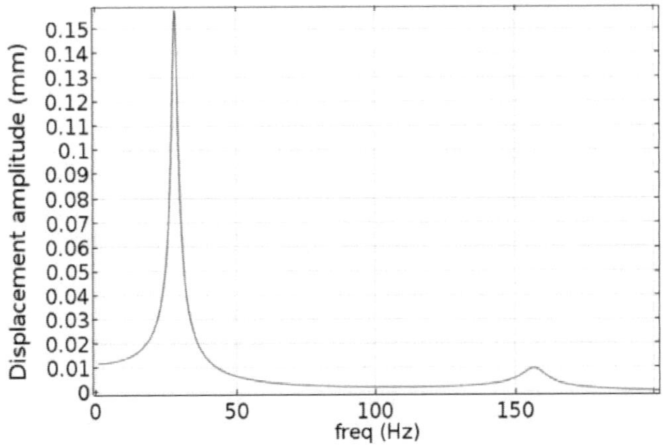

Figure 5. the simulation of the displacement amplitude of the fiber tip from 10Hz to 200Hz.

3. Fabrication and Measurement

The manufacture of coil is important to the fiber endoscope, the driven microcoil should be soft enough to be wrapped on the tube, and be easy to fabricate by the traditional MEMS planar manufacture technology. The polyimide film is used as flexible substrates because of its stability and compatibility in fabrication[7]. The coil is fabricated on the ployimide film that has been fixed on the planar substrate by traditional MEMS planar fabrication process, and to protect the coil, the processed coil is encapsulated by the solidified polyimide. Figure 6 shows the fabrication process of the coil and the optical image.

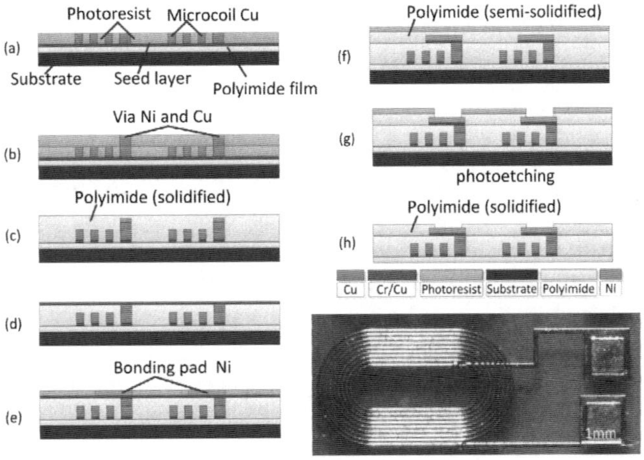

Figure 6. the main fabrication process of the driven microcoil and the photo of the wrapped coil.

To measure the scanning angle of the fiber scanner that was discussed in previous section. The measurement system is built as shown in figure 7. The driven alternating current is generated by the function generator and current amplifier, and the vibration situation of fiber vibration system can be observed by optic camera. figure. 8 shows the photograph of the first two order frequency 24Hz and 157Hz. weight and magnet have the same phase at the first order frequency, and arrive the maximum at the same time. but they have the opposite phase under the second order frequency, the fiber has a large bending deformation and has larger scanning angle. The measured maximum scanning angle of the first two resonant frequency is 2.98° and 9.47° .

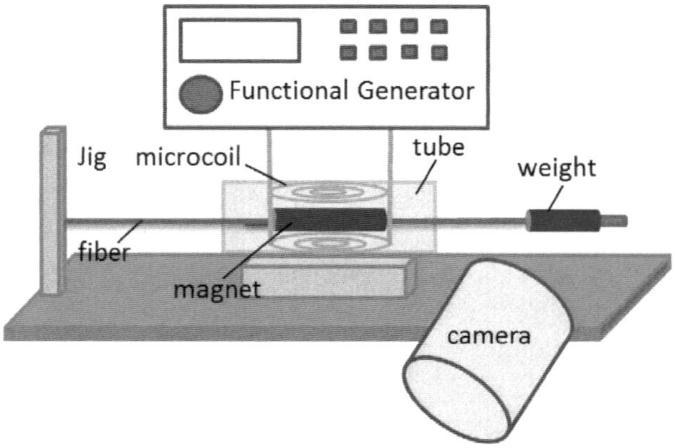

Figure 7. Measurement device for fiber scanner.

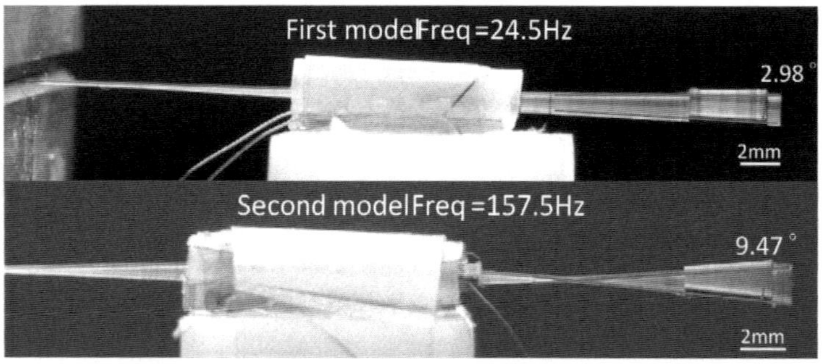

Figure 8. Measured corresponding scanning angles.

4. Conclusions

This paper presents a design of a fiber-optical endoscope with electromagnetic driving flexible microcoil, the electromagnetic and dynamic property has been discussed and simulated, the driving coil is fabricated on the polyimide film by planar MEMS process. The fiber scanner system vibrates at its own the second order resonant frequency can provide a larger scanning angle, and the scanning angle is 9.47° by optical camera.

5. References

[1] Jiang H, 2009 *Computed Tomography: Principles, Design, Artifacts, and Recent Advances, Second Edition*

[2] Farman A G, Scarfe W C 2006 Development of imaging selection criteria and procedures should precede cephalometric assessment with cone-beam computed tomography. *American Journal of Orthodontics & Dentofacial Orthopedics.* **130** 257

[3] Gannot I, 2010 *Optical Fibers and Sensors for Medical Diagnostics and Treatment Applications VII.*

[4] Huo L, Xi J, Wu J, 2010 Forward-viewing resonant fiber-optic scanning endoscope of appropriate scanning speed for 3D OCT imaging. *Optics Express.* **14** 14375-84

[5] Sun B, Sawada R, Yang Z, Zhang Y 2014 Design and fabrication of driving microcoil with large tilt-angle for medical scanner application. *Design, Test, Integration and Packaging of Mems/moems* **3** 1-6

[6] Judy J W, Muller R S 1997 Magnetically actuated, addressable microstructures. *Journal of Microelectromechanical Systems.* **3** 249-256

[7] Liaw D J, Wang K L, Huang Y C, Lee K R, Lai J Y 2012 Advanced polyimide materials: syntheses, physical properties and applications. *Progress in Polymer Science*, **37** 907-974.

Acknowledgments
This work was supported in part by the National Natural Science Foundation of China (No. 61571287), the Advanced Research Ministry of Education Joint Foundation (6141A02033403), the Hi-Tech Research and Development Program of China (2015AA042701) and the SJTU-SMC Young Scholar Program (16X100080044).

Design, Simulation and Fabrication of A MEMS-based Double-layer Spiral Planar Inductor with Patterned Permalloy as Magnetic Layers

Xiaomin Zhu,[1] Ping Cheng,[1] Mingming Chen,[1] and Guifu Ding[1]

[1]National Key Laboratory of Science and Technology on Micro/Nano Fabrication, Department of Micro/Nano Electronics, School of Electronic Information and Electrical Engineering, Shanghai Jiao Tong University, Shanghai 200240, CHINA

xmin_zhu@163.com

Abstract. There have been significant efforts to boost the inductance value by adopting the sandwich structures using permalloy magnetic shielding layers. However, this structure will introduce high ac conductor losses and high eddy currents. In order to solve these problems, patterned permalloy can solve this problem effectively. According to the simulation results based on the application of finite element method in the frequency domain, the optimum permalloy pattern is which the blank of the permalloy are perpendicular to the coil inside. The double-layer planar inductor has a size of 1.5×1.5×0.1mm consisted of 13-turn spiral Cu coil for each layer and a 20-μm-thick patterned permalloy magnetic shielding layer. The inductor shows a higher inductance than the traditional planar inductor. The patterned permalloy made the inductor more stable in high frequency than the none-patterned. And the inductor has an inductance of 1.3μH and quality factor of 2.8 at 1.5MHz, with an inductance per unit of 578nH/mm^2, which is much higher than that in the reported literatures.

1. Introduction

On-chip spiral inductors have played an increasingly important role in integrated circuits, such as converters, filters, mixers, and oscillators [1]. Due to the widespread use in the wireless communications technology and the strongly desire of smaller size for integrated packaging, the planar spiral inductors have gained interest in this area. The challenge of integrating inductors on-chip have stood the way of fully integration of many electronic systems such as power converter devices [2]. In the current inductive elements market, the conventional large size winding coils are still dominant. However, there have been increased efforts to reduce the size of inductor without sacrificing the performance. Planar spiral structures are often desirable for inductors because the parallelized fabrication processes can make it easy for packaging while the spiral configurations can ensure the number of turns in a very small size.

The size of the inductor can be reduced by MEMS technology. The development of MEMS technology make it possible to fabricate the planar spiral micro-inductors by some micro-fabrication processes like photolithography, wet etching, electroplating technique, etc. And the performance should be insured while used in the packaging. In recent years, there have been significant efforts to boost the inductance value by adopting the sandwich structures using magnetic shielding layers. Permalloy is a kind of soft magnetic materials with high relative permeability. It shows great compatibility with MEMS technology and plays significant role in increasing the inductance and quality factor of planar spiral inductors [3]. The permalloy magnetic films are often deposited above and below the spiral plane using the photolithography and electroplating techniques.

Now the plane sandwich structure are greatly applied in fabricating micro-inductor due to its stability and small size and compatibility with processing procedure. The metal layers of the sandwich structure can prevent the magnetic field lines from spreading outside and thus keep the magnetic field loops around the conductor lines. However, this structure will introduce high ac conductor losses and high eddy currents. Patterned permalloy can solve this problem effectively. As the magnetic shielding layers, appropriate patterned method can also reduce the eddy currents while it play a role in shielding the magnetic field lines outside.

Many researches have focused on the packaging micro-inductor applied in the DC/DC power conversion. Many electronic systems have the module of DC/DC power conversion. Figure 1 (a) shows the typical circuits of the principal components of a DC/DC converter [2]. Some electronic components such as PWM logic, OSC, Slope comp and UVLO have already integrated in the Control IC part. However, the inductor is the most difficult to be integrated in the power converter chip. At present, the series of XCL205 with a size of 2mm*3mm fabricated by TOREX have integrated the inductor on chip successfully. The X-ray scanning picture of the device is shown in figure 1 (b). The inductor is pressed on the chip part and connect with each other inside by lead wire. In addition, the inductance value at 1MHz is 1.5μH. It can be seen that the inductor is a handmade device and thus is poor at mass production.

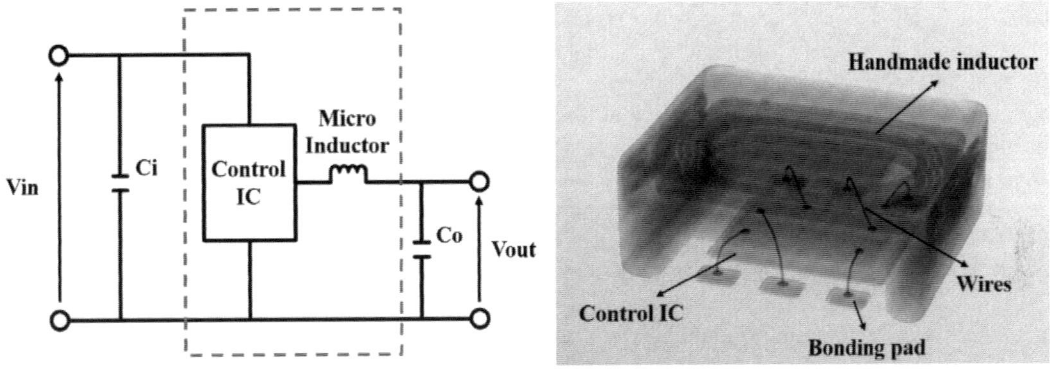

(a) (b)

Figure 1. (a) Schematic illustration of the DC/DC converter circuits. The components in the dashed box would be integrated together; (b) The X-ray scanning picture of the XCL205 fabricated by TOREX.

Mishra et al. [4] developed a toroid-based inductor using multilayered ferromagnetic-polymer composites as magnetic cores, which was fabricated by laminating NiFeMo magnetic foils and BCB polymer adhesive layer by layer. The inductor had a size of 7mm × 4mm × 1.5mm, with an inductance of 1.96nH/mm^2. Fukuda et al. [5] designed a single-layer planar inductor using screen printed NiZnCu /resin ferrite composite layer to improve its inductance. The inductor had a size of 6mm × 6mm × 0.2mm. At 1.5MHz, the inductance was 1.5μH with an inductance value per unit of 42nH/mm^2. Sadler D J et al. [6] developed a spiral-type inductor with 4.52μH using AZ-4000 series photoresist as dielectric layer and using Ni/Fe permalloy as core. The inductor had a size of 4.2mm × 4.2mm, with an inductance value per unit of 256nH/mm^2. Kowase et al. [7] developed a large-current planar inductor with a sandwich structure, which consisted of two-turn Cu square spiral coil, top and bottom ferrite magnetic cores. The inductor had a size of 15mm × 15mm × 1.4mm, and an inductance of 140nH, with an inductance value per unit of $0.62nH/mm^2$. For the above reported inductor, the size of device was too big and the inductance value per area was relatively small.

This study aimed to get a high inductance value in a small device size. A double-layer planar inductor with up and below patterned permalloy magnetic shielding layers have been fabricated. The patterned permalloy figure which the blank of the permalloy are perpendicular to the coil inside have been proved better than other figures. The inductor with a small device size of 1.5×1.5×0.1mm can been packaged easily which oversize is satisfied with the application DT8515 buck DCDC converter. And the inductor has an inductance of 1.3μH and quality factor of 2.8 at 1.5MHz, with an inductance per unit of $578nH/mm^2$.

2. Inductor Design

2.1. The Structure of the Inductor

Planar, rectangular spiral geometry was chosen as a well-tested technique for maximizing inductance per area, and a nominal inductor size of 1.5mm × 1.5mm was selected. within these parameters, The number of turns was varied as an inductance vs. coil resistance tradeoff. However, a 13-turn design was qualitatively selected as a most likely candidate for fabrication and integration and all extended characterization was performed with 13-turn inductors by simulation. Single-layer coils are easy to fabricate, but the multi-layer coils have more coil turns without increasing the device size [9]. All the above factors and the microfabrication processes considered, figure 2 (a) shows the illustrated structure designed of the planar spiral inductors [8]. Double-layer structure is adopted and each coil layer consists of 13 turns. Each turn is 20μm in width and each coil spacing is 20μm. The area of the inductor is 1.5mm×1.5mm except the two metal test pads. The part of via connects the coil layers effectively above and below to form a double-layer structure. The thickness of each layer is 20±2μm and the two layers are 10±2μm apart from each other. And the above and below layers are permalloy magnetic layers which play a role in shielding the magnetic field lines outside the inductor structure. To find the optimum permalloy pattern and bring some laws of the geometric figures, eight different patterned permalloy figures were designed and simulated. Figure 2 (b) illustrated these eight geometric figures. Type7 is a reference and other pattern were started from Type8. There are two basic categories. Type3 and Type5 are which permalloy blank are perpendicular to the coil inside. And Type2 and Type4 are which permalloy blank are parallel to the coil inside. It can be seen obviously from the figure 2 (b).

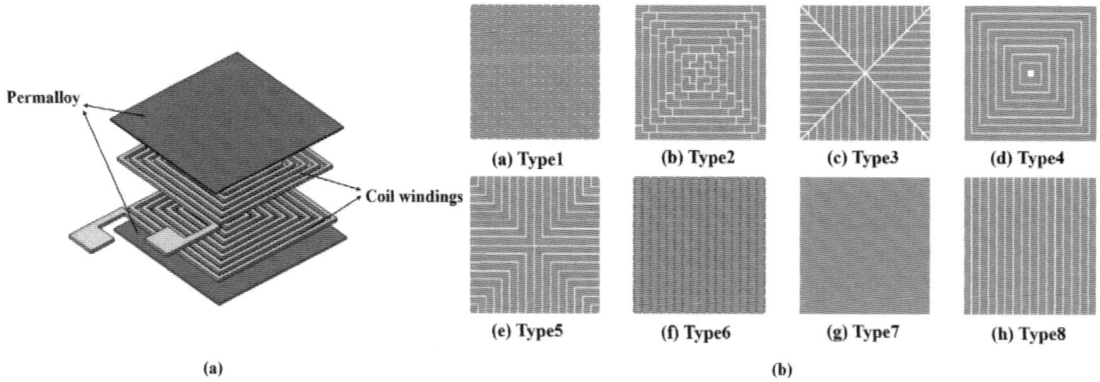

Figure 2. (a) The illustrated structure of the inductor; (b) Eight geometric figures of the patterned permalloy designed as magnetic layers.

IOP Conf. Series: Journal of Physics: Conf. Series **986** (2018) 012036 doi:10.1088/1742-6596/986/1/012036

2.2. The simulation results of the inductor designed above

2.2.1. The method of the analysis and the mathematical description. Inductances are a fundamental component in electronic field, thus it is important to know how it can be modeled. For arbitrary complex structures, numerical procedures are developed either in time domain or in frequency domain and among them, then the Finite Element Method (FEM) can be used to computed. FEM analysis requires a convenient mesh grid study to obtain a correct and optimized solution of the model. The optimization of the mesh is related to some specific physics characteristics, like the skin depth [9]. This work was developed based on the application of finite element method in the frequency domain, using the software COMSOL Multiphysics with the ACDC module [10]. The simulations have made in a frequency domain with the ACDC module and the Magnetic Field Interface. The equations (1) are computed by COMSOL Multiphysics:

$$\begin{cases} (j\omega\sigma - \omega^2\epsilon)A + \nabla \times H = J_e \\ B = \nabla \times A \end{cases} \qquad (1)$$

where: ω – the angular frequency of a magnetic field, σ – the electrical conductivity of materials, ϵ – permittivity of materials, A – a magnetic vector potential, H – magnetic field intensity, B – magnetic flux density, J_e – externally generated current density.

This system of equations is computed to each finite element of the model. Solving the given equations allows us to obtain the magnetic flux distribution. In order to find a close solution it must be imposed a constrain in the outer boundaries:

$$n \times A = 0 \qquad (2)$$

This equation fixes the tangential component of the magnetic potential A to zero.

Additionally, the winding is supply by a constant voltage, this imposes the following equations:

$$J_e = \sigma \frac{V_{coil}}{L} \qquad (3)$$

where: V_{coil} – the applied voltage specified, J_e – current density in windings, σ – the conductivity of materials, L is equal to the physics interface's thickness [12] [19] [20].

2.2.2. The Simulation Results from COMSOL Multiphysics. The simulation results of the inductance value L and the quality factor Q are shown in figure 3. Figure 3 (a) shows that the inductor with the patterned permalloy of Type 2, Type 4 and Type 6 do not show a better inductance value than the others. In addition, the inductor of Type 1 and Type 7 are unstable in high frequency. At 1.5MHz, the comparision of the inductance value L is Type5>Type3>Type8, and the Type3 shows a better stability than the Type5. Figure 3 (b) shows that the maximum value of the quality factor Q are in the comparision of Type3 > Type5 > Type2 > Type6 > Type8 > Type4 > Typel > Type7. And the corresponding frequency f of the maximum value Q are in the comparision of Type4 > Type2 > Type6 > Type3 > Type5 > Type8 > Typel > Type7.

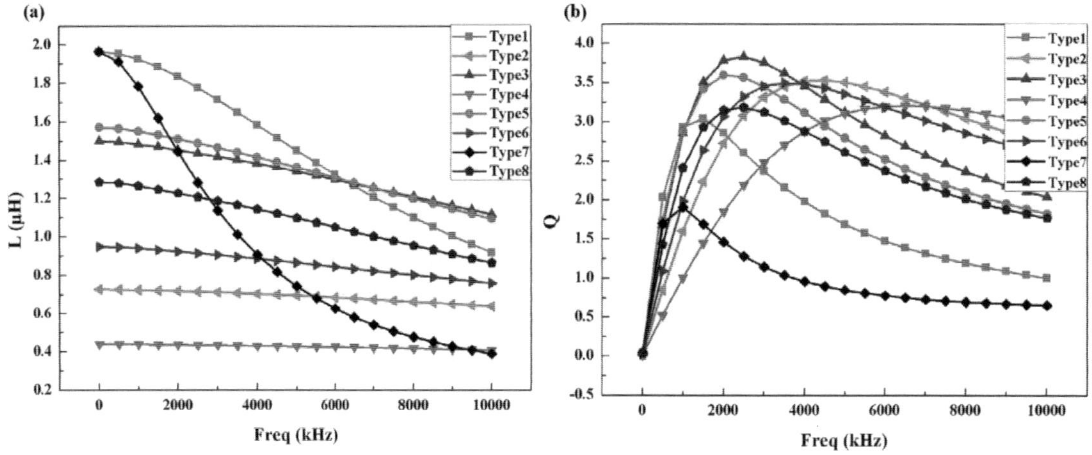

Figure 3. Simulated results of the eight types. (a) Variation of the inductance value with the frequency; (b) Variation of the quality factor with the frequency.

According to the simulation results above, the final designed inductor structure with patterned permalloy is shown in figure 4. The Type3 and Type 5 pattern are chosen due to their excellent properties in the simulation results. And the Type 7 is chosen to be a reference object. The coil inside was a double-layer spiral planar inductor and each coil layer consists of 13 turns. The size of the inductor is 1.5mm×1.5mm.

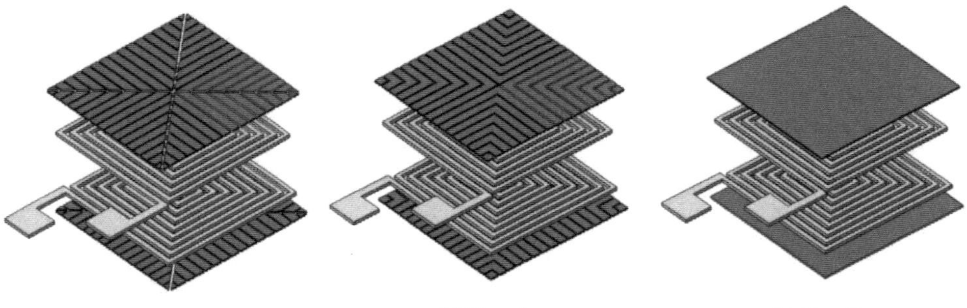

Figure 4. Schematic diagram of the experimental inductors. Type3, Type5 and Type7 are chosen according to the simulated results.

3. Fabrication Procedure

Firstly, the Cu film was prepared on a 1mm thickness glass substrate by electroplating process. The photoresist was spin coated on the electroplated Cu film, and patterned using photolithography and wet etching [11]. Then the spiral Cu conductor line was obtained. The main fabrication processes of the double-layer spiral planar inductor are illustrated in the figure 5. The double-layer structure was realized by a through-hole filled with Cu connecting two layers of Cu coil [6]. The second coil layer is electroplated after the spin-coating. The via connected the upper and lower coil and the distance between the two coil layers was designed as 10μm. The permalloy magnetic films are deposited above and below the spiral plane using the photolithography and electroplating techniques. The pictures of the inductor samples fabricated by some MEMS technology are shown in figure 6.

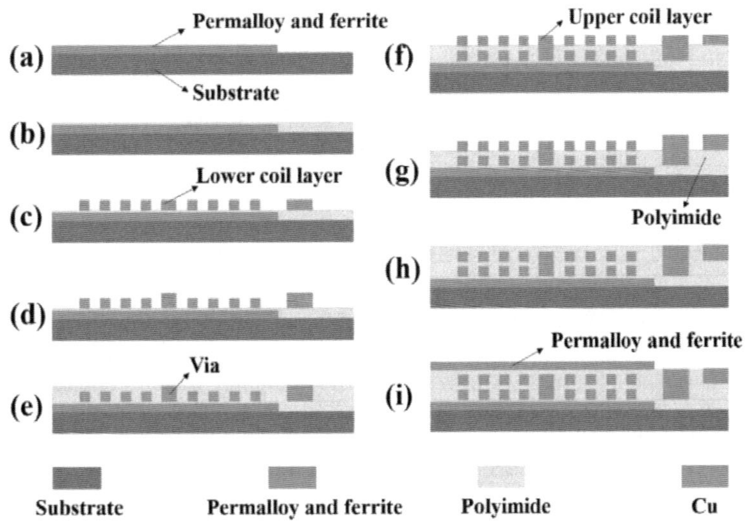

Figure 5. Main fabrication processes (a)-(i).

Figure 6. Fabricated inductor samples.

4. Results and Discussion

Figure 7 showed the relationship between the inductance L and the quality factor Q with the frequency at a testing voltage of 5V. The testing frequency range was 10kHz-10MHz. All the testing results are from the inductor with the patterned permalloy of Type 3, Type 5 and Type 7. Figure 7 (a) shows that the inductor of Type 7 are unstable in high frequency. At 1.5MHz, the comparision of the inductance value L is Type 7>Type 5>Type 3, and the Type3 shows a better stability than the Type5. Figure 7 (b) shows that the maximum value of the quality factor Q are in the comparision of Type 3>Type 5>Type 7. And the corresponding frequency f of the maximum value Q are in the comparision of Type 3>Type 5>Type 7. The trend are achieved agreeable to the simulated results.

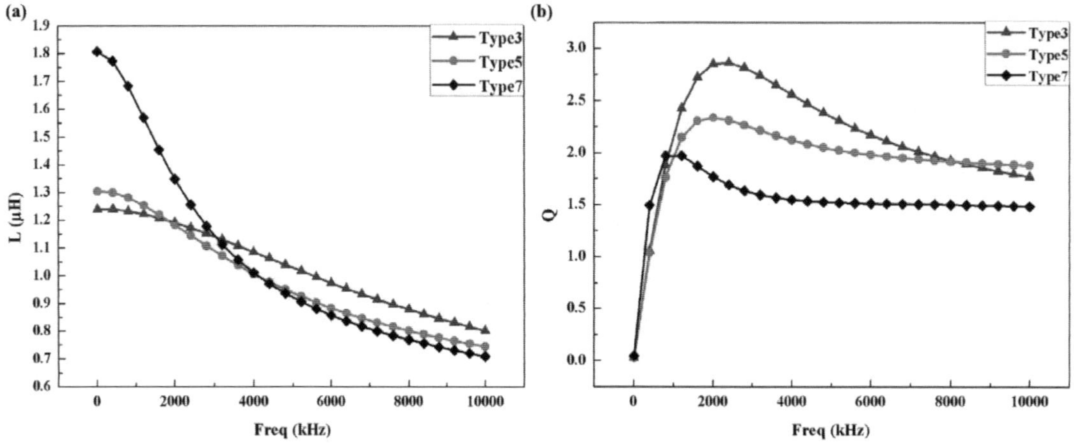

Figure 7. Experimental data of the chosen three types of the planar inductors with patterned permalloy magnetic layers filled with MnZn ferrite/PI composite materials. (a) Relation between inductance L and frequency; (b) Relation between the quality factor Q and frequency.

5. Conclusions

In conclusion, the MEMS-based spiral-type double-layer planar inductor was designed, simulated, fabricated and tested. The patterned permalloy inductor structure can reduce the eddy currents and improve the quality factor effectively at the same time boosting the inductance value. The optimum permalloy pattern is which the blank of the permalloy are perpendicular to the coil inside. In this study, the novel structure inductor have been fabricated by some MEMS technologies. The double-layer planar inductor has a size of 1.5×1.5×0.1mm consisted of 13-turn spiral Cu coil for each layer and a 20-μm-thick patterned permalloy magnetic shielding layer. And the inductor has an inductance of 1.3μH and quality factor of 2.8 at 1.5MHz, with an inductance per unit of 578nH/mm^2. The results indicated that the patterned permalloy made the inductor more stable in high frequency than the none-patterned. It will have many potential applications in magnetic sensors and actuators, and electronic circuit components.

6. Acknowledgments

This work is sponsored by National Basic Research Program of China (973 Program) and National State Key Laboratory Development Program.

References

[1] Ni W, Kim J and Kan E C. Permalloy patterning effects on RF inductors[J]. *IEEE transactions on magnetics,* 2006, 42(10): 2827-2829.
[2] Galle P, Wu X, Milner L, et al. Ultra-compact power conversion based on a CMOS-compatible microfabricated power inductor with minimized core losses[C]//*Electronic Components and Technology Conference,* 2007. ECTC'07. Proceedings. 57th. IEEE, 2007: 1889-1894.
[3] Pulijala V and Syed A. Comparison of the effects of 60nm and 96nm thick patterned permalloy thin films on the performance of on-chip spiral inductors[J]. *Journal of Magnetism and Magnetic Materials,* 2016, 419: 245-248.
[4] Mishra D, Raj P M and Tummala R. Design, fabrication and characterization of thin power inductors with multilayered ferromagnetic-polymer composite structures[J]. *Microelectronic Engineering,* 2016, 160: 34-38.
[5] Fukuda Y, Inoue T, Mizoguchi T, et al. Planar inductor with ferrite layers for DC-DC converter[J]. *IEEE Transactions on Magnetics,* 2003, 39(4): 2057-2061.
[6] Sadler D J, Gupta S and Ahn C H. Micromachined spiral inductors using UV-LIGA techniques[J]. *IEEE Transactions on Magnetics,* 2001, 37(4): 2897-2899.

IOP Conf. Series: Journal of Physics: Conf. Series **986** (2018) 012036 doi:10.1088/1742-6596/986/1/012036

[7] Kowase I, Sato T, Yamasawa K, et al. A planar inductor using Mn-Zn ferrite/polyimide composite thick film for low-Voltage and large-current DC-DC converter[J]. *IEEE Transactions on Magnetics,* 2005, 41(10): 3991-3993.
[8] Rawat M and Yadav K L. Electrical, magnetic and magnetodielectric properties in ferrite-ferroelectric particulate composites[J]. *Smart Materials and Structures,* 2015, 24(4): 045041.
[9] Giovanardi F. Impedance Analysis of a Pot Core Inductor[J].
[10] "ACDC Module Users Guide", COMSOL, www.comsol.com.
[11] Kim J, Herrault F, Yu X, et al. Microfabrication of air core power inductors with metal-encapsulated polymer vias[J]. *Journal of Micromechanics and Microengineering,* 2013, 23(3): 035006.

Research on subsurface deformed layer in ultra-precision cutting of single crystal copper by focused ion beam etching method

Y Chen[1], X J Huang[1] and J X Kong[1]

[1] Institute of Mechanical Manufacturing Technology, China Academy of Engineering Physics, 64 Mianshan Road, Mianyang, CHINA

E-mail: chen_yioh@sina.cn

Abstract. In this paper, the focused ion beam was used to study the subsurface deformed layer of single crystal copper caused by the nanoscale single-point diamond fly cutting, and the possibility of using nanometer ultra-precision cutting to remove the larger deformation layer caused by traditional rough cutting process was explored. The maximum cutting thickness of single-point diamond cutting was about 146 nm, and the surface of the single-crystal copper after cutting was etched and observed by using the focused ion beam method. It was found that the morphology of the near-surface layer and the intermediate layer of the copper material were larger differences: the near-surface of the material was smaller and more compact, and the intermediate material layer of the material was more coarse sparse. The results showed that the traditional precision cutting would residual significant subsurface deformed layer and the thickness was on micron level. Even more, the subsurface deformed layer was obviously removed from about 12μm to 5μm after single-point diamond fly cutting in this paper. This paper proved that the large-scale subsurface deformed layer caused by traditional cutting process could be removed by nanometer ultra-precision cutting. It was of great significance to further establish the method that control of the deformation of weak rigid components by reducing the depth of the subsurface deformed layers.

1. Introduction

Weak rigid components, which required high accuracy, were widely used in aerospace and other industries [1-3]. So far, the deformation of the weak rigid components was difficult to control due to the residual stress [4]. And the stress injury caused by residual stress was the main factor to form the subsurface deformed layers [5]. So, it was one of the most effective methods to study the deformation mechanism and control method of weak rigid components by studying subsurface deformed layers.

The depth of subsurface deformed layers was an important parameter affecting the quality of manufacturing quality. Nowadays, the measurements of the depth of subsurface deformed layers mostly were studied by computer simulation, which was proved to be a kind of useful method [6-10]. Verified by the molecular dynamics simulation method, Luo et al. reported that the use of nanoscale diamond tools to produce nanostructures results in the transferability of shapes [11]. Uezakia et al. used a method of local compressive stress to successfully suppress the plastic flow of material and improve the surface quality of the product [12]. Fang et al. used molecular dynamics to study the phase transition of germanium during the nano-cutting process [13].

Compared with the results obtained by the simulation method, although there were some researchers using direct experimental methods, such as transmission electron microscopy (TEM) [14][15], X-ray diffraction [16][17], Raman scattering [18], to study subsurface metamorphic layers. However, there was still no validated experimental method to observe the subsurface deformed layer during nano-cutting process. For manufacturing engineers, finding the optimum machining parameters that produce minimum subsurface deformed layers in materials, especially in metal materials, remained a difficult issue.

In this paper, the focused ion beam (FIB) method was used to study the subsurface deformed layer of single crystal copper caused by the traditional precision cutting and the nanoscale single-point diamond fly cutting. The possibility of using nanometer ultra-precision cutting to remove the larger deformation layer caused by traditional precision cutting process was discussed.

2. Materials and Methods

The material used in this paper was commercial pure single crystal copper. The pure copper was first processed by cold forging method at room temperature, and then, combined with 400℃×8-hour annealing, the grain size of the pure copper material will be reduced to about 10~20 μm.

The fine grain size copper samples were first rough cut using a precision machining machine. Then, it was cut by a homemade ultra-precision single-point diamond fly cutting machine. The fly-cutting machine used an air-bearing tooling spindles and a hydrostatic support rail which driven by a linear motor. A single-point diamond tools with a radius of 5 mm were used to cut the pure copper on fly-cutting machine.

The main process parameters of ultra-precision cutting were shown in Table 1.

Table 1. the main process parameters of ultra-precision cutting

Depth of cut	Speed of horizontal rail	Spindle speed
a_p	V	ω
3 μm	12 mm/min	280 r/min

The principle of single-point diamond flying cut process was shown in figure 1. The parameter R, equal to 5 mm, represents the radius of the diamond tool. The parameter f, equal to V/ω, was the cutting feed in horizontal direction.

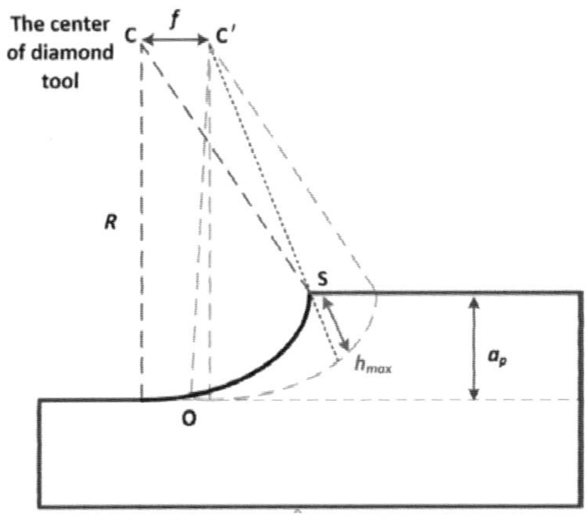

Figure 1. Principle of single-point diamond flying cut

As shown in the figure 1, the center of the diamond tool moved from C to C′ after a single feed. The parameter h_{max}, represented the maximum cutting thickness of flying cut, can be calculated by the formula (1) as follow.

$$h_{\max} = R - \sqrt{R^2 + f^2 - 2f\sqrt{2Ra_p - a_p^{\,2}}} \tag{1}$$

Through the parameters of fly-cutting process used in this paper, the maximum cutting thickness of flying cut can be calculated by formula (1) as 146 nm. Therefore, the thickness of ultra-precision cutting process in this paper was on the nanometer scale.

After the two times of the single point diamond fly cutting process, one side of the pure copper sample's surface were processed to the mirror as shown in figure 2. The surface roughness of the pure copper was measured at about 80 nm using a dynamic interferometer.

Figure 2. View of single-crystal copper cutting by single-point diamond flying cutting

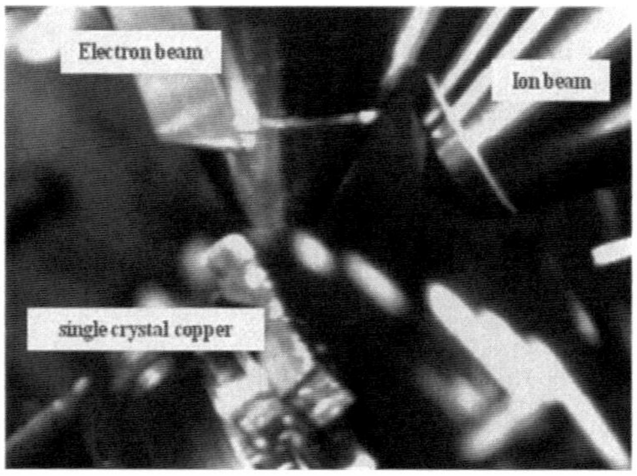

Figure 3. FIB etched and observed the copper cut by single-point diamond fly cutting tools

The FIB (focused ion beam) method was used to etch the surface of the material and observe the subsurface deformed layer in situ after the copper sample's cutting. The FIB used in this paper had a gallium ion source. When high energy gallium ions collided with the samples, they sputter the atoms from the surface and remove the material. Therefore, the FIB method was used as a micro processing tool in this paper. The etching process was direct etching without auxiliary gas. As figure 3 shown, a small piece of copper sample was sampled by wire cutting and placed into the FIB chamber for etching.

3. Results and Discussion

The material morphology of the pure copper surface after ultra-precision cutting and FIB etching was shown in figure 4. The FIB etching was using oblique etching method. The initial location of the FIB etch was located at the long side of the trapezoid (about 60μm) in the figure. From the initial position, the etching depth would gradually increase. Through the process optimization, the final etching depth of the copper, located at the short side of the trapezoid (about 12μm) in the figure, was about 20 μm by adjusting the etching time.

IOP Conf. Series: Journal of Physics: Conf. Series **986** (2018) 012037 doi:10.1088/1742-6596/986/1/012037

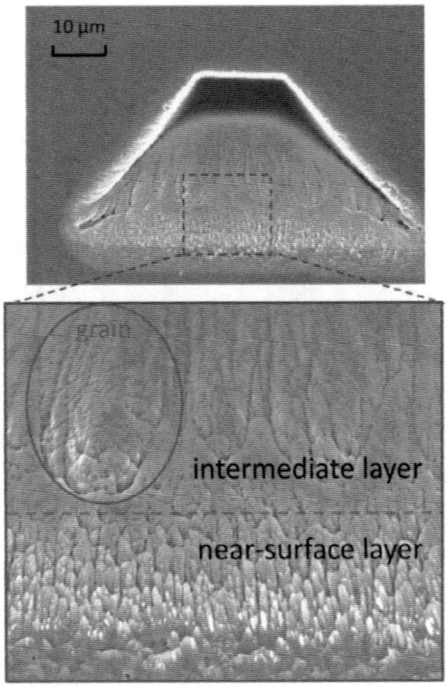

Figure 4. View of the near-surface and the intermediate material layer of the etched copper

The subsurface material morphology was observed in-situ along the Z direction. It could be found that there was a big difference between the material morphology of the pure copper material near the surface layer and the intermediate layer. Due to the role of pre-forging and heat treatment, the material grain size of intermediate layer was about 15μm. The grain size conforms to the experience.

The morphology of material near the surface changed after the cutting process. Obviously, the material morphology became more small and compact. The scale was reduced to the submicron level. This shows that the properties of near-surface pure copper materials had changed, forming a subsurface deformed layer. On the other hand, according to the tilt angle of the sample during the etching of the FIB, the depth of the subsurface deformed layer of the pure copper, obtained through geometric conversion and measured by the calibrated FIB algorithm, was about 5 μm.

The near-surface materials were further etched in order to determine whether the changes in the near-surface material were caused by the gallium ion sputter of the FIB. In this etching process, the pure copper sample was first etched down overall about 20 microns. Then, on the inner material, the pure copper material was removed obliquely using the etching method. The etch results shown in Figure 5.

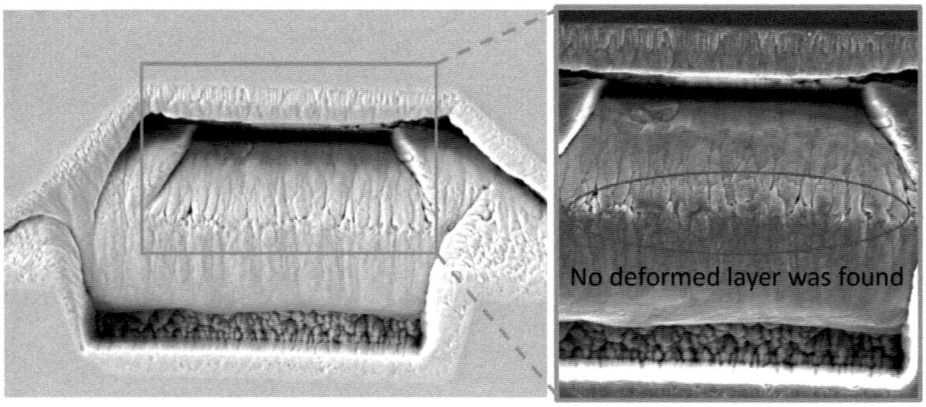

Figure 5. View of multiple etching on inner copper materials

As shown on figure 5, no smaller and more compact grains of the material were found. The results indicated that the smaller and more compact material deformed layer existed only on the near surface of the material and was not caused by gallium ion sputtering of FIB. Therefore, the subsurface deformed layer of pure copper in this paper was caused by the mechanical and thermal effects of cutting. This result was consistent with the findings of other scholars.

In order to identify whether the deformed layer was formed by rough cutting or ultra-precision cutting, FIB etching was performed on the other side (backside) of the pure copper sample that had not undergone single point diamond fly-cutting process. The etching result was shown as figure 6.

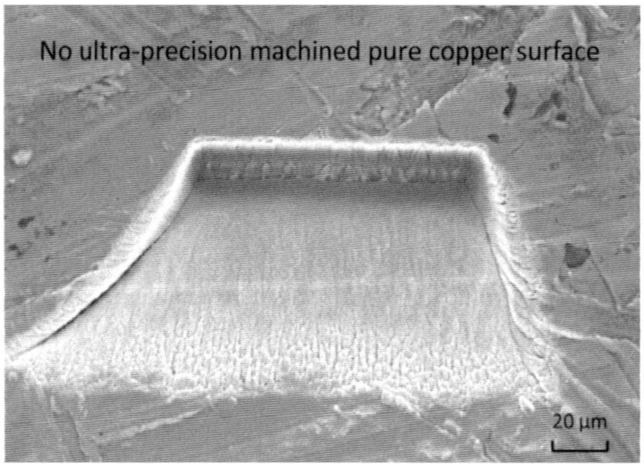

Figure 6. the FIB etching results of copper surface without ultra-precision cutting

The subsurface deformed layer near the surface of no ultra-precision machined pure copper material was also found in the figure 5. And on this time, the depth of the subsurface deformed layer, also measured by the calibrated FIB algorithm, was about 12 μm. Therefore, the rough cutting process would surely course the subsurface deformed layer.

Compared with the depth of subsurface deformed layer after ultra-precision cutting process, which was 5 μm, the depth of the subsurface deformed layer without ultra-precision fly cutting process was larger. The difference between the two depths was approximately equal to the sum of the depths of the two ultra-precision fly cutting processes ($2 \cdot a_p = 6$ μm). Furthermore, it could be found that the morphology and size of the subsurface deformed layer's grain in the two experiments were basically the same. Therefore, the subsurface deformed layer found in this paper was mainly caused by rough cutting, while the nanoscale ultra-precision cutting process could accurately remove them.

4. Conclusions

In this paper, the focused ion beam was used to study the subsurface deformed layer of single crystal copper caused by the nanoscale single-point diamond fly cutting. By etching the rough cutting surface and the ultra-precision cutting surface of the pure copper material respectively, the following conclusions could be drawn:

(1) The subsurface deformed layer of single-crystal copper was mostly formed by rough cutting. The grain of subsurface deformed layer material was smaller and more compact, whose scale was in the sub-micron level.

(2) It had been proved that the nanoscale single point diamond fly cutting process could obviously reduce the subsurface deformed layer formed in the previous rough cutting process. The subsurface deformed layer caused by nanoscale fly cutting was not significant. This result provides a basis for the design and optimization of ultra-precision cutting processes.

The paper proves that the large-scale subsurface deformed layer caused by traditional rough cutting process could be removed by nanoscale ultra-precision cutting. The research results were instructive for the machining of materials sensitive to subsurface damage. At the same time, it was of great significance to further establish the method that control of the deformation of weak rigid components by reducing the depth of the subsurface deformed layers.

Acknowledgments

Supported by Science Challenge Project, No. JCKY2016212A506-0107.

References

[1] Huang P L, Li J F, Sun J and Jia X M 2016 *Int. J. Adv. Manuf. Tech.* **84** 2461-69.
[2] Wu Q, Li D P, Ren L and Mo S 2016 *Int. J. Adv. Manuf. Tech.* **85** 1291-1302.
[3] Ezugwu E O 2005 *Int. J. Mach. Tool Manu.* **45** 1353-67.
[4] Hu Y, Yang R, Wang D and Yao Z 2018 *J. Mater Process Tech.* **251** 197-204.
[5] Wang Q, Bai Q, Chen J, Sun Y, Guo Y and Liang Y 2015 *Appl. Surf. Sci.* **344** 38-46.
[6] Maiti R, Gerhardt L C, Lee Z S, Byers R A, Woods D, Sanz-Herrera J A and Carré M J 2016 *J. Mech. Behav. Biomed.* **62** 556 69.
[7] Zhao H, Shi C, Zhang P, Zhang L, Huang H and Yan J 2012 *Appl. Surf. Sci.* **259** 66-71.
[8] Lai M, Zhang X and Fang F 2013 *Nanoscale res. lett.* **8** 353.
[9] Tarasov S, Rubtsov V and Kolubaev A 2010 *Wear* **268** 59-66.
[10] Outeiro J C, Campocasso S, Denguir L A, Fromentin G, Vignal V and Poulachon G 2015 *CIRP Ann-Manuf. Techn.* **64** 53-6.
[11] Luo X, Tong Z and Liang Y 2014 *Appl. Surf. Sci.* **321** 495-502.
[12] Uezaki K, Shimizu J and Zhou L 2014 *Precis. Eng.* **38** 371-8.
[13] Lai M, Zhang X, Fang F, Wang Y, Feng M and Tian W 2013 *Nanoscale res. lett.* **8** 13.
[14] Yan J, Asami T, Harada H and Kuriyagawa T 2012 *CIRP Ann-Manuf. Techn.* **61** 131-4.

[15] M'Saoubi R, Larsson T, Outeiro J, Guo Y, Suslov S, Saldana C and Chandrasekar S 2012 *CIRP Ann-Manuf. Techn.* **61** 99-102.
[16] Tanaka H, Shimada S and Anthony L 2007 *CIRP Ann-Manuf. Techn.* **56** 53-6.
[17] Mahmoodi M, Sedighi M and Tanner D A 2012 *Mater Design* **40** 516-20.
[18] Conti C, Colombo C, Realini M, Zerbi G and Matousek P 2014 *Appl. spectrosc.* **68** 686-91.

Error correcting circuit design with carbon nanotube field effect transistors

Xiaoqiang Liu[1], Li Cai[1], Xiaokuo Yang[1], Baojun Liu[2] and Zhongyong Liu[1]

[1]Graduate School, Air Force Engineering University, No.1 East Changle Road, Xi'an 710051, Shaanxi, China
[2]First Aviation University of Air Force, No.23 Aviation Road, Xinyang 464000, Henan, China

Email: qianglicai@163.com

Abstract. In this work, a parallel error correcting circuit based on (7, 4) Hamming code is designed and implemented with carbon nanotube field effect transistors, and its function is validated by simulation in HSpice with the Stanford model. A grouping method which is able to correct multiple bit errors in 16-bit and 32-bit application is proposed, and its error correction capability is analyzed. Performance of circuits implemented with CNTFETs and traditional MOSFETs respectively is also compared, and the former shows a 34.4% decrement of layout area and a 56.9% decrement of power consumption.

1. Introduction

As the feature size decreases, performance of electronic devices working in intense radiation environment is severely deteriorated. Energetic particles such as protons and alpha particles affect the charge distribution in sensitive areas of devices, causing single event upset(SEU) in memory devices. Error correcting circuit is an effective method to prevent SEU from happening [1].

The basic idea of error correcting is increasing the minimum hamming distance of the code word set by inserting redundant check bits into original data bits. Therefore, errors could be detected and corrected. Among several error correcting codes, Hamming code [2] is the most widely used due to its simplicity and validity, not only because its code word is shorter than repetition code and other codes [3], but also it's easy to encode and decode. A Hamming encoder or decoder can be serial or parallel [4], serial encoder/decoder can be implemented by software in a general-purpose processor, and it is cost saving but time-consuming. Parallel encoder/decoder is usually implemented in ASIC, however, it's time-saving.

Considering the layout areas and power consumption of encoding and decoding circuits, a parallel error correcting circuit orienting to (7,4) Hamming code is implemented with carbon nanotube field effect transistors(CNFET) in this paper. Compared with MOSFETs, CNFETs has a series of advantages such as smaller size, lower working current, higher carrier mobility and exemption from short-channel effect [5], CNFETs is considered as the most promising device to replace MOSFETs. At present, several models for CNTFETs are available [6]. Considering the model precision and simulation time cost, Stanford model for CNFETs [7, 8] is utilized in this work.

The rest of this work is organized as follows. Section 2 is the implementation of (7, 4) Hamming encoder and decoder with modulo-2 adder. Section 3 verifies the function of the proposed circuit, and discusses the error detecting and correcting capability of grouping method in 16-bit and 32-bit application. In section 4, performance of circuits designed respectively with CNTFETs and MOSFETs is compared in terms of layout area and power consumption. Section 5 is a brief summary.

2. Designing Error Correcting Circuit with CNTFETs

2.1. Hardware Framework

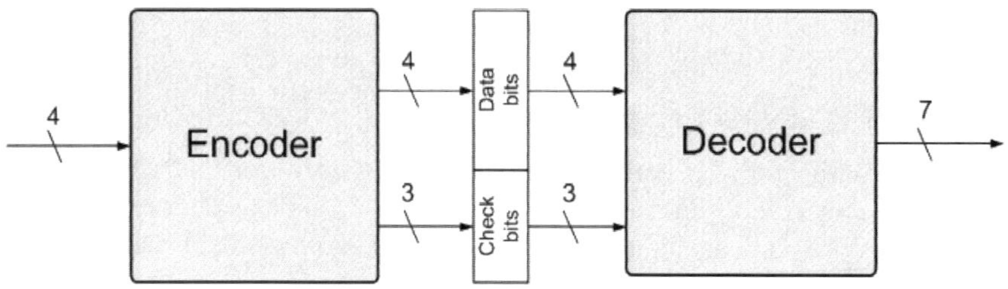

Figure 1. Hardware framework of proposed error correcting circuit

The framework of the error correcting circuit is shown in figure 1, it consists of three parts: encoder, decoder and memory. Original data is encoded before storing into the memory and decoded before reading out. In this circuit, the (7, 4) Hamming code is applied, 4-bit data is input into the encoder and then 3-bit check bits are generated and stored into the memory together with the original data bits. The generator matrix and parity matrix that we use are shown as follows.

$$G = \begin{pmatrix} 1 & 0 & 0 & 0 & 0 & 1 & 1 \\ 0 & 1 & 0 & 0 & 1 & 0 & 1 \\ 0 & 0 & 1 & 0 & 1 & 1 & 0 \\ 0 & 0 & 0 & 1 & 1 & 1 & 1 \end{pmatrix}, H = \begin{pmatrix} 0 & 0 & 0 & 1 & 1 & 1 & 1 \\ 0 & 1 & 1 & 0 & 0 & 1 & 1 \\ 1 & 0 & 1 & 0 & 1 & 0 & 1 \end{pmatrix} \quad (1)$$

With generator matrix, all allowable code words can be calculated, as it shows in Table.1.

Table 1. Allowable code word set of (7, 4) Hamming code

0000 000	0100 101	1000 011	1100 110
0001 111	0101 010	1001 100	1101 001
0010 110	0110 011	1010 101	1110 000
0011 001	0111 100	1011 010	1111 111

2.2. The Modulo-2 Adder

From the generator matrix and the parity matrix shown in (1), we can derive the linear relationship between the data bits, the check bits and the syndromes. In the encoder circuit, check bits are generated by the linear combination of data bits and in the decoder circuit, syndrome is generated by the linear combination of data bits and check bits,

$$\begin{cases} c_2 = d_2 + d_1 + d_0 \\ c_1 = d_3 + d_1 + d_0 \\ c_0 = d_3 + d_2 + d_0 \end{cases}, \begin{cases} s_2 = d_0 + c_2 + c_1 + c_0 \\ s_1 = d_2 + d_1 + c_1 + c_0 \\ s_0 = d_3 + d_1 + c_2 + c_0 \end{cases} \quad (2)$$

From equation (2) we can see that the critical component of the encoder and the decoder is modulo-2 adder, which can be built with exclusive-or gates. In figure 2, the exclusive-or gate is implemented with CNTFETs, and the 4-bit modulo-2 adder consists of 3 exclusive-or gates, table 2 lists parameters of the Stanford model for CNTFETs.

IOP Conf. Series: Journal of Physics: Conf. Series **986** (2018) 012038 doi:10.1088/1742-6596/986/1/012038

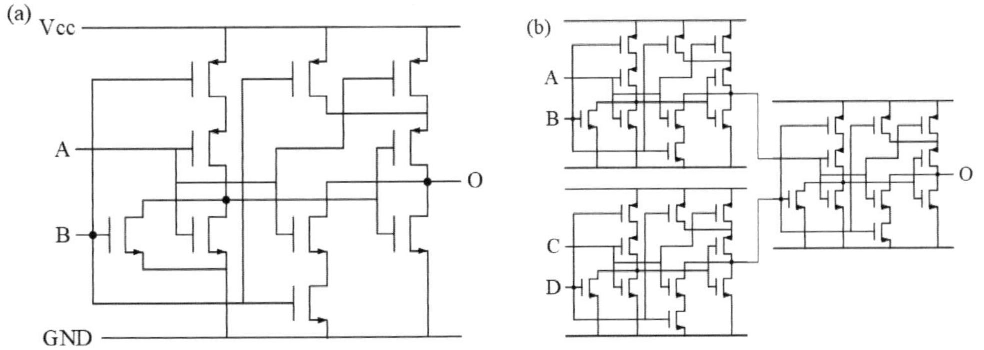

Figure 2. (a) Exclusive-or gate and (b) 4-bit modulo-2 adder implemented with CNTFETs

Table 2. Parameters of CNFETs

Parameter	Value	
L_g	32nm	Length of channel
L_{dd}	32nm	Length of drain extension
L_{ss}	32nm	Length of source extension
T_{ox}	4nm	Thickness of gate insulator
K_{ox}	16	Relative permittivity of gate
N_t	3	Number of tubes
(n,m)	(19,0)	Chirality vector of carbon
Pitch	20nm	Distance between neighboring

2.3. Encoder Circuit

The structure of encoder circuit depends on the generator matrix G. The structure of encoder circuit is built according to equation (2) and shown in figure 3.

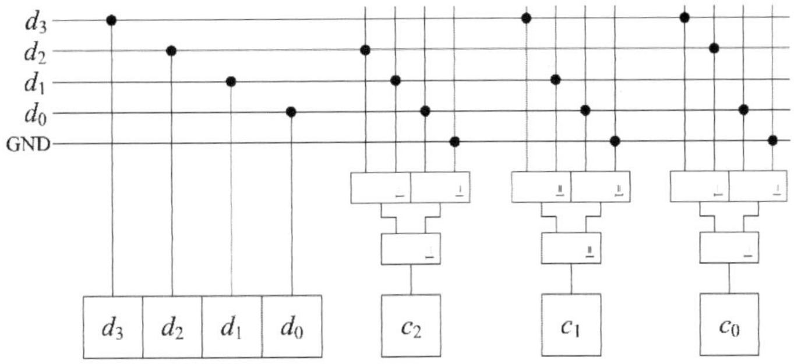

Figure 3. Structure of encoder circuit

2.4. Decoder Circuit

The structure of decoder circuit depends on the parity check matrix H. According to equation (3), the decoder circuit is built and shown in figure 4. The syndrome S is input into the error correcting logic to correct errors in the code word. For (7, 4) Hamming code, the error correcting logic is a 3-8 decoder.

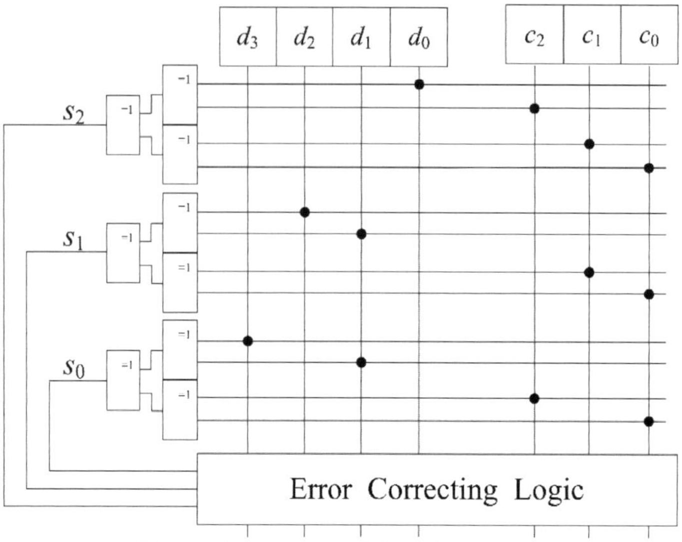

Figure 4. Structure of decoder circuit

3. Function Verification of Error Correcting Circuit

3.1. Encoder

The encoder circuit has 4 inputs (d_3,d_2,d_1,d_0) and 7 outputs ($d_3,d_2,d_1,d_0,c_2,c_1,c_0$). To validate its function, we run simulations in HSpice with the Stanford model for CNFETs. In the simulation, the input data varies from $(0000)_b$ to $(1111)_b$, and the waveform of output data is shown in figure5.

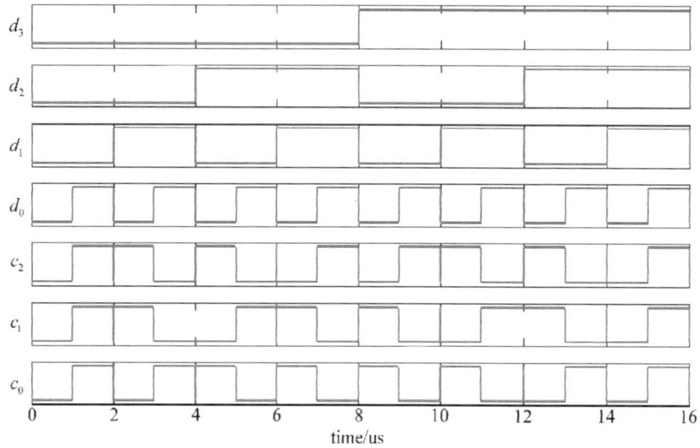

Figure 5. Waveform of encoder circuit

Comparing the waveforms in figure 5 with the allowable code words listed in table 2, we can see that the input data is encoded correctly by the encoder implemented with CNTFET.

3.2. Decoder

The decoder circuit has 7 inputs and 7 outputs, data stored in the memory is input into the decoder and then corrected data is output. To test the error correcting capability of the decoder, several error code words are needed. In the simulation, error code words listed in table 3 is input successively into the decoder, and the waveform of the syndrome is obtained and shown in figure 6. Compared with table 3, we can see that the ordinal number of the 1-bit error (e_1~e_6) is equal to its syndrome, so the syndrome can be used to correct 1-bit error. However, 2-bit error (e_7, e_8) cannot be corrected.

IOP Conf. Series: Journal of Physics: Conf. Series **986** (2018) 012038 doi:10.1088/1742-6596/986/1/012038

Table 3. List of error code words used in the simulation and their error bits

Error code word	Error bit	Error code word	Error bit
e_1=0000100	5	e_5=1100011	2
e_2=0000110	3	e_6=0010101	1
e_3=0100111	6	e_7=1101010	4,5
e_4=0110010	7	e_8=1100010	3,6

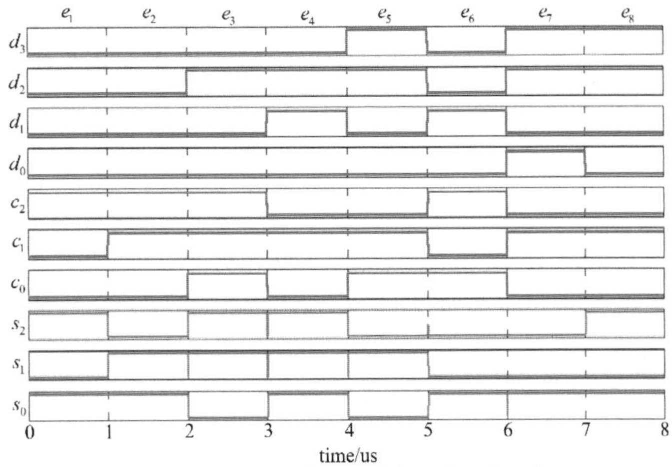

Figure 6. Waveform of decoder circuit

3.3. Error correcting capability

Error correction capability of Hamming code is very limited, to correct more than 1-bit errors in 16-bit data or 32-bit data, we can apply the grouping method. The original data are equally divided into several groups, each of the group contains 4-bits of data, then we apply (7, 4) Hamming code to these groups respectively. If each group contains no more than 1-bit error, the error can be corrected, otherwise, errors can only be detected. Table 4 lists the probabilities of correctable errors in 16-bit/32-bit application.

Table 4. Probabilities of correctable errors in 16-bit/32-bit application

Number of error bits	probability of correctable	
	16-bit data	32-bit data
1	100%	100%
2	77.78%	89.09%
3	41.88%	69.29%

From table 4, we can conclude that all 1-bit error can be corrected, most of the 2-bit errors can be corrected with the probability of 77.78% and 89.09% in 16-bit and 32-bit applications respectively, for 3-bit error, probability of correctable errors is vastly decreased in 16-bit application, but still acceptable in 32-bit application.

4. Performance Comparison with MOSFETs

4.1. Layout Area

Area of the layout is an important performance index for integrated circuit, smaller circuit layout area means higher integration. In this part, areas of circuits implemented with CNTFETs and 32-nm MOSFETs respectively are compared in this section.

Carriers in p-type MOSFETs and n-type MOSFETs have different motilities. With the same drain voltage and relative gate voltage, carriers in n-type moves faster than that in p-type, therefore, p-type MOSFETs in an integrated circuit have wider channels than n-type MOSFETs. In this work, width of p-type MOSFET is twice of that of n-type. For CNTFETs, carriers in p-type and n-type CNTFETs have similar motilities due to ballistic transport in carbon nanotubes, therefore, two types of CNTFETs have equal channel width, which means that size of CNTFETs is smaller than that of MOSFETs. Layout of single exclusive-or gate implemented with two types of FETs are compared in figure 7.

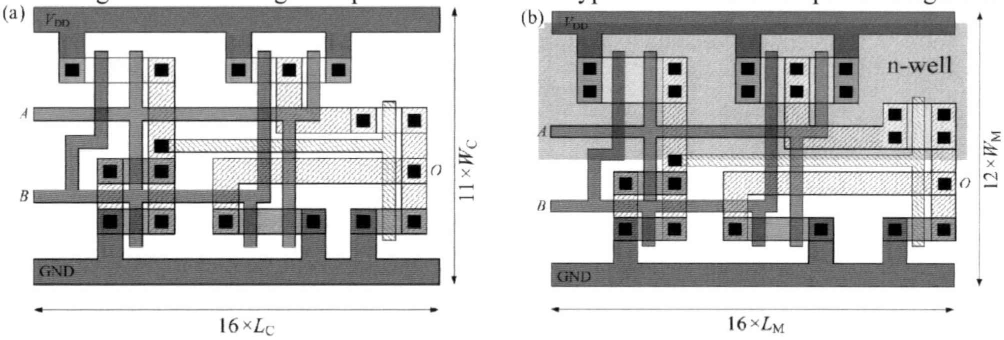

Figure 7. Layout of exclusive-xor gate implemented with (a) CNTFETs and (b) MOSFETs

Area of CNTFETs exclusive-xor gate can be calculated by parameters listed in table 2, the length of a CNTFET's channel is L_C=32nm and its width is W_C=60nm, which is derived from CNTFET's pitch. The area of CNTFETs exclusive-xor gate is $A_{CNT}=16\times L_C\times 11\times W_C=337920\text{nm}^2$. In the MOSFETs exclusive-xor gate, the length/width ratio of n-type MOSFET is 5, so L_M=32nm and W_M=160nm.The area of MOSFETs exclusive-xor gate is $A_{MOS}=16\times L_M\times 12\times W_M=337920\text{nm}^2$. From the result, we can see that A_{CNT} is about 34.4% of A_{MOS}, which means that CNTFETs have much higher integration than MOSFETs.

4.2. Power Consumption

Power consumption of two types of encoder is calculated and compared in this section. In the simulation, input data of the two encoders are the same as it shows in figure 5, and the power consumption is measured by the total working current provided by voltage source, since both of the power supply voltage in these two circuits are V_{DD}=0.9V. Waveforms of the total working current of these two encoders is shown in figure 8.

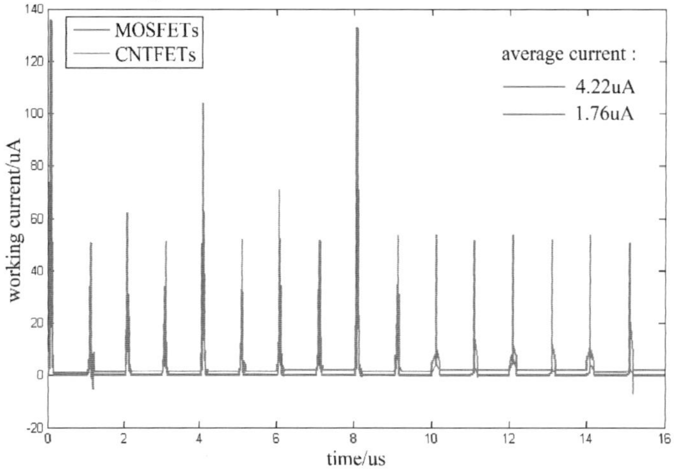

Figure 8. Working current of encoders implemented with two types of FETs

As it shows in figure 8, working current of the encoder implemented with CNTFETs is lower than MOSFETs, it shows a 56.93% decrement of average working current.

5. Conclusion

A parallel error correcting circuit based on (7, 4) Hamming code is implemented with CNFETs in this work. We choose a generator matrix and its parity check matrix to build the encoder and decoder circuits. In the simulation, 4-bit data can be encoded correctly by the encoder, and the decoder is able to correct 1-bit error. Regarding the limitations of (7, 4) Hamming code, a grouping method for 16-bit and 32-bit application is proposed and its error correction capability is analysed, the result shows that it could correct most of the 2-bit errors. At last, performances of circuits implemented respectively with CNTFETs and MOSFETs are compared, and the former shows a 34.4% decrement of layout area and a 56.9% decrement of power consumption.

Acknowledgements

This work was supported by the National Natural Science Foundation of China (Grant No. 11405270).

References

[1] Bentoutou Y IEEE Trans. Aerosp. Electron. Syst. 2012 48 648–57
[2] Hamming R W Bell System Technical Journal 1950 29 147–60
[3] Sadia A, Minahil Z and Salma Z F Int. Conf. Aero. Sci. Eng. 2013 Islamabad: 1–5
[4] Sridhar K P, Agalya R and Narmatha D International Conference on Circuit Power and Computing Technology 2014, Nagercoil: 1094–98
[5] Zhang Z and Delgado-Frias J G IEEE Trans. Nanotechnol. 2012 11 788–98
[6] Al-Shaggah A, Rjoub A and Khasawneh M. IEEE Jordan Conf. Appl. Elec. Eng. Comput. Technol. 2013 Amman: 1–6
[7] Deng J and Wong H S P IEEE Trans. Electron Devices 2007 54 3186–94.
[8] Deng J and Wong H S P IEEE Trans. Electron Devices 2007 54 3195–205